Information Security and Cryptography

Information Security – protecting information in potentially hostile environments – is a crucial factor in the growth of information-based processes in industry, business, and administration. Cryptography is a key technology for achieving information security in communications, computer systems, electronic commerce, and in the emerging information society.

Springer's Information Security & Cryptography (IS&C) book series covers all relevant topics, ranging from theory to advanced applications. The intended audience includes students, researchers and practitioners.

Jacques Patarin · Emmanuel Volte · Benoît Cogliati

Guide to Feistel Ciphers

Security Proofs and Cryptanalysis

Second Edition

Springer

Jacques Patarin
Laboratoire de Mathématiques de Versailles
University Paris-Saclay
Versailles, Yvelines, France

Emmanuel Volte
Lycée Stanislas
Cannes, France

Benoît Cogliati
Meudon, France

ISSN 1619-7100 ISSN 2197-845X (electronic)
Information Security and Cryptography
ISBN 978-3-031-99849-2 ISBN 978-3-031-99850-8 (eBook)
https://doi.org/10.1007/978-3-031-99850-8

Originally published with the title: Feistel Ciphers

This Springer imprint is published by the registered company Springer Nature Switzerland AG
The registered company address is: Gewerbestrasse 11, 6330 Cham, Switzerland

Preface to the Second Edition

Since the first edition of this book (eight years ago), AES is still the dominant secret key algorithm used in the world, and AES is not a Feistel cipher.

However, Feistel schemes are still a very hot cryptographic subject, both from a theoretical and practical point of view, with many publications in international cryptographic conferences. This is due to the flexibility and simplicity of the Feistel construction that facilitate cryptographic constructions for some specific purposes, and security analysis. During these eight years, more focus has also been done around quantum attacks (despite the fact that it is not yet clear when a dangerous for cryptography quantum computer may exist).

The main differences from the first edition of this book are:

- A new chapter about quantum attacks,
- Some explanations about the very famous linear and differential attacks on the DES,
- We have completely rewritten the proofs in Chap. 15 about the Pi XOR Pj theorem.

We have also improved many parts in various chapters.

I would like to give special thanks to Benoît Cogliati who has joined us in this second edition. Many improvements in this book from the first edition were done thanks to him.

I am also happy to thank the Springer team for their decision to make a second edition of this book.

Versailles, France
April 2025

Jacques Patarin

Preface to the First Edition

Feistel ciphers take an important part in secret key cryptography from both theoretical and practical point of view. From a theoretical point of view, it is from these ciphers that Luby and Rackoff proved in 1989 their famous theorem. This subsequently leads to a very large number of research papers in cryptography. This theorem gave a very innovative and powerful method to obtain security proof for "generic" ciphers. It was then possible to prove that one can obtain pseudo-random permutations (i.e., permutations easily generated by computers that are indistinguishable from truly random permutations) using pseudo-random functions. More recently (2008–2011), again from Feistel ciphers, it was possible to prove the equivalence between the Random Oracle Model and the Ideal Cipher Model, a famous problem that was left open for many years.

From a practical point of view, Feistel ciphers had their days of glory with the DES algorithm and it variants (3DES with 2 or 3 keys, XDES,...) that were the most widely used secret key algorithms around the world between 1977 and 2000. Since then, the AES algorithm, which is not a Feistel cipher, became the standard for secret key encryption. However, 3DES is still used in many applications, like in banking applications. Notice that the replacement of DES by AES is due to the fact that the parameters used in DES (in particular the size of the key) or in 3DES (in particular the size of the inputs and the outputs) have become too small for many modern applications, whereas the principle of Feistel ciphers stay very strong.

Versailles, France
January 2015

Jacques Patarin

Contents

Part I
Definitions and First Security Results

Part I
Definitions and First Security Results

Chapter 1
Introduction General Definitions

Abstract In this chapter, we present the general concept of cryptography and introduce the notation and definitions used throughout this book.

1.1 Introduction

Cryptology is the science of secure communications. It includes cryptography which designates defense techniques and cryptanalysis which designates attack techniques. Some authors use the word cryptography instead of cryptology. Cryptography is particularly interested in encryption, signature, and authentication techniques. We talk about confidentiality for encryption, integrity for signatures, and authentication. Since the paper by Diffie and Helman (1976), we know that there exist two kinds of cryptography: secret-key cryptography (also known as symmetric cryptography), which is the traditional cryptography where the secret key is shared by both parties, and public key cryptography where the secret key is owned only by the person who encrypts or signs the message. Feistel ciphers, which are the topic of this book are most of the time used for encryption (see Fig. 1.1). However, it is common to use them for secret-key signature (we talk about MAC for Message Authentication Code), and it is also possible to use them for secret-key authentication.

Feistel ciphers do not provide directly public key algorithm but they are sometimes used in standard public key algorithm protocols where they play an important role for security (for example in the RSA-OAEP norm). Moreover, generally public key encryption is used to send the secret key of a secret-key algorithm which will do most of the encryption work: this is called `bulk encryption`. Indeed secret-key algorithm (and particularly Feistel ciphers based algorithms) are generally much faster to compute than public key algorithms.

Let us come back to the situation where two parties, traditionally named Alice and Bob want to use a secret-key algorithm. Generally, we can distinguish two kinds of secret-key algorithm: block cipher encryption algorithms and stream cipher encryption algorithms, where the message to be sent is combined with pseudo-random bit generator.

J. Patarin et al., *Guide to Feistel Ciphers*, Information Security and Cryptography,
https://doi.org/10.1007/978-3-031-99850-8_1

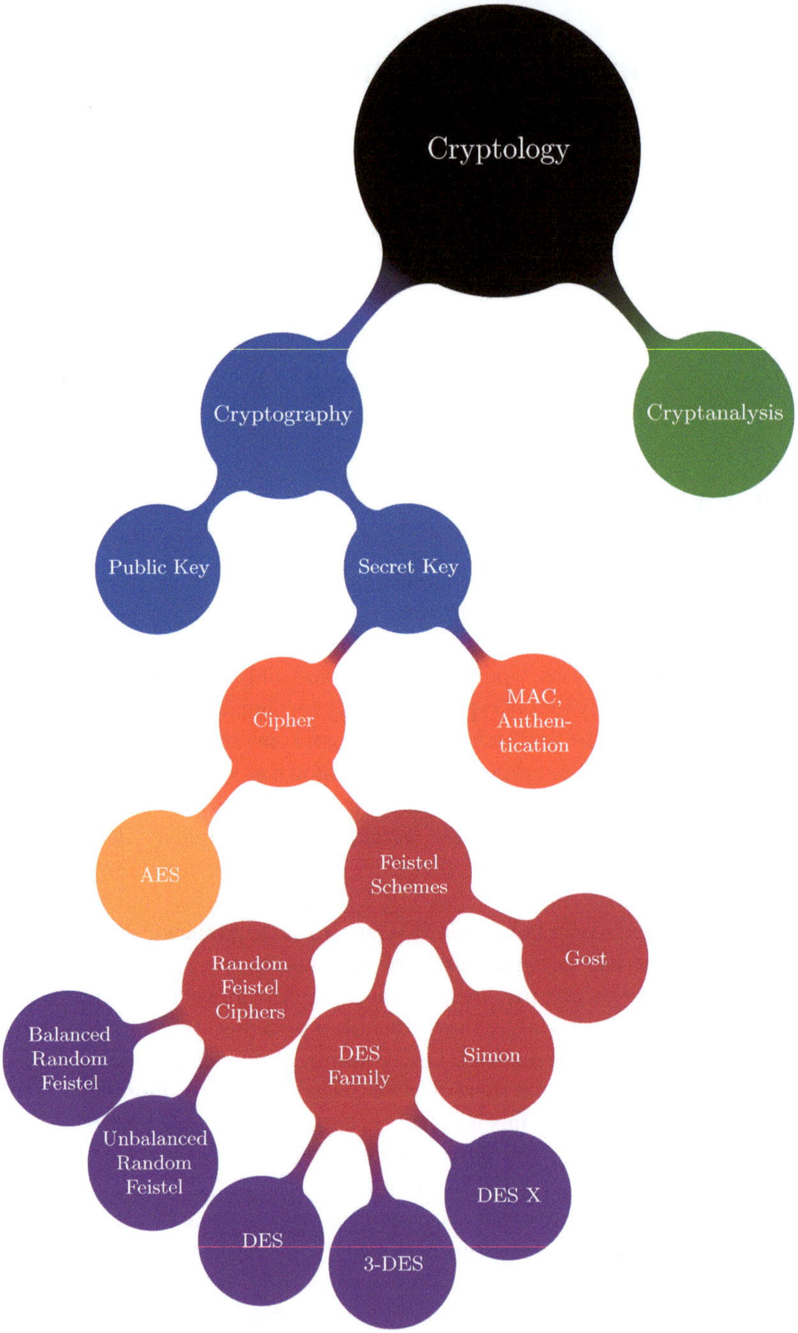

Fig. 1.1 Encryption with Feistel ciphers in cryptology

Feistel ciphers can be seen as examples of block ciphers (even though they can be used for other purposes like MAC or pseudo-random bit generator). Moreover, a block cipher encryption algorithm can be defined as a pseudo-random permutation generator. The word `permutation` means `bijection on finite set`. The word `pseudo-random` means that the bijections generated by the generator can be easily computed by a computer program from a secret key, but an attacker who do not know the key will not be able to distinguish, in practice, permutations generated by the generator from truly random permutations.

In this section, we are a bit informal and we present here some notions very quickly in order to introduce Feistel ciphers. Thus, `indistinguishable inpractice` can be defined in a very precise way (see Goldreich 2007 for example), by the means of probabilistic algorithms that are polynomial or not.

Let us suppose that we want to generate pseudo-random permutations taking as input a 100-bit string and giving as output a 100-bit string. How is it possible to handle this? Horst Feistel had the idea to proceed the way we are going to describe in Sect. 1.3. His method, which was not the only possible method has proven very useful in cryptography and led to many security results. Mainly, using functions from n bits to n bits, he constructed a bijection from $2n$ bits to $2n$ bits (here $n = 50$). In order to generate a pseudo-random permutation, Feistel's construction shows that the bijection property is not a problem. However, the pseudo-random property will lead to many studies.

1.2 General Notation

We will use the following notation.

- The set of all integers is denoted $\mathbb{N}$.
- For $n \in \mathbb{N}$, $n \geq 1$, $\{0, 1\}^n$ denotes the set of binary strings of length n.
- If S and T are two sets, the set of all functions from S to T will be denoted $\mathscr{F}(S, T)$. The set of all functions from $\{0, 1\}^m$ to $\{0, 1\}^n$ is denoted $\mathscr{F}_{m,n}$. When $m = n$, the set of all functions from $\{0, 1\}^n$ to $\{0, 1\}^n$ is simply denoted $\mathscr{F}_n$.
- If S is a set, the set of all permutations of S will be denoted $\mathscr{P}(S)$. The set of all permutations of $\{0, 1\}^n$ is simply denoted $\mathscr{P}_n$.
- If L and R are two elements of $\{0, 1\}^n$, we will denote $[L, R]$ the element of $\{0, 1\}^{2n}$ which is the concatenation of L and R.
- The XOR (*eXclusive OR*) operator is denoted $\oplus$, i.e., if $L \in \{0, 1\}^n$ and $R \in \{0, 1\}^n$, $L \oplus R$ is the n-bit string obtained by bitwise addition modulo 2 of L and R.
- We denote σ the permutation of $\{0, 1\}^{2n}$ that swaps the two n-bit halves of its argument, i.e., for all $L, R \in \{0, 1\}^n$, $\sigma([L, R]) = [R, L]$.
- The composition of two functions f and g is denoted $f \circ g$, i.e., $(f \circ g)(x) = f(g(x))$.
- The *alternating group*, i.e., the subgroup of $\mathscr{P}_n$ whose elements have even signature, is denoted $\mathscr{A}_n$.

- If A is as finite set, $|A|$ is the cardinal of A.
- For $a, b \in \mathbb{R}^+$, $a \geq b$ means $a \geq b$ or $a \simeq b$. Similarly, $a \gtrsim b$ means $a \leq b$ or $a \simeq b$.
- For $a, b \in \mathbb{R}^+$, $a \ll b$ means that a is much less than b. Similarly, $a \gg b$ means that a is much greater than b.
- $\mathbb{E}$ denotes the expectation function.
- KP (respectively CP, ACP) stands for Known Plaintext (respectively, Chosen Plaintext and Adaptive Chosen Plaintext).

1.3 Block Ciphers

Definition 1.1 (*Block cipher*) Let $\mathscr{K}$ and $\mathscr{M}$ be two non-empty sets. A block cipher with key space $\mathscr{K}$ and message space $\mathscr{M}$ is a mapping $E : \mathscr{K} \times \mathscr{M} \to \mathscr{M}$ such that for all $K \in \mathscr{K}$, $M \mapsto E(K, M)$ is a permutation of the message space $\mathscr{M}$. We will denote E_K the mapping $M \mapsto E(K, M)$, and E_K^{-1} the inverse mapping.

Note that this definition is purely "syntactical". A "practical" block cipher usually comes with two efficient procedures for computing $E_K(M)$ and $E_K^{-1}(C)$ for any key K, any message M, and any ciphertext C.

1.4 Attack Models

Informally, a good block cipher, when used with a key drawn uniformly at random, should behave as a perfectly random permutation of its message space. This is formally captured with the abstract notion of *distinguishers*. A distinguisher can be thought as the most basic and fundamental attacker one might consider. It can make queries to a black box or "permutation oracle" which can be either the block cipher under a uniformly random key, or a perfectly random permutation of the message space. Its goal is to decide which world it is in, solely based on the answers of the permutation oracle. We give formal definitions below.

Definition 1.2 (*Distinguisher*) Let $\mathscr{M}$ be some non-empty set. A distinguisher is a (potentially randomized) algorithm D with access to a permutation oracle $P \in \mathscr{P}(\mathscr{M})$. Namely, it can make queries of the form $(+, x)$ and receive the value $P(x)$, or queries of the form $(-, y)$ and receive the value $P^{-1}(y)$. The distinguisher eventually outputs a single bit $b \in \{0, 1\}$, which we denote $\mathsf{D}^P \to b$.

Let $\mathscr{D}$ and $\mathscr{D}'$ be two distributions over the set of permutations of $\mathscr{M}$. The advantage of D in distinguishing $\mathscr{D}$ and $\mathscr{D}'$ is defined as

$$\mathscr{A}_{\mathscr{D},\mathscr{D}'}(\mathsf{D}) \stackrel{\text{def}}{=} \left|\Pr\left[P \leftarrow \mathscr{D} : \mathsf{D}^P \to 1\right] - \Pr\left[P' \leftarrow \mathscr{D}' : \mathsf{D}^{P'} \to 1\right]\right|.$$

When the distinguisher is randomized, the two probabilities are taken over the random draw of P (resp. P') and the random coins of the distinguisher.

As we will see shortly, whether a distinguisher is successful or not will be captured by a quantity called its advantage. Before that, we must specify more precisely what we call the *attack model*, which describes how the distinguisher is allowed to query the permutation oracle.

We classify attacks according to the way the distinguisher can query the permutation oracle.

- Cipher-Only Attacks (here no plaintext is available)
- Know-Plaintext Attacks (KPA)
- Non-adaptive Chosen-Plaintext Attacks (NCPA)
- (adaptive) Chosen-Plaintext Attacks (CPA)
- Non-adaptive Chosen-Ciphertext Attacks
- (adaptive) Chosen-Ciphertext Attacks
- Non-adaptive Chosen-Plaintext and Ciphertext Attacks (NCCA)
- (adaptive) Chosen-Plaintext and Ciphertext Attacks (CCA).

Alternative Notation
In the cryptographic community:

- NCPA is sometimes denoted CPA-1.
- CPA is sometimes denoted CPA-2.
- NCCA is sometimes denoted CCA-1 or CPCA-1.
- CCA is sometimes denoted CPCA or CCA-2 or CPCA-2.

There is no classical notation for Chosen-Ciphertext Attack with no possibility to choose a plaintext, since instead of speaking of Chosen-Ciphertext security for a block cipher E, we will speak of CPA for E^{-1}. We mention these Chosen-Ciphertext Attacks (with no possibility to choose a plaintext) for completeness, but we will not use them.

Definition 1.3 (*Advantage*) Let E be a block cipher with key space $\mathscr{K}$ and message space $\mathscr{M}$, and let D be a distinguisher. The *advantage* of D against E is defined as its advantage in distinguishing the probability distribution over $\mathscr{P}(\mathscr{M})$ induced by the uniformly random draw of a key for the block cipher, and the uniform distribution over $\mathscr{P}(\mathscr{M})$, viz.

$$\mathscr{A}_E(\mathsf{D}) \stackrel{\text{def}}{=} \left|\Pr\left[K \leftarrow_{\$} \mathscr{K} \,:\, \mathsf{D}^{E_K} \to 1\right] - \Pr\left[P \leftarrow_{\$} \mathscr{P}(\mathscr{M}) \,:\, \mathsf{D}^{P} \to 1\right]\right|.$$

For any attack model ATK, and any integers t and q, we define

$$\mathscr{A}_E^{\text{ATK}}(t, q) \stackrel{\text{def}}{=} \max_{\mathsf{D}} \{\mathscr{A}_E(\mathsf{D})\},$$

where the maximum is taken over all distinguishers of the class ATK running in time at most t and making at most q oracle queries. When considering computationally unbounded distinguishers, we denote $\mathscr{A}_E^{\text{ATK}}(q)$ as a shorthand for $\mathscr{A}_E^{\text{ATK}}(\infty, q)$.

There is a simple relation between the advantage and the probability to distinguish:

$$\text{Probability to distinguish } = \frac{1}{2} + \frac{\text{Advantage}}{2}$$

Most results in this book will deal with computationally unbounded distinguishers, hence we highlight some classical results about them. First, for computationally unbounded distinguishers, one can always restrict our attention without loss of generality to *deterministic* distinguishers.

Lemma 1.1 *Let E be a block cipher,* ATK *be an attack model, and q be a non-negative integer. Then there exists a deterministic distinguisher* D *of the class* ATK *making q queries such that*

$$\mathscr{A}_E(\mathsf{D}) = \mathscr{A}_E^{\text{ATK}}(q).$$

Proof Let D' be the probabilistic distinguisher with the best advantage. Consider now the computationally unbounded distinguisher D which enumerates all possible random tapes and computes the corresponding advantage of D' when run with this random tape. Then D fins the random tape which maximizes this advantage, and simply runs D' with this random tape. □

Computationally unbounded distinguishers have the nice property that their advantage can be simply linked with the statistical distance between the answers of the block cipher and the uniform distribution over all possible answers. This will be developed in Chap. 3 about the H-coefficients technique. We recall the definition of statistical distance (also called total variation distance).

Definition 1.4 (*Statistical Distance*) Let X be some non-empty set and μ and ν be two probability distributions over X. The statistical distance between μ and ν, denoted $\|\mu - \nu\|$, is defined as

$$\|\mu - \nu\| \stackrel{\text{def}}{=} \frac{1}{2} \sum_{x \in X} |\mu(x) - \nu(x)|.$$

1.5 Kerckhoffs's Principle

In cryptography, Kerckhoffs's principle says that "A crypto-system should be secure even if everything about the system, except the key, is public knowledge." This principle was written in 1883 by Auguste Kerckhoffs, a Dutch cryptographer working in France. In conformity with this principle, we will always assume that the design of the ciphers, but not the secret key, are known by the attackers.

References

Diffie, W., and M. Helman. 1976. New directions in cryptogaphy. *IEEE Transactions on Information Theory* 22: 644–654.
Goldreich, O. 2007. *Foundations of Cryptography*. Cambridge: Cambridge University Press.

References

[illegible]

Chapter 2
Balanced Feistel Ciphers, First Properties

Abstract Feistel ciphers are named after Horst Feistel who studied these schemes in the 1960s. In this chapter, we will only present classical Feistel ciphers, i.e., balanced Feistel ciphers with the $\oplus$ group law (Xor). In Chaps. 8, 9, and 10, we will see that there are many variants of these ciphers.

2.1 Definition of Classical Feistel Ciphers

Classical Feistel ciphers are also known as *balanced* Feistel ciphers. We start with the definition of the 1-round Feistel transformation.

Definition 2.1 Let $f \in \mathscr{F}_n$. The *1-round balanced Feistel network* associated with f, denoted $\Psi(f)$, is the function from $\{0, 1\}^{2n}$ to $\{0, 1\}^{2n}$ defined by (see also Fig. 2.1):

$$\forall (L, R) \in (\{0, 1\}^n)^2,\ \Psi(f)([L, R]) = [S, T] \iff \begin{cases} S = R \\ T = L \oplus f(R). \end{cases}$$

It is quite easy to see that for any function f, $\Psi(f)$ is actually a permutation of $\{0, 1\}^{2n}$, as we show in the following proposition. Recall that σ denotes the permutation of $\{0, 1\}^{2n}$ that swaps the two n-bit halves of its argument.

Proposition 2.1 *For any function $f \in \mathscr{F}_n$, $\Psi(f)$ is a permutation of $\{0, 1\}^{2n}$ and its inverse is $\Psi(f)^{-1} = \sigma \circ \Psi(f) \circ \sigma$.*

Proof

$$\Psi(f)([L, R]) = [S, T] \iff \begin{cases} R = S \\ L = T \oplus f(S). \end{cases}$$

J. Patarin et al., *Guide to Feistel Ciphers*, Information Security and Cryptography,
https://doi.org/10.1007/978-3-031-99850-8_2

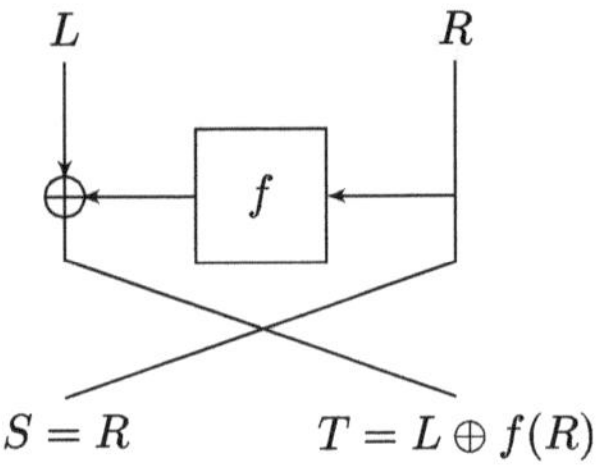

Fig. 2.1 The basic (1-round) balanced Feistel network associated with round function f

Therefore, for all $[S, T]$ we have exactly one solution $[L, R]$ and $\Psi(f)$ is a permutation of $\{0, 1\}^{2n}$. Moreover, its inverse is given by

$$\begin{aligned}\Psi(f)^{-1}[S, T] &= [T \oplus f(S), S] \\ &= \sigma([S, T \oplus f(S)]) \\ &= \sigma(\Psi(f)([T, S])) \\ &= \sigma(\Psi(f)(\sigma([S, T]))),\end{aligned}$$

hence the result. □

Before going further in the description of Feistel ciphers, we will make comments on this first round. Notice that $\Psi(f_1)$ is always a permutation even though f_1 is not bijective. This is an important property of Feistel ciphers. In contrast, in some other ciphers like AES for example, designers manage to have bijective transformations. Here the choice for f_1 is much larger since we do not have to take into account the bijective feature of f_1. However, clearly one round of a Feistel cipher is not enough to obtain a pseudo-random permutation: indeed the left-hand part of the output is exactly the right-hand part of the input. It was not encrypted at all. However, if we compose several bijections, we still get a bijection. This is what we are going to do below. Thus even though one round of a Feistel cipher is not good to hide the inputs, this will not be the case anymore after several rounds as we will see. An architect who builds a tower with one floor that collapses will not consider the possibility of constructing a tower by adding several floors of the same kind and hope that the tower will be solid. However, this is what we will do, but this construction will be justified by the security results we will obtain. Cryptography with bijections does not behave like the architecture of towers!

Definition 2.2 Let $r \geq 1$ and let $f_1, f_2, \ldots, f_r$ be r functions in $\mathscr{F}_n$. The *r-round balanced Feistel network* associated with $f_1, \ldots, f_r$, denoted $\Psi^r(f_1, \ldots, f_r)$, is the function from $\{0, 1\}^{2n}$ to $\{0, 1\}^{2n}$ defined by (see also Fig. 2.2)

$$\Psi^r(f_1, \ldots, f_r) = \Psi(f_r) \circ \cdots \circ \Psi(f_2) \circ \Psi(f_1).$$

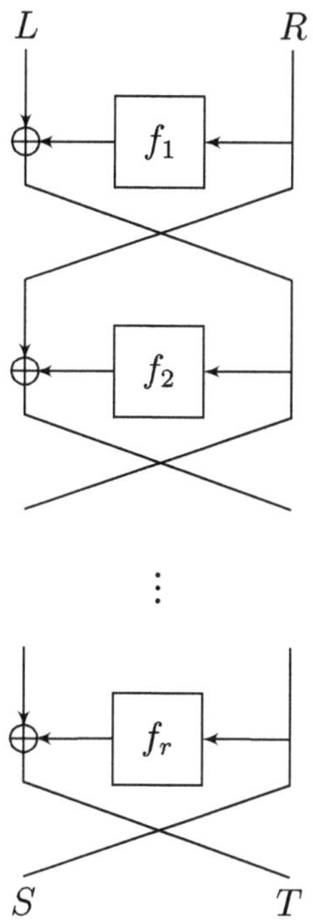

Fig. 2.2 The r-round balanced Feistel network associated with round functions $f_1, \ldots, f_r$

Theorem 2.1 *For any functions* $f_1, \ldots, f_r \in \mathscr{F}_n$, $\Psi^r(f_1, \ldots, f_r)$ *is a permutation of* $\{0, 1\}^{2n}$ *and*

$$(\Psi^r(f_1, \ldots, f_r))^{-1} = \sigma \circ \Psi^r(f_r, \ldots, f_1) \circ \sigma.$$

Proof $\Psi^r(f_1, \ldots, f_r)$ is a permutation of $\{0, 1\}^{2n}$ since it is the composition of r permutations of $\{0, 1\}^{2n}$. Moreover, by Proposition 2.1, and since σ^2 is the identity function, one has

$$\begin{aligned}(\Psi^r(f_1, \ldots, f_r))^{-1} &= (\Psi(f_1))^{-1} \circ \cdots \circ (\Psi(f_r))^{-1} \\ &= \sigma \circ \Psi(f_1) \circ \sigma \circ \cdots \circ \sigma \circ \Psi(f_r) \circ \sigma \\ &= \sigma \circ \Psi(f_1) \circ \cdots \circ \Psi(f_r) \circ \sigma,\end{aligned}$$

from which the result follows. □

Up to the initial and final application of the "swapping" function σ, the inverse of an r-round balanced Feistel network is simply another r-round Feistel network where the round functions f_1, ..., f_r are used in the reverse order. Since the computation of σ is very fast, we see that the computation of $\Psi^r(f_1, \ldots, f_r)$ should take about the same time in the forward or backward direction (i.e., when encrypting or decrypting).

Definition 2.3 Let $r \geq 1$. The *r-round Feistel transformation*, denoted Ψ^r, maps a tuple of functions $(f_1, \ldots, f_r) \in (\mathscr{F}_n)^r$ to the permutation $\Psi^r(f_1, \ldots, f_r)$ of $\{0, 1\}^{2n}$ as defined by Definition 2.2.

Remark 2.1 Balanced Feistel networks can be defined on any group $(G, *)$, not only $(\{0, 1\}^n, \oplus)$.

From Feistel networks, we can finally define Feistel ciphers, by letting round functions depend on secret keys.

Definition 2.4 Let $r \geq 1$ and let $F = (f_K)$ be a family of functions in $\mathscr{F}_n$ indexed by a set $\mathscr{K}$. The *r-round balanced Feistel cipher* associated with F is the block cipher with key space $\mathscr{K}^r$ and message space $\{0, 1\}^{2n}$ which maps a key $(K_1, \ldots, K_r) \in \mathscr{K}^r$ and a plaintext $[L, R] \in \{0, 1\}^{2n}$ to the ciphertext $\Psi^r(f_{K_1}, \ldots, f_{K_r})([L, R])$. In other words, the permutation of $\{0, 1\}^{2n}$ associated with key $(K_1, \ldots, K_r)$ is the Feistel network $\Psi^r(f_{K_1}, \ldots, f_{K_r})$.

2.2 Signature of Balanced Feistel Networks

Theorem 2.2 (Patarin 2025) *When $n \geq 2$, the signature of a Feistel permutation is even, i.e.,*

$$\forall f_1, f_2, \ldots, f_r \in \mathscr{F}_n, \ \Psi^r(f_1, \ldots, f_r) \in \mathscr{A}_{2n}.$$

Proof Let f_1 be a function of F_n. Let $\Psi'(f_1)([L, R]) = [L \oplus f_1(R), R]$. We will show that the signature of both σ and $\Psi'(f_1)$ is even. Since $\Psi(f_1) = \sigma \circ \Psi'(f_1)$, $\Psi(f_1)$ has an even signature as well, and by composition, any Feistel permutations has an even signature.

Consider σ: All its cycles have 1 or 2 elements since $\sigma \circ \sigma$ is the identity. There are exactly 2^n cycles with 1 element since $\sigma([L, R]) = [L, R]$ if and only if $L = R$ (and a cycle with 1 element has an even signature). Hence, there are $(2^{2n} - 2^n)/2$ cycles with 2 elements, which is even for $n \geq 2$.

Consider now $\Psi'(f_1)$: All the cycles have 1 or 2 elements since $\Psi'(f_1) \circ \Psi'(f_1)$ is the identity. Moreover $\Psi'(f_1)([L, R]) = [L, R]$ if and only if $f_1(R) = 0$, so the number of cycles with 2 elements is $k \cdot 2^n/2$, with k being the number of values R such that $f_1(R) \neq 0$. So when $n \geq 2$ the signature of $\Psi'(f_1)$ is even. □

The fact that Feistel ciphers have always an even signature is not in general a cryptographic security problem. Indeed, this property has influence only when you know the images of all inputs (except maybe 2 which can be deduced from the others). Thus this property is mathematically interesting but it has a small cryptographic impact.

2.3 Random Feistel Ciphers

As we have seen, when the functions $f_1, \ldots, f_d$ are randomly and independently chosen in $\mathscr{F}_n$ (or when they are generated from a pseudo-random generator), $\Psi^d(f_1, \ldots, f_d)$ is called a `Random Feistel Cipher`, or a `Luby-Rackoff` construction, since we will see in Chap. 4 some very famous security results on these ciphers proved by Luby and Rackoff (1988). On the contrary, many important Feistel ciphers are designed with functions $f_1, \ldots, f_d$ which are not random or pseudo-random, for example DES variants as we will see in Part III.

From the Luby-Rackoff theorem that we will see in Chap. 4, it is possible to prove that random Feistel ciphers provide a PRPG (Pseudo-random Permutation Generator) from a PRFG (Pseudo-Random Function Generator). Moreover, in cryptography, it is also proved that is possible to generate a PRFG from a PGNG (Pseudo-Random Number Generator) and a PRNG from any one-way function (see Fig. 2.3). However, this is not the topic of this book. The interested reader is referred to Goldreich et al. (1986) and Hastad et al. (1993). Since a proof of the existence of a one-way function will provide a proof of the famous theoretical open computer science problem $P \neq NP$, this design of a PRPG is interesting but do not provide the existence of a PRPG. Moreover, Feistel ciphers that are based on functions $f_1, \ldots, f_d$ that are not pseudo-random (like 3DES) are often much more computationally efficient than random Feistel ciphers as constructed in Fig. 2.3.

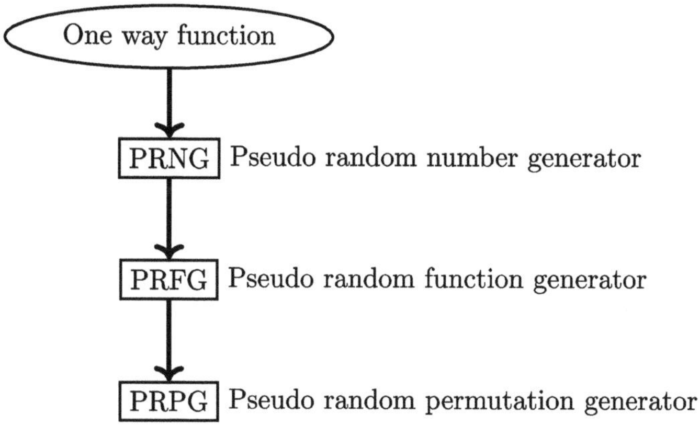

Fig. 2.3 A possible construction of PRPG from any one-way function

2.4 Efficient Attacks for One, Two, and Three Rounds

We show that for one, two, and three rounds, balanced Feistel ciphers can be broken very efficiently with a *constant* number of queries, independently of the size parameter n. These attacks are generics, i.e., they work for any round functions.

2.4.1 KPA for One Round with $q = 1$

Consider the following KPA-distinguisher D:

1. D makes a query to the oracle, and receives a random plaintext $[L, R]$ together with $[S, T] = \mathcal{O}([L, R])$;
2. if $S = R$, D outputs 1, otherwise it outputs 0.

Clearly, when $\mathcal{O} = \Psi^1(f_1)$, D *always* outputs 1 since

$$\Psi^1(f_1)([L, R]) = [S, T] \iff \begin{cases} S = R \\ T = L \oplus f_1(R). \end{cases}$$

On the other hand, when $\mathcal{O}$ is a random permutation of $\{0, 1\}^{2n}$, then $[S, T]$ is uniformly random in $\{0, 1\}^{2n}$, and the probability that $S = R$ (and hence that D outputs 1) is exactly 2^{-n}. Therefore, by definition of the advantage (cf. Definition 1.3) we have

$$\mathcal{A}_{\Psi^1}(\mathsf{D}) = 1 - \frac{1}{2^n}.$$

Since $\mathcal{D}$ makes exactly one query, it follows that

$$\mathcal{A}_{\Psi^1}^{\mathrm{KPA}}(D) \geq 1 - \frac{1}{2^n}.$$

Hence, there is a very efficient known-plaintext attack against Ψ^1, making only one query and distinguishing Ψ^1 from a random permutation with probability negligibly close to one.

2.4.2 NCPA for Two Rounds with $q = 2$

Consider the following NCPA-distinguisher D:

1. D chooses $L, L', R \in \{0, 1\}^n$, with $L \neq L'$, and queries $[S, T] := \mathcal{O}([L, R])$ and $[S', T'] := \mathcal{O}([L', R])$;
2. D checks whether $S \oplus S' = L \oplus L'$; if this holds, D outputs 1, otherwise D outputs 0.

Note that D chooses L, L', and R *before* making any query to the oracle, hence it is non-adaptive. By definition of Ψ^2, we have

$$\Psi^2(f_1, f_2)([L, R]) = [S, T] \iff \begin{cases} S = L \oplus f_1(R) \\ T = R \oplus f_2(L \oplus f_1(R)). \end{cases}$$

Hence, when $\mathscr{O} = \Psi^2(f_1, f_2)$, one has

$$S \oplus S' = L \oplus f_1(R) \oplus L' \oplus f_1(R) = L \oplus L',$$

so that D always outputs 1.

On the other hand, when $\mathscr{O}$ is a random permutation of $\{0, 1\}^{2n}$, then $[S', T']$ is uniformly random in $\{0, 1\}^{2n} \setminus \{[S, T]\}$. Since there are exactly 2^n possible values of $[S', T']$ in $\{0, 1\}^{2n} \setminus \{[S, T]\}$ such that $S' = S \oplus L \oplus L'$ (because $L \oplus L' \neq 0$), D outputs 1 with probability

$$\frac{2^n}{2^{2n} - 1}.$$

Hence, we have, by definition of the advantage,

$$\mathscr{A}_{\Psi^2}(\mathsf{D}) = 1 - \frac{2^n}{2^{2n} - 1}.$$

Since D makes exactly two queries, this implies

$$\mathscr{A}_{\Psi^2}^{\mathrm{NCPA}}(\mathsf{D}) \geq 1 - \frac{2^n}{2^{2n} - 1}.$$

Hence, there is a very efficient non-adaptive chosen-plaintext attack against Ψ^2, making only two queries and distinguishing Ψ^2 from a random permutation with probability negligibly close to one.

2.4.3 CCA for Three Rounds with $q = 3$

We consider the following CCA-distinguisher D:

1. D chooses $L, L', R \in \{0, 1\}^n$, with $L \neq L'$, and queries $[S, T] := \mathscr{O}([L, R])$ and $[S', T'] := \mathscr{O}([L', R])$;
2. D asks for the value $[L'', R''] := \mathscr{O}^{-1}([S', T' \oplus L \oplus L'])$.
3. D checks if $R'' = S' \oplus S \oplus R$; if this holds, D outputs 1. Otherwise D outputs 0.

If $\mathcal{O}$ is a permutation randomly chosen, the probability that D outputs 1 is $\simeq 1/2^n$. Now assume that $\mathcal{O} = \Psi^3(f_1, f_2, f_3)$.

Then $\mathcal{O}([L, R]) = [S, T] \Leftrightarrow \begin{cases} S = R \oplus f_2(L \oplus f_1(R)) \\ T = L \oplus f_1(R) \oplus f_3(R \oplus f_2(L \oplus f_1(R)). \end{cases}$

And $\mathcal{O}^{-1}[S, T] = [L, R] \Leftrightarrow \begin{cases} L = T \oplus f_3(S) \oplus f_1(S \oplus f_2(T \oplus f_3(S)))) \\ R = S \oplus f_2(T \oplus f_3(S)) \end{cases}$

Thus $\mathcal{O}^{-1}[S', T' \oplus L \oplus L'] = [L'', R''] \Rightarrow R'' = \underbrace{S' \oplus f_2(\underbrace{T' \oplus L \oplus L' \oplus f_3(S')}_{L \oplus f_1(R)})}_{S \oplus R}$

Therefore $R'' = S' \oplus S \oplus R$.

Thus the probability that D outputs 1 when $\mathcal{O} = \Psi^3(f_1, f_2, f_3)$ is 1 and we obtain that

$$\mathcal{A}_{\Psi^3}^{CCA}(\mathsf{D}) \geq 1 - \frac{1}{2^n}.$$

This attack is able to distinguish $\Psi(f_1, f_2, f_3)$ when f_1, f_2, and f_3 are randomly and independently chosen in $\mathcal{F}_n$ from a truly random permutation of $\mathcal{P}_{2n}$ with a high probability when we can choose 2 plaintext/ciphertext pairs and obtain the corresponding ciphertexts, and then choose 1 ciphertext and obtain the plaintext. This attack is a CCA with $q = 3$ plaintext/ciphertext pairs.

This attack can be found as follows. The idea is to create a “circle” in R, S, X, as in Fig. 2.4, where $X_i = L_i \oplus f_1(R_i)$, i.e., to have $R_2 = R_1$, $S_3 = S_2$ and $X_3 = X_1$. We always have

$$R_i = R_j \Rightarrow L_i \oplus L_j = X_i \oplus X_j \tag{2.1}$$

$$X_i = X_j \Rightarrow R_i \oplus R_j = S_i \oplus S_j \tag{2.2}$$

$$S_i = S_j \Rightarrow X_i \oplus X_j = T_i \oplus T_j\ . \tag{2.3}$$

First, we choose $R_2 = R_1$ and $L_2 \neq L_1$. So from (2.1), we have

$$X_2 \oplus X_1 = L_1 \oplus L_2\ . \tag{2.4}$$

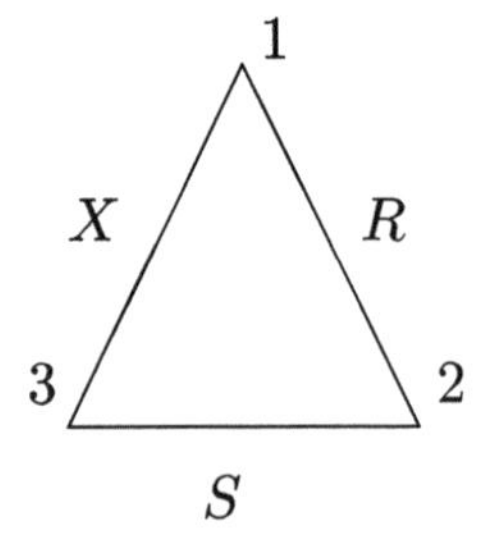

Fig. 2.4 A “circle” in R, S, X (here it looks more as a triangle)

Second, we choose $S_3 = S_2$. So from (2.3), we have

$$X_2 \oplus X_3 = T_2 \oplus T_3 \,. \tag{2.5}$$

So from (2.4) and (2.5) we can impose $X_3 = X_1$ by choosing $T_3 = T_2 \oplus L_1 \oplus L_2$. Then from (2.2) we will have $R_3 = R_1 \oplus S_1 \oplus S_3$ $(= R_1 \oplus S_1 \oplus S_2)$.

2.5 Conclusion

We have seen that it is possible to mount attacks on Feistel ciphers with 1, 2, and 3 rounds even when the round functions are perfect. This shows that these ciphers are not secure if we apply only 1, 2, or 3 rounds. Several questions arise. They will be the topic of the following chapters. Do there exist similar attacks for 4 rounds or more? This is studied in Chap. 6. On the contrary, is it possible to obtain security results? For 3 rounds, was it unavoidable to use a more complex attack (CCA instead of KPA or NCPA)? We will get an answer with Luby–Rackoff Theorems in Chap. 4. Can we design more general Feistel ciphers? Examples will be given in Chaps. 8, 9, and 10.

Problems

2.1 Is $\Psi^3(f, f, f)$ secure in CPA? Here $f_1 = f_2 = f_3$. Similarly, is $\Psi^7(f_1, f_2, f_3, f_4, f_3, f_2, f_1)$ secure in CPA?

2.2 Let $G = F_2 \circ F_1$ where F_1 is a Feistel cipher with a key k_1 of 40 bits, and F_2 is also a Feistel cipher with a key k_2 of 40 bits. Can G be a secure cipher? Here F_1 and F_2 have many rounds and we assume that the computation of F_1 and F_2 is not very slow.

2.3 Let F be a Feistel cipher with 10 rounds and a secret key of 256 bits that generates permutations on 40 bits. Can F be a secure cipher?

2.4 In a foreign country, the law asks for all cryptographic permutations to have a key with a maximal length of 50 bits. How can we build an efficient and secure permutation according to such a law?

References

Goldreich, O., S. Goldwasser, and S. Micali. 1986. How to construct random functions. *Journal of the ACM* 33: 792–807.

Hastad, J., R. Impagliazzo, L. Levin, and M. Luby. 1993. Construction of a pseudo-random generator from any one-way function. *SIAM Journal on Computing* 28: 12–24.

Luby, M., and C. Rackoff. 1988. How to construct pseudo-random permutations from pseudo-random functions. *SIAM Journal on Computing* 17: 373–386.

Patarin, J. 2025. Generic attacks on feistel schemes. In *Cryptology ePrint Archive: Report 2008/036*.

Chapter 3
The H-Coefficient Method

Abstract The "H-coefficient technique" was introduced in 1990 and 1991 in Patarin (1991, 2025). Since then, it has been used many times to prove various results on pseudo-random functions and pseudo-random permutations (Chen et al. 2014; Gilbert and Minier 2001; Pieprzyk 1991; Yun et al. 2011). Recently, it has also been used on key-alternating ciphers (Even-Mansour), cf. Chen and Steinberger (2014) for example. We will use this technique in Chap. 4 for the specific cases of Ψ^3, Ψ^4, and then in many proofs of security of this book. In this chapter, in Sect. 3.1, we will present the "H-coefficient technique", in a general way (not only for Ψ^k), with different formulations when we study different cryptographic attacks (known-plaintext attacks, chosen-plaintext attacks, etc.). In Sect. 3.4, we will present an example with the exact values of the H coefficient on Ψ^k with $q = 2$ plaintext/ciphertext pairs. Finally, in Sect. 3.5, we will present two simple and powerful composition theorems based on H-coefficient method in CCA.

3.1 Six "H-Coefficient" Theorems

We will formulate six theorems. These theorems are the basis of a general proof technique called the "H-coefficient technique", that allows to prove security results for function generators and permutation generators (and thus applies for random and pseudo-random Feistel ciphers). Five of these theorems were mentioned and proved in Patarin (1991, 2003, 2009). One (Theorem 3.3) was proved in Cogliati et al. (2014). The "Expectation Method" of Hoang and Tessaro (2016) can also be seen as another related theorem.

3.1.1 Notation Definition of H

- K will denote a set of values that we will sometimes call "keys". In this chapter, we will generally consider that K is a set of r-uples of functions $(f_1, \ldots, f_r)$ of

J. Patarin et al., *Guide to Feistel Ciphers*, Information Security and Cryptography,
https://doi.org/10.1007/978-3-031-99850-8_3

$\mathscr{F}_n$. (However generally only $|K|$ will be important, not the nature of the elements of K). We will say that G is a "function generator".
- G is an application of $K \to \mathscr{F}_N$. For example, with a balanced Feistel cipher, we have $N = 2n$.
- $\mathbb{E}$ denotes the expectation function.

Definition 3.1 Let q be an integer (q will be the number of queries). Let $a = (a_i)_{1 \le i \le q}$ be a sequence of pairwise distinct elements of $\{0, 1\}^N$. Let $b = (b_i)_{1 \le i \le q}$ be a sequence of elements of $\{0, 1\}^N$. We denote by $H(a, b)$ or simply by H if the context of the a_i and b_i is clear, the number of $(f_1, \ldots, f_r) \in K$ such that

$$\forall i,\ 1 \le i \le q,\ G(f_1, \ldots, f_r)(a_i) = b_i.$$

Therefore, H is the number of "keys" (i.e., elements of K) that send all the a_i inputs to the exact values b_i.

3.1.2 Theorem in KPA

Theorem 3.1 (H-coefficient technique, sufficient condition for security against KPA) *Let α and β be real numbers, $\alpha > 0$ and $\beta > 0$*

If:

(1) For random values a_i, b_i, $1 \le i \le q$, of $\{0, 1\}^N$ such that the a_i are pairwise distinct, with probability $\ge 1 - \beta$ we have

$$H \ge \frac{|K|}{2^{Nq}}(1 - \alpha).$$

Then

(2) For every KPA with q (random) known plaintexts we have $\mathscr{A}^{KPA} \le \alpha + \beta$, where $\mathscr{A}^{KPA}$ denotes the advantage to distinguish $G(f_1, \ldots, f_r)$ when $(f_1, \ldots, f_r) \in_R K$ from a function $f \in_R \mathscr{F}_N$

Proof Let D be a distinguisher (with no limitations in the number of computations) that takes the (a_i, b_i), $1 \le i \le q$, as input and outputs 0 or 1. Let P_1 be the probability that D outputs 1 when $\forall i,\ 1 \le i \le q,\ b_i = G(f_1, \ldots, f_r)(a_i)$ when $(f_1, \ldots, f_r) \in_R K$. Let P_1^* be the probability that D outputs 1 when $b_i = f(a_i)$ when $f \in_R \mathscr{F}_N$. We want to prove that $|\mathbb{E}(P_1 - P_1^*)| \le \alpha + \beta$. Let M be the set of all pairwise distinct a_i, $1 \le i \le q$, (so $|M| \simeq 2^{Nq}(1 - \frac{q(q-1)}{2 \cdot 2^N})$). When the a_i, $1 \le i \le q$, are fixed, let $W(a)$ be the set of all $b_1, \ldots, b_q$, such that the distinguisher D outputs 1 on the input (a_i, b_i), $1 \le i \le q$. When the a_i, $1 \le i \le q$, are fixed in M, then we have

$$P_1^* = \frac{|W(a)|}{2^{Nq}} \tag{3.1}$$

and

$$P_1 = \frac{1}{|K|} \sum_{b \in W(a)} [Numbers\ of\ (f_1, \dots, f_r) \in K/$$

$$\forall i,\ 1 \leq i \leq q,\ G(f_1, \dots, f_r)(a_i) = b_i]$$

so

$$P_1 = \frac{1}{|K|} \sum_{b \in W(a)} H(a, b). \tag{3.2}$$

Moreover, by hypothesis, we have that the number $\mathcal{N}$ of (a, b) such that

$$H(a, b) \geq \frac{|K|}{2^{Nq}}(1 - \alpha) \text{ satisfies} : \mathcal{N} \geq |M| \cdot 2^{Nq}(1 - \beta). \tag{3.3}$$

When the (a_i), $1 \leq i \leq q$, are fixed, let $\mathcal{N}(a)$ be the set of all b such that

$$H(a, b) \geq \frac{|K|}{2^{Nq}}(1 - \alpha).$$

From (3.3) we have

$$\sum_{a \in M} |\mathcal{N}(a)| \geq |M| \cdot 2^{Nq}(1 - \beta). \tag{3.4}$$

From (3.2), we have

$$P_1 \geq \frac{1}{|K|} \sum_{b \in W(a) \cap \mathcal{N}(a)} H(a, b)$$

so

$$P_1 \geq \frac{(1 - \alpha)}{2^{Nq}} |W(a) \cap \mathcal{N}(a)|$$

and

$$P_1 \geq \frac{(1 - \alpha)}{2^{Nq}} (|W(a)| - |\mathcal{N}'(a)|) \tag{3.5}$$

where $\mathcal{N}'(a)$ is the set of all b such that $b \notin \mathcal{N}(a)$. $|\mathcal{N}'(a)| = 2^{Nq} - |\mathcal{N}(a)|$, so

$$\sum_{a \in M} |\mathcal{N}'(a)| = |M| 2^{Nq} - \sum_{a \in M} |\mathcal{N}(a)|$$

then from (3.4), we have

$$\sum_{a \in M} |\mathcal{N}'(a)| \leq \beta \cdot |M| \cdot 2^{Nq}, \text{ and } \mathbb{E}(|\mathcal{N}'(a)|) \leq \beta \cdot 2^{Nq} \tag{3.6}$$

(where the expectation is computed when the (a_i), $1 \le i \le q$, are randomly chosen in M). From (3.5) and (3.1), we have

$$P_1 \ge (1-\alpha)(P_1^* - \frac{|\mathscr{N}'(a)|}{2^{Nq}})$$

$$P_1 \ge (1-\alpha)P_1^* - \frac{|\mathscr{N}'(a)|}{2^{Nq}}$$

so from (3.6) we get:

$$\mathbb{E}(P_1) \ge (1-\alpha)\mathbb{E}(P_1^*) - \beta$$

and

$$\mathbb{E}(P_1) \ge E(P_1^*) - \alpha - \beta. \tag{3.7}$$

Now if we consider the distinguisher D' that outputs 1 if and only if D outputs 0, we have $P_1' = 1 - P_1$ and $P_1^{'*} = 1 - P_1^*$ and from (3.7) we get: $\mathbb{E}(P_1') \ge \mathbb{E}(P_1^{'*}) - \alpha - \beta$ (because (3.7) is true for all distinguisher D, so it is true for D'). This gives

$$E(1-P_1) \ge E(1-P_1^*) - \alpha - \beta$$

and

$$E(P_1) - E(P_1^*) \le \alpha + \beta. \tag{3.8}$$

From (3.7) and (3.8) we get $|E(P_1 - P_1^*)| \le \alpha + \beta$ and this implies that $\mathscr{A}^{KPA} \le \alpha + \beta$. □

3.1.3 Theorems in NCPA

Theorem 3.2 (H-coefficient technique, sufficient condition for security against NCPA) *Let α and β be real numbers, $\alpha > 0$ and $\beta > 0$*

If:

(1) For all sequences $a = (a_i)$, $1 \le i \le q$, of pairwise distinct elements of $\{0,1\}^N$ there exists a subset $E(a)$ of $\{0,1\}^{qN}$ such that $|E(a)| \ge (1-\beta)\cdot 2^{Nq}$ and such that for all sequences $b = (b_i)$, $1 \le i \le q$, of $E(a)$ we have

$$H \ge \frac{|K|}{2^{Nq}}(1-\alpha).$$

Then

(2) For every NCPA with q chosen plaintexts, we have $\mathscr{A}^{NCPA} \le \alpha + \beta$ where $\mathscr{A}^{NCPA}$ denotes the advantage to distinguish $G(f_1, \ldots, f_r)$ when $(f_1, \ldots, f_r) \in_R K$ from a function $f \in_R \mathscr{F}_N$.

Proof The proof can be obtained as below for Theorem 3.4. □

Theorem 3.3 *We have*

$$\mathscr{A}^{NCPA}_{permutation} = \max_a \left(\frac{1}{2} \sum_b \frac{|H(a,b) - \tilde{H}|}{|K|} \right)$$

where

- $\mathscr{A}^{NCPA}_{permutation}$ *denotes the advantage to distinguish* $G(f_1, \ldots, f_r)$ *when* $(f_1, \ldots, f_r) \in_R K$ *from a random permutation* $f \in_R \mathscr{P}_n$.
- $a = (a_i)$, $1 \le i \le q$, *is a sequence of pairwise distinct elements of* $\{0, 1\}^N$.
- $b = (b_i)$, $1 \le i \le q$, *is a sequence of pairwise distinct elements of* $\{0, 1\}^N$.
- $\tilde{H}$ *is the mean value of* H, *i.e.,* $\tilde{H} = \frac{|K|}{2^N(2^N-1)\ldots(2^N-q+1)}$ *with* q *queries.*

Proof The proof of theorem 3.3 is given in Cogliati et al. (2014) p.13, 14. □

3.1.4 Theorem in CPA

Theorem 3.4 (H-coefficient technique, sufficient condition for security against CPA) *Let* α *and* β *be real numbers,* $\alpha > 0$ *and* $\beta > 0$. *Let* E *be a subset of* $\{0, 1\}^{Nq}$ *such that* $|E| \ge (1-\beta) \cdot 2^{Nq}$.

If:

(1) For all sequences a_i, $1 \le i \le q$, *of pairwise distinct elements of* $\{0, 1\}^N$ *and for all sequences* b_i, $1 \le i \le q$, *of* E *we have*

$$H \ge \frac{|K|}{2^{Nq}}(1-\alpha).$$

Then

(2) For every CPA with q *chosen plaintexts, we have* $\mathscr{A}^{CPA} \le \alpha + \beta$ *where* $\mathscr{A}^{CPA}$ *denotes the probability to distinguish* $G(f_1, \ldots, f_r)$ *when* $(f_1, \ldots, f_r) \in_R K$ *from a function* $f \in_R \mathscr{F}_N$.

Proof Let D be a (deterministic) distinguisher which tests a function f of $\mathscr{F}_N$. D can test any function f of $\mathscr{F}_N$. D can use f at most q times, that is to say that D can ask for the values of some $f(C_i)$, $C_i \in \{0, 1\}^N$, $1 \le i \le q$. The value C_1 is chosen by D, then D receive $f(C_1)$, then D can choose any $C_2 \ne C_1$, then D receive $f(C_2)$, etc. Here we have adaptive chosen plaintexts. (If $i \ne j$, C_i is always different from C_j). After a finite but unbounded amount of time, D outputs "1" or "0". The output is denoted by D(f).

We will denote by P_1^*, the probability that D outputs 1 when f is chosen randomly in $\mathscr{F}_N$. Therefore

$$P_1^* = \frac{\text{Number of functions } f \text{ such that } \mathsf{D}(f) = 1}{|\mathscr{F}_N|}$$

where $|\mathscr{F}_N| = 2^{N \cdot 2^N}$.

We will denote by P_1, the probability that D outputs 1 when $(f_1, \ldots, f_r) \in_R K$ and $f = G(f_1, \ldots, f_r)$. Therefore

$$P_1 = \frac{\text{Number of } (f_1, \ldots, f_r) \in K \text{ such that } \mathsf{D}(G(f_1, \ldots, f_r)) = 1}{|K|}.$$

We will prove

Lemma 3.1 *For all such distinguisher* D,

$$|P_1 - P_1^*| \leq \alpha + \beta.$$

Then Theorem 3.4 will be an immediate corollary of this lemma since $\mathscr{A}^{PRF}$ *is the best* $|P_1 - P_1^*|$ *that we can get with such* D *distinguishers.*

Proof (Evaluation of P_1^*) Let f be a fixed function, and let $C_1, \ldots, C_m$ be the successive values that the distinguisher D will ask for the values of f (when D tests the function f). We will note $\sigma_1 = f(C_1), \ldots, \sigma_q = f(C_q)$. D$(f)$ depends **only** of the outputs $\sigma_1, \ldots, \sigma_q$. This means that if f' is another function of $\mathscr{F}_N$ such that $\forall i, 1 \leq i \leq q, f'(C_i) = \sigma_i$, then D$(f)$ = D(f'). Since for $i < q$, the choice of C_{i+1} depends only of $\sigma_1, \ldots, \sigma_i$. Also the distinguisher D cannot distinguish f from f', because D will ask for f and f' exactly the same inputs, and will obtain exactly the same outputs. Conversely, let $\sigma_1, \ldots, \sigma_q$ be q elements of $\{0, 1\}^N$. Let C_1 be the first value that D chooses to know $f(C_1)$, C_2 the value that D chooses when D has obtained the answer σ_1 for $f(C_1)$, …, and C_q the qth value that D presents to f, when D has obtained $\sigma_1, \ldots, \sigma_{q-1}$ for $f(C_1), \ldots, f(C_{q-1})$. Let D$(\sigma_1, \ldots, \sigma_q)$ be the output of D (0 or 1). Then

$$P_1^* = \sum_{\substack{\sigma_1, \ldots, \sigma_q \\ \mathsf{D}(\sigma_1, \ldots \sigma_q) = 1}} \frac{\text{Number of functions } f \text{ such that } \forall i,\ 1 \leq i \leq q,\ f(C_i) = \sigma_i}{2^{N \cdot 2^N}}.$$

Since the C_i are all distinct the number of functions f such that $\forall i,\ 1 \leq i \leq q,\ f(C_i) = \sigma_i$ is exactly $|\mathscr{F}_N|/2^{Nq}$. Therefore

$$P_1^* = \frac{\text{Number of outputs } (\sigma_1, \ldots, \sigma_q) \text{ such that } \mathsf{D}(\sigma_1, \ldots \sigma_q) = 1}{2^{Nq}}$$

Let $\mathcal{N}$ be the number of outputs $\sigma_1, \ldots, \sigma_q$ such that $\mathsf{D}(\sigma_1, \ldots \sigma_q) = 1$. Then $P_1^* = \frac{\mathcal{N}}{2^{Nq}}$.

Evaluation of P_1

With the same notation $\sigma_1, \ldots, \sigma_q$, and $C_1, \ldots C_q$:

$$P_1 = \frac{1}{|K|} \sum_{\substack{\sigma_1,\ldots,\sigma_q \\ \mathsf{D}(\sigma_1,\ldots\sigma_q)=1}} [\text{Number of } (f_1, \ldots, f_r) \in K \text{ such that} \quad \forall i,\ 1 \le i \le q,\ G(f_1, \ldots, f_r)(C_i) = \sigma_i]. \tag{3.9}$$

Now (by definition of β) we have at most $\beta \cdot 2^{Nq}$ sequences $(\sigma_1, \ldots, \sigma_q)$ such that $(\sigma_1, \ldots, \sigma_q) \notin E$. Therefore, we have at least $(\mathcal{N} - \beta \cdot 2^{Nq})$ sequences $(\sigma_1, \ldots, \sigma_q)$ such that $\mathsf{D}(\sigma_1, \ldots \sigma_q) = 1$ and $(\sigma_1, \ldots, \sigma_q) \in E$. Therefore, from this, hypothesis (1) of Theorems 3.4 and 3.9, we have

$$P_1 \ge \frac{(\mathcal{N} - \beta \cdot 2^{Nq}) \cdot \frac{|K|}{2^{Nq}}(1-\alpha)}{|K|}.$$

Therefore

$$P_1 \ge \Big(\frac{\mathcal{N}}{2^{Nq}} - \beta\Big)(1-\alpha)$$

$$P_1 \ge (P_1^* - \beta)(1-\alpha).$$

Thus

$$P_1 \ge P_1^* - \alpha - \beta. \tag{3.10}$$

We now have to prove the inequality in the other side. For this, let P_0^* be the probability that $\mathsf{D}(f) = 0$ when $f \in_R \mathcal{F}_N$. Then $P_0^* = 1 - P_1^*$. Similarly, let P_0 be the probability that $\mathsf{D}(f) = 0$ when $(f_1, \ldots, f_r) \in_R K$ and $f = G(f_1, \ldots, f_r)$. Then $P_0 = 1 - P_1$. We will have $P_0 \ge P_0^* - \alpha - \beta$. Then we can proceed as in the proof of Theorem 3.1 and we obtain $1 - P_1 \ge 1 - P_1^* - \alpha - \beta$, i.e.,

$$P_1^* \ge P_1 - \alpha - \beta. \tag{3.11}$$

Finally, from (3.10) and (3.11), we have $|P_1 - P_1^*| \le \alpha + \beta$, as claimed. This ends the proof of Lemma 3.1 and Theorem 3.4. □

3.1.5 *Theorems in CCA*

Theorem 3.5 (H-coefficient technique, sufficient condition for security against CCA) *Let α be a real number, $\alpha > 0$.*

If:

(1) For all sequences of pairwise distinct elements a_i, $1 \le i \le q$, *and for all sequences of pairwise distinct elements* b_i, $1 \le i \le q$, *we have*

$$H \ge \frac{|K|}{2^{Nq}}(1-\alpha).$$

Then

(2) For every CCA with q *queries (i.e.,* q *chosen plaintexts or ciphertexts) we have* $\mathscr{A}^{CCA} \le \alpha + \frac{q(q-1)}{2 \cdot 2^N}$ *where* $\mathscr{A}^{CCA}$ *denotes the probability to distinguish* $G(f_1, \ldots, f_r)$ *when* $(f_1, \ldots, f_r) \in_R K$ *from a permutation* $f \in_R \mathscr{P}_N$.

We first introduce some definition and prove several lemmas.

Let D be a (deterministic) distinguisher which tests a permutation f of $\mathscr{P}_N$ with an adaptive chosen plaintext and chosen ciphertext way. This means that D will proceed like this:

1. D chooses first a value γ_1, and asks either for the value $f(\gamma_1)$, or for the value $f^{-1}(\gamma_1)$. The wanted value is given to D.
2. After some computations, D chooses a second value γ_2 and D asks either for the value $f(\gamma_2)$, or for the value $f^{-1}(\gamma_2)$. Without loss of generality for our security results, we can assume that these values were not already known by D.
3. Then, after some computations, D chooses a value γ_3 and D asks either for the value $f(\gamma_3)$, or for the value $f^{-1}(\gamma_3)$, etc., with q values $\gamma_1, \gamma_2, \ldots, \gamma_q$.
4. Then, after some computations, D stops and outputs 0 or 1.

We denote by D(f) this output.

By definition, we will denote

$$P_1^{**} = \frac{\text{Number of } f \in \mathscr{P}_N \text{ that } \mathsf{D}(f) = 1}{|\mathscr{P}_N|}$$

and (as before)

$$P_1 = \frac{\text{Number of } (f_1, \ldots, f_r) \in K \text{ such that } \mathsf{D}(G(f_1, \ldots, f_r)) = 1}{|K|}.$$

We denote $\gamma(f)$ the values $(\gamma_1, \gamma_2, \ldots, \gamma_q)$ chosen by D when D tests the permutation f.

We denote $\delta(f)$ the values $(\delta_1, \delta_2, \ldots, \delta_q)$ given to D from these queries $\gamma_1, \gamma_2, \ldots, \gamma_q$.
$\forall i,\ 1 \le i \le q$, if D asks for $f(\gamma_i)$ (i.e., "direct query"), we define $a_i = \gamma_i$ and $b_i = \delta_i$.
$\forall i,\ 1 \le i \le q$, if D asks for $f^{-1}(\gamma_i)$ (i.e., "inverse query"), we define $a_i = \delta_i$ and $b_i = \gamma_i$.
We denote by $a(f)$ the values $(a_1, \ldots, a_q)$ and by $b(f)$ the values $(b_1, \ldots, b_q)$.

Then: $\forall i, 1 \leq i \leq q,\ f(a_i) = b_i$. Notice that these values $a_1, \ldots, a_q, b_1, \ldots, b_q,$ $\gamma_1, \ldots, \gamma_q$ depend only of the answers $\delta_1, \ldots, \delta_q$ given to D and do not depend on all the function f. Similarly, the output (0 or 1) of D is fixed when $\delta_1, \ldots, \delta_m$ are fixed.

Definition 3.2 Let $d = (d_1, \ldots, d_q) \in \{0, 1\}^{qN}$. If there is at least one permutation $f \in \mathscr{P}_N$ such that $\delta(f) = d$, we will say that "d is compatible with a permutation".

Definition 3.3 Let $d = (d_1, \ldots, d_q)$ compatible with a permutation. We will denote by $a(d)$ (respectively $b(d)$ and $\gamma(d)$) the value $a(f)$ (respectively $b(f)$ and $\gamma(f)$) where f is a permutation such that $\delta(f) = d$.

Remark 3.1 $a(d)$ depends only on d and not on a specific f such that $\delta(f) = d$. For example $\gamma(d)$ is the value chosen by D when D receives the answers $\delta_1, \ldots, \delta_q$ to its queries.

Definition 3.4 Let $\delta = (\delta_1, \ldots, \delta_q)$ be compatible with a permutation. We will denote $\mathsf{D}(\delta)$ the output of $\phi(f)$ where f is a permutation such that $\mathsf{D}(f) = \delta$.

For example, $\gamma(\delta)$ depends only on δ and not on the specific f chosen (because $\mathsf{D}(f)$ is the output given by D when it receives the answers $\delta_1, \ldots, \delta_q$).

Remark 3.2 Let $\gamma(\delta) = (\gamma_1, \ldots, \gamma_q)$. Then γ_1 is a constant when D is fixed, γ_2 depends only on δ_1, γ_3 depends only on δ_1 and δ_1, etc. Finally, γ_q depends on $\delta_1, \delta_2, \ldots, \delta_{q-1}$.

Thus $\gamma(\delta)$ does not depend on δ_q but only on $\delta_1, \delta_2, \ldots, \delta_{q-1}$ but $\mathsf{D}(\delta)$ depends on $\delta_1, \delta_2, \ldots, \delta_{q-1}$ and δ_q.

Lemma 3.2 *Let $d = (d_1, \ldots, d_q)$ be an element of $\{0, 1\}^{qN}$ compatible with a permutation. Then the number of $f \in \mathscr{P}_N$ such that $\delta(f) = d$ is exactly $(2^N - q)!$*

Proof Let $a(d) = (a_1, \ldots, a_q)$ and $b(d) = (b_1, \ldots, b_q)$. Then f is a permutation such that $\delta(f) = d$ if and only if $\forall i,\ 1 \leq i \leq q,\ f(a_i) = b_i$. Moreover, by definition of D, since D has chosen the a_i pairwise distinct, the b_i are pairwise distinct. Thus, f is here exactly fixed on q points and we have $(2^N - q)!$ possibilities for f. □

Lemma 3.3 *For all distinguishers of permutations* D, *the number of values $(d_1, \ldots, d_q)$ compatible with a permutation is exactly $2^N(2^N - 1)\ldots(2^N - q + 1)$.*

Proof On $\mathscr{P}_N$ let $\mathscr{R}$ be this relation

$$f \mathrel{\mathscr{R}} g \Leftrightarrow \delta(f) = \delta(g)$$

$\mathscr{R}$ is an equivalence relation. Moreover, by Lemma 3.2, each equivalence class contains exactly $(2^N - q)!$ permutations. Therefore we have exactly $\frac{|\mathscr{P}_N|}{(2^n-q)!}$ equivalence classes, i.e., $2^N(2^N - 1)\ldots(2^N - q + 1)$ equivalence classes. Each d compatible with a permutation characterizes exactly one such class: the class of all $f \in \mathscr{P}_N$ such that $\delta(f) = d$. □

Definition 3.5 We will denote by Σ the set of all values δ compatible with a permutation such that $\mathsf{D}(\delta) = 1$.

Lemma 3.4 *We have*

$$P_1^{**} = \frac{|\Sigma|}{2^N(2^N - 1)\dots(2^N - q + 1)}.$$

Therefore

$$\frac{|\Sigma|}{2^{Nq}} \le P_1^{**} \le \frac{|\Sigma|}{2^{Nq}(1 - \frac{q(q-1)}{2^{N+1}})}.$$

Proof By definition

$$P_1^{**} = \frac{\text{Number of } f \in \mathscr{P}_N \; such that \; \mathsf{D}(f) = 1}{(2^N)!}.$$

Let A be the set of all permutations $f \in \mathscr{P}_N$ such that $\mathsf{D}(f) = 1$. $\forall d \in \Sigma$, let A_d be the set of all permutations $f \in A$ such that $\delta(f) = d$. We have $A = \cup_{d\in\Sigma} A_d$ and the sets A_d are pairwise disjoint. Therefore, $|A| = \sum_{d\in\Sigma} |A_d|$. From Lemma 3.2, we have $|A_d| = (2^N - q)!$. So $|A| = |\Sigma|(2^n - q)!$. Now since $P_1^{**} = \frac{|A|}{(2^N)!}$, we obtain

$$P_1^{**} = \frac{|\Sigma|}{2^N(2^N - 1)\dots(2^N - q + 1)}.$$

Moreover, if $\epsilon_1 > 0, \dots, \epsilon_q > 0$, we have $1 \ge (1 - \epsilon_1)(1 - \epsilon_2)\dots(1 - \epsilon_q) \ge 1 - \epsilon_1 - \epsilon_2 \dots - \epsilon_q$ (proof by induction on q). Thus

$$\frac{|\Sigma|}{2^{Nq}} \le P_1^{**} \le \frac{|\Sigma|}{2^{Nq}(1 - \frac{q(q-1)}{2^{N+1}})}$$

□

Lemma 3.5

$$P_1 = \sum_{\delta\in\Sigma} \frac{H(a(\delta), b(\delta))}{|K|}.$$

Proof By definition

$$P_1 = \frac{\text{Number of } k \in K \text{ such that } \mathsf{D}(G(k) = 1)}{|K|}.$$

Let C be the set of $k \in K$ such that $\mathsf{D}(G(k)) = 1$. $\forall\delta \in \Sigma$, let C_δ be the set of all $k \in K$ such that $\forall i,\ 1 \le i \le q,\ G(k)(a_i) = b_i$ where $(a_1, \dots, a_q)$ is $a(\delta)$ and

$(b_1, \ldots, b_q)$ is $b(\delta)$. We have $C = \cup_{\delta \in \Sigma} C_\delta$, and the sets C_δ are pairwise distinct. Now, by definition of the H coefficients, $|C_\delta|$ is exactly $H(a(\delta), b(\delta))$. Thus,

$$P_1 = \frac{|C|}{|K|} = \sum_{\delta \in \Sigma} \frac{H(a(\delta), b(\delta))}{|K|}.$$

□

We are now ready to prove Theorem 3.5.

Proof

$$P_1 = \sum_{\delta \in \Sigma} \frac{H(a(\delta), b(\delta))}{|K|}$$

So

$$P_1 \geq \frac{|\Sigma|}{2^{Nm}}(1 - \alpha) \geq \frac{|\Sigma|}{2^{Nm}} - \alpha.$$

We have seen that

$$P_1^{**} \leq \frac{|\Sigma|}{2^{Nq}(1 - \frac{q(q-1)}{2^{N+1}})}.$$

Therefore

$$P_1 \geq P_1^{**}(1 - \frac{q(q-1)}{2^{N+1}}) - \alpha. \tag{3.12}$$

This relation is valid for all D and thus for all D' such that D' outputs 0 when D outputs 1 and D' outputs 1 when D outputs 0. So we have

$$1 - P_1 \geq 1 - P_1^{**} - \alpha - \frac{q(q-1)}{2^{N+1}}. \tag{3.13}$$

With (3.12) and (3.13) we get:

$$|P_1 - P_1^{**}| \leq \alpha + \frac{q(q-1)}{2^{N+1}}.$$

□

Theorem 3.6 (H-coefficient technique, a more general sufficient condition for security against CCA) *Let α and β be real numbers, $\alpha > 0$ and $\beta > 0$*

If: There exists a subset E of $(\{0, 1\}^{qN})^2$ such that

(1a) For all $(a, b) \in E$, we have

$$H \geq \frac{|K|}{2^{Nq}}(1 - \alpha) \mathring{1}$$

with

$$\overset{\circ}{1} \overset{déf}{=} \frac{1}{(1-\frac{1}{2^N})(1-\frac{2}{2^N})\dots(1-\frac{q-1}{2^N})}.$$

(1b) *For all CCA acting on a random permutation f of $\mathscr{P}_N$, the probability that $(a, b) \in E$ is $\geq 1 - \beta$ where (a, b) denotes here the successive $b_i = f(a_i)$ or $a_i = f^{-1}(b_i)$, $1 \leq i \leq q$, that will appear.*
Then

(2) *For every CCA with q queries (i.e., q chosen plaintexts or ciphertexts) we have $\mathscr{A}^{PRP} \leq \alpha + \beta$ where $\mathscr{A}^{PRP}$ denotes the probability to distinguish $G(f_1, \dots, f_r)$ when $(f_1, \dots, f_r) \in_R K$ from a permutation $f \in_R \mathscr{P}_N$.*

Proof The proof is very similar to the proof of Theorem 3.5, we are just a bit more precise. We have

$$P_1 = \sum_{\delta \in \Sigma} \frac{H(a(\delta), b(\delta))}{|K|}$$

Let

$$\Sigma_E = \{\delta \in \Sigma \text{ such that } (a(\delta), b(\delta)) \in E\}.$$

We have

$$P_1 \geq \sum_{\delta \in \Sigma_E} \frac{H(a(\delta), b(\delta))}{|K|} \geq \overset{\circ}{1} \frac{|\Sigma_E|}{2^{Nq}}(1-\alpha). \tag{3.14}$$

Let Q be the set of all the values γ compatible with a permutation such that $(a(\gamma), b(\gamma)) \notin E$. We have $|\Sigma_E| \geq |\Sigma| - |Q|$. The probability that $(a(\gamma), b(\gamma)) \in E$ when γ is randomly and uniformly chosen in the set of values of $\{0, 1\}^{qN}$ compatible with a permutation is equal to $1 - \overset{\circ}{1} \frac{|Q|}{2^{Nq}}$. So we have

$$1 - \overset{\circ}{1} \frac{|Q|}{2^{Nq}} \geq \beta.$$

Therefore

$$|\Sigma_E| \geq |\Sigma| - \frac{\beta 2^{Nq}}{\overset{\circ}{1}}.$$

From (3.14), we obtain

$$P_1 \geq \overset{\circ}{1} \frac{|Q|}{2^{Nq}}(1-\alpha) - \beta(1-\alpha) \geq \overset{\circ}{1} \frac{|Q|}{2^{Nq}} - \alpha - \beta.$$

Now since $P_1^{**} = \frac{\overset{\circ}{1}|\Sigma|}{2^{Nq}}$, we obtain

$$P_1 \geq P_1^{**} - \alpha - \beta. \tag{3.15}$$

Finally, by considering D' such that

$$\mathsf{D}'(f) = 1 \Leftrightarrow \mathsf{D}(f) = 0$$

and

$$\mathsf{D}'(f) = 0 \Leftrightarrow \mathsf{D}(f) = 1$$

we obtain

$$1 - P_1 \geq 1 - P_1^{**} - \alpha - \beta \tag{3.16}$$

and from (3.15) and (3.16):

$$|P_1 - P_1^{**}| \leq \alpha + \beta$$

as claimed. □

3.1.6 Comments About These Theorems

There are a lot of variants and generalizations of these theorems. For example, in all these theorems 1, 2, 3, 4, 5, the results are still true if we change $H \geq \frac{|K|}{2^{Nq}}(1 - \alpha)$ by $H \leq \frac{|K|}{2^{Nq}}(1 + \alpha)$. However, for cryptographic uses, lower bounds for H are much more practical since often it will be easier to evaluate the exceptions where H is smaller than the mean value than the exceptions when H is greater than the mean value.

There are two noticeable points about these theorems.

1. They create a connection between the security of a generic cipher and the "H-coefficients", i.e., the transition probabilities, or the number of keys that sends some inputs on some outputs.
2. They show that we just here have to prove H greater than or equal to a certain value (near the mean value of H) for some sets of inputs/outputs, but we do not need to prove H less than a certain value.

Remark 3.3 Recently, in Hoang and Tessaro (2016) another H-coefficient theorem (or a generalization of the H-coefficient theorem in CCA) was given. Essentially, the idea is that, instead of introducing some sets E with good or bad properties, a computation of the mean value (computed with the probability on random permutations) is introduced. This is called the "Expectation Method" by the authors.

3.2 How to Distinguish Random Functions from Random Permutations

Theorem 3.7 (Pseudo-random function/pseudo-random permutation switching lemma) *When we want to distinguish a random function of $\mathscr{F}_N$ from a random permutation of $\mathscr{P}_N$, we have*

$$\mathscr{A}^{CPA} \leq \frac{q(q-1)}{2 \cdot 2^N}$$

Proof Let G be a perfect permutation generator, i.e., for all a_i pairwise distinct, $1 \leq i \leq q$, and for all b_i pairwise distinct, $1 \leq i \leq q$, we have $H = \frac{K}{2^{Nq}} \overset{\circ}{1}$. Let E be the set of all pairwise distinct values $(a_1, \ldots, a_q)$. Then

$$|E| = 2^N(2^N - 1)\ldots(2^n - q + 1) \geq 2^{Nq}(1 - \frac{q(q-1)}{2 \cdot 2^N})$$

From Theorem 3.4, we obtain (with $\alpha = 0$, $\beta = \frac{q(q-1)}{2}$ and since $\overset{\circ}{1} \geq 1$): $Adv^{CPA} \leq \frac{q(q-1)}{2 \cdot 2^N}$.

This means that when $q \ll \sqrt{2^N}$, the probability to distinguish a random permutation from a random function adaptive chosen plaintext attack is negligible. This result is sometimes called the "permutation switching lemma", and has been proved independently by many authors, with different proof techniques. When we study permutations versus functions, the bound $\sqrt{2^N}$ is called the "birthday bound" in analogy with the famous "birthday paradox", i.e., below this bound, the probability to have a collision (and therefore to detect a collision) is negligible. □

3.3 Triangular Evaluation on Generic Designs

Let denote by $\mathscr{A}(A \leftrightarrow B)$ the advantage to distinguish between two constructions A and B. In this chapter, we were looking for a bound $\mathscr{A} \leq \alpha$ between $G(k)$, with $k \in K$, and $f \in_R \mathscr{P}_N$, or $f \in_R \mathscr{F}_N$. This was an Advantage between a perfect (ideal) construction ($f \in_R \mathscr{P}_N$, or $f \in_R \mathscr{F}_N$) and a generator $G(k)$ that we call a "generic design". However, in cryptography, generally we do not use a space for the keys as large as K, but much smaller keys (of typical length 80, 128, or 256 bits for example in secret key cryptography). Therefore how analysis about $G(k)$ with a huge "key" space can be useful? In fact, in cryptography, generally k is not chosen as $k \in_R K$ with a huge set K, but k is generated from a generator A, from a smaller key. Very often, this generator is well known and it is already known or assumed that the Advantage to distinguish it from a perfect generator $k \in_R K$ is less than or equal to ϵ (Fig. 3.1).

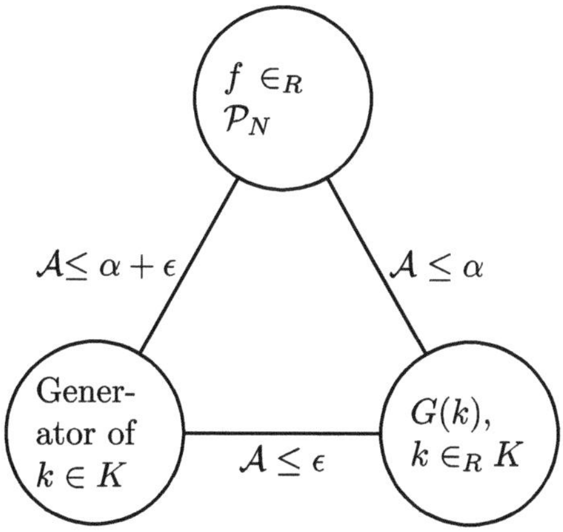

Fig. 3.1 Triangular evaluation

From any constructions A, B, C, we have $\mathscr{A}(A \leftrightarrow B) \leq \mathscr{A}(A \leftrightarrow C) + \mathscr{A}(C \leftrightarrow B)$. Here this gives: $\mathscr{A}(A \leftrightarrow f \in_R \mathscr{P}_N) \leq \epsilon + \alpha$. Therefore, when ϵ is known, only the evaluation of α (between generic designs) is needed to obtain the wanted bound on $\mathscr{A}(A \leftrightarrow f \in_R \mathscr{P}_N)$.

3.4 Example: Exact Values of H for Ψ^r and $q = 2$

Let D be a distinguisher which tests a function f with q queries. Here D take a function f as input and gives 0 or 1 as output.

We will denote by $\mathsf{D}(f)$ the output (1 or 0) on the function f.

We will denote by P_1^* the probability that $\mathsf{D}(f) = 1$ when f is randomly chosen in $\mathscr{F}_{2n}$.

Then

$$P_1^* = \frac{\text{Number of functions } f \text{ such that } \mathsf{D}(f) = 1}{(2^{2n})^{(2^{2n})}}.$$

And we will denote by P_1 the probability that $\mathsf{D}(f) = 1$ when $f_1, \ldots, f_r$ are r functions randomly chosen in $\mathscr{F}_n$, and $f = \Psi^r(f_1, \ldots, f_r)$. So :

$$P_1 = \frac{\text{Number of } (f_1, \ldots, f_r) \text{ such that } \mathsf{D}(\Psi^r(f_1, \ldots, f_r)) = 1}{|\mathscr{F}_n|^r}.$$

In Chap. 4, we will see that Luby and Rackoff proved (Luby and Rackoff 1988) that for Ψ^3 (or for Ψ^r, $r \geq 3$), for all CCA with q queries, we have

$$|P_1 - P_1^*| \leq \frac{q(q-1)}{2^n}$$

i.e.,

$$Adv^{CCA}(\Psi^3) \leq \frac{q(q-1)}{2^n}.$$

Here we will just consider the case $q = 2$. Of course, this case is less important than the cases where q is great. But when q is small ($q = 0, 1, 2$ or 3) it is possible to study the problem completely, and to obtain the exact values of the maximum of $|P_1 - P_1^*|$, and this is done for each r. Then this case will show with high precision how our generator of permutations "better and better pseudo-random" becomes when the number of rounds increases.

Remark 3.4 For $q = 0$ and $q = 1$ we have $|P_1 - P_1^*| = 0$ if $r \geq 2$. So the real problem begins when $q \geq 2$.

Let $L_1, R_1, L_2, R_2, S_1, T_1, S_2, T_2$ be elements of $\{0, 1\}^n$ such that $[L_1, R_1] \neq [L_2, R_2]$ and $[S_1, T_1] \neq [S_2, T_2]$.

The key property is that we are able to find the exact number H_r of r -tuples of functions $(f_1, \ldots, f_r)$ such that : $\forall i, 1 \leq i \leq q, \Psi^r(f_1, \ldots, f_r)[L_i, R_i] = [S_i, T_i]$ when q is very small ($q = 2$ here). Then, the values H_r will give us the maximum of $|P_1 - P_1^*|$.

For $q = 2$, an explicit calculus gives the values H_r. (The main idea is to make an induction on r). When r is even, these values are

Theorem 3.8 *Let* $a_0 = \dfrac{1}{1 - \frac{1}{2^{2n}}} \cdot \dfrac{|\mathscr{F}_n|^r}{2^{4n}}$, *where* $|\mathscr{F}_n| = 2^{n.2^n}$.

Then for Ψ^r, *when* r *is even and* $r \geq 2$, *we have :*

Case 1 : $R_1 \neq R_2$ ***and*** $S_1 \neq S_2$. *Then* $H_r = a_0 \left(1 - \dfrac{1}{2^{rn}}\right)$.

Case 2 : $R_1 \neq R_2$, $S_1 = S_2$ *and* $R_1 \oplus R_2 \neq T_1 \oplus T_2$.
or : $R_1 = R_2$, $S_1 \neq S_2$ *and* $S_1 \oplus S_2 \neq L_1 \oplus L_2$

$$\text{Then } H_r = a_0 \left(1 - \frac{1}{2^{(\frac{r}{2}-1)n}} - \frac{1}{2^{(\frac{r}{2})n}} + \frac{1}{2^{(r-1)n}}\right).$$

Case 3 : $R_1 \neq R_2$, $S_1 = S_2$ ***and*** $R_1 \oplus R_2 = T_1 \oplus T_2$
or : $R_1 = R_2$, $S_1 \neq S_2$ ***and*** $S_1 \oplus S_2 = L_1 \oplus L_2$

$$\text{Then } H_r = a_0 \left(1 + \frac{1}{2^{(\frac{r}{2}-2)n}} - \frac{1}{2^{(\frac{r}{2}-1)n}} - \frac{2}{2^{\frac{rn}{2}}} + \frac{1}{2^{(r-1)n}}\right).$$

Case 4 : $R_1 = R_2$ ***and*** $S_1 = S_2$

$$\text{Then } H_r = a_0 \left(1 - \frac{1}{2^{(r-2)n}}\right).$$

Proof This is a sketch of the proof. The main idea is to make an induction on r. Let

$$h_i = \frac{H_r \cdot 2^{4n}}{|\mathscr{F}_n|^r},$$

and

$$h'_i = \frac{H_{r+2} \cdot 2^{4n}}{|\mathcal{F}_n|^{r+2}},$$

where i, $1 \le i \le 4$ denotes the case number i. Then:

$$\begin{cases} h'_1 = \left(1 - \frac{1}{2^n} + \frac{1}{2^{2n}}\right) h_1 + \left(\frac{1}{2^n} - \frac{2}{2^n}\right) h_2 + \frac{h_3}{2^n} \\ h'_2 = \left(1 - \frac{1}{2^n}\right) h_1 + \frac{h_2}{2^n} \\ h'_3 = \left(1 - \frac{1}{2^n}\right) h_1 + \frac{h_3}{2^n} \\ h'_4 = \left(1 - \frac{2}{2^n}\right) h_2 + \frac{h_3}{2^n} + \frac{h_4}{2^n}. \end{cases}$$

□

When r is odd and $r \ge 3$, the computation of H_r is also possible. There are then five cases as stated in the Theorem 3.9.

Theorem 3.9 *Let* $a_0 = \dfrac{1}{1 - \frac{1}{2^{2n}}} \cdot \dfrac{|\mathcal{F}_n|^r}{2^{4n}}$, *where* $|\mathcal{F}_n| = 2^{n.2^n}$.
Then for Ψ^r, *when* r *is odd and* $r \ge 3$, *we have*

Case 1 : $R_1 \ne R_2$, $S_1 \ne S_2$ ***and*** $R_1 \oplus R_2 \ne S_1 \oplus S_2$

Then $H_r = a_0 \left(1 - \dfrac{1}{2^{(\frac{r}{2}-\frac{1}{2})n}} - \dfrac{1}{2^{(\frac{r+1}{2})n}} + \dfrac{1}{2^{rn}}\right)$.

Case 2 : $R_1 \ne R_2$ ***and*** $R_1 \oplus R_2 = S_1 \oplus S_2$ *(so* $S_1 \ne S_2$*).*

Then $H_r = a_0 \left(1 + \dfrac{1}{2^{(\frac{r}{2}-\frac{3}{2})n}} - \dfrac{1}{2^{(\frac{r}{2}-\frac{1}{2})n}} - \dfrac{2}{2^{(\frac{r}{2}+\frac{1}{2})n}} + \dfrac{1}{2^{rn}}\right)$.

Case 3 : $R_1 = R_2$ ***and*** $S_1 \ne S_2$ *(**or*** $R_1 \ne R_2$ ***and*** $S_1 = S_2$*)*

Then $H_r = a_0 \left(1 - \dfrac{1}{2^{(r-1)n}}\right)$.

Case 4 : $R_1 = R_2$ ***and*** $S_1 = S_2$ ***and*** $L_1 \oplus L_2 \ne T_1 \oplus T_2$

Then $H_r = a_0 \left(1 - \dfrac{1}{2^{(\frac{r}{2}-\frac{3}{2})n}} - \dfrac{1}{2^{(\frac{r}{2}-\frac{1}{2})n}} + \dfrac{1}{2^{(r-2)n}}\right)$.

Case 5 : $R_1 = R_2$ ***and*** $S_1 = S_2$ ***and*** $L_1 \oplus L_2 = T_1 \oplus T_2$

Then $H_r = a_0 \left(1 + \dfrac{1}{2^{(\frac{r}{2}-\frac{5}{2})n}} - \dfrac{1}{2^{(\frac{r}{2}-\frac{3}{2})n}} - \dfrac{2}{2^{(\frac{r}{2}-\frac{1}{2})n}} + \dfrac{1}{2^{(r-2)n}}\right)$.

Proof This value for r odd can easily be obtained from the values obtained for r even (see Patarin 1991 for details). □

Examples. As above, let $h_i = \frac{H_r \cdot 2^{4n}}{|\mathscr{F}_n|^r}$ where i denotes the case number i.

- One round: for Ψ, we have $h = 0$, or $h = 2^{2n}$, or $h = 2^{3n}$.
- Two rounds: for Ψ^2, we have $h = 0$, or $h = 1$, or $h = 2^n$.
- Three rounds: for Ψ^3, we have $h_1 = 1 - \frac{1}{2^n}$, $h_2 = 2 - \frac{1}{2^n}$, $h_3 = 1$, $h_4 = 0$, $h_5 = 2^n$.
- Four rounds: for Ψ^4, we have $h_1 = 1 + \frac{1}{2^{2n}}$, $h_2 = 1 - \frac{1}{2^n}$, $h_3 = 2 - \frac{1}{2^n}$, $h_4 = 1$.
- Five rounds: for Ψ^5, we have $h_1 = 1 - \frac{1}{2^{3n}}$, $h_2 = 1 + \frac{1}{2^n} - \frac{1}{2^{3n}}$, $h_3 = 1 + \frac{1}{2^{2n}}$, $h_4 = 1 - \frac{1}{2^n}$, $h_5 = 2 - \frac{1}{2^n}$.
- Six rounds: for Ψ^6, we have $h_1 = 1 + \frac{1}{2^{2n}} + \frac{1}{2^{4n}}$, $h_2 = 1 - \frac{1}{2^{3n}}$, $h_3 = 1 + \frac{1}{2^n} - \frac{1}{2^{3n}}$, $h_4 = 1 + \frac{1}{2^{2n}}$.
- When the number of rounds tends to infinity, all the values h_i tend to $\frac{1}{1-\frac{1}{2^{2n}}} = 1 + \frac{1}{2^{2n}} + \frac{1}{2^{4n}} + \frac{1}{2^{6n}} + \ldots$.

3.5 Two Simple Composition Theorems in CCA

3.5.1 A Simple Mathematical Property

Theorem 3.10 *Let $x_1, \ldots, x_n$ and $y_1, \ldots, y_n$ be real numbers and let α and β be real numbers, $\alpha \geq 0$, $\beta \geq 0$ such that*

- $\sum_{i=0}^{n} x_i = 0$.
- $\sum_{i=0}^{n} y_i = 0$.
- $\forall i,\ 1 \leq i \leq n,\ x_i \geq -\alpha$.
- $\forall i,\ 1 \leq i \leq n,\ y_i \geq -\beta$.

Then: $\sum_{i=1}^{n} x_i y_i \geq -n\alpha\beta$.

Proof First note that, since $x_i \geq -\alpha$ and $y_i \geq -\beta$ for $i = 1, \ldots, n$, one has

$$\sum_{i=1}^{n} (x_i + \alpha)(y_i + \beta) \geq 0.$$

Thus

$$\sum_{i=1}^{n} (x_i + \alpha)(y_i + \beta) = \sum_{i=1}^{n} x_i y_i + \alpha \sum_{i=1}^{n} y_i + \beta \sum_{i=1}^{n} x_i + n\alpha\beta \geq 0.$$

Using the fact that $\sum_{i=0}^{n} x_i = 0$ and $\sum_{i=0}^{n} y_i = 0$ yields $\sum_{i=1}^{n} x_i y_i \geq -n\alpha\beta$, as claimed. □

3.5.2 A Composition Theorem in CCA with H-Coefficients

Theorem 3.11 *Let G_1 and G_2 two permutation generators (with the same key space K) such that*

(1) *For all sequences of pairwise distinct elements a_i, $1 \le i \le q$, and for all sequences of pairwise distinct elements b_i, $1 \le i \le q$, we have $H_1 \ge \frac{|K|}{2^N(2^N-1)\dots(2^N-q+1)}(1-\alpha_1)$ and similarly $H_2 \ge \frac{|K|}{2^N(2^N-1)\dots(2^N-q+1)}(1-\alpha_2)$ where H_1 denotes the H coefficient for G_1 and H_2 the H coefficient for G_2. Then:*

(2) *If we compose 2 such generators G_1 and G_2 with random independent keys, for the composition generator $G' = G_2 \circ G_1$, we have for all sequences of pairwise distinct elements a_i, $1 \le i \le q$, and for all sequences of pairwise distinct elements b_i, $1 \le i \le q$, $H' \ge \frac{|K|^2}{2^N(2^N-1)\dots(2^N-q+1)}(1-\alpha_1\alpha_2)$, where H' denotes the H coefficient for G'.*

Proof Let $\tilde{H}_1$ (respectively $\tilde{H}_2$) denotes the mean value of H_1. (respectively H_2). We have

$$\tilde{H}_1 = \tilde{H}_2 = \frac{|K|}{2^N(2^N-1)\dots(2^N-q+1)}.$$

Let us denote by $\tilde{H}'$ the mean value of H for $G' = G_2 \circ G_1$. We have

$$\tilde{H}' = \frac{|K|^2}{2^N(2^N-1)\dots(2^N-q+1)}$$

Let $a = (a_1, \dots, a_q)$ be q pairwise distinct plaintexts, and $b = (b_1, \dots, b_q)$ be q ciphertexts of G'. Let J be the set of all $(t_1, \dots, t_q)$ pairwise distinct values of $\{0,1\}^N$. We have $|J| = 2^N(2^N-1)\dots(2^N-q+1)$. For $G' = G_2 \circ G_1$, we have

$$H(a,b) = \sum_{t \in J} H_1(a,t) H_2(t,b).$$

We also have $\sum_{t\in J} H_1(a,t) = |K|$ and $\sum_{t\in J} H_2(t,b) = |K|$ since each key sends a value a to a specific value t. We also have $|K| = \tilde{H}_1 \cdot |J| = \tilde{H}_2 \cdot |J|$. By hypothesis, we also have

$$\forall t \in J,\ H_1(a,t) \ge \tilde{H}_1(1-\alpha_1) \text{ and } H_2(a,t) \ge \tilde{H}_2(1-\alpha_2)$$

$\forall t \in J$, let $x_t = \frac{H_1(a,t)}{\tilde{H}_1} - 1$ and $y_t = \frac{H_2(a,t)}{\tilde{H}_2} - 1$. $\forall t \in J$, we have $x_t \ge -\alpha_1$, and $y_t \ge -\alpha_2$, $\sum_{t\in J} x_t = 0$ and $\sum_{t\in J} y_t = 0$. Therefore, from Theorem 3.10, we have $\sum_{t\in J} x_t y_t \ge -|J|\alpha_1\alpha_2$. For $G' = G_2 \circ G_1$, we have

$$\begin{aligned} H(a,b) &= \sum_{t\in J} H_1(a,t)\cdot H_2(t,b) \\ &= \sum_{t\in J} \left(\tilde{H}_1 x_t - \tilde{H}_1\right)\left(\tilde{H}_2 y_t - \tilde{H}_2\right) \\ &= \sum_{t\in J} \tilde{H}_1\tilde{H}_2 x_t y_t - \tilde{H}_1\tilde{H}_2 y_t - \tilde{H}_1\tilde{H}_2 x_t + \tilde{H}_1\tilde{H}_2 \\ &\geq -\tilde{H}_1\tilde{H}_2|J|\alpha_1\alpha_2 + |J|\tilde{H}_1\tilde{H}_2. \end{aligned}$$

Moreover $\tilde{H}' = \frac{|K|^2}{|J|} = |J|\tilde{H}_1\tilde{H}_2$. We have proved: $H(a,b) \geq \tilde{H}'(1-\alpha_1\alpha_2)$ as claimed. □

Corollary 3.1 *From Theorem 3.6 (H-coefficients in CCA) with $\beta = 0$, we see that we have $\mathscr{A}^{PRP} \leq \alpha_1\alpha_2$ where $\mathscr{A}^{PRP}$ denotes the advantage in CCA to distinguish $G_2 \circ G_1$ (when the keys are independently and randomly chosen) from a permutation $f \in_R \mathscr{P}_n$.*

By induction, we see

Theorem 3.12 *Let q and k be two integers. Let $\alpha_1, \ldots, \alpha_k$ be k real values. Let $G_1, \ldots, G_k$ be k permutation generators such that for all sequences of pairwise distinct elements a_i, and for all sequences of pairwise distinct elements b_i, $1 \leq i \leq q$, we have*

$$H \geq \frac{|K|}{2^N(2^N-1)\ldots(2^N-q+1)}(1-\alpha_j)$$

If we compose k such generators $G_1, \ldots, G_k$ with random and independent keys, for the composition generator $G' = G_k \circ \cdots \circ G_1$, we have for all sequences of pairwise distinct elements a_i, $1 \leq i \leq q$ and for all sequences of pairwise distinct elements b_i, $1 \leq i \leq q$, $H \geq \frac{|K|}{2^N(2^N-1)\ldots(2^N-q+1)}(1-\alpha_1\ldots\alpha_k)$. Therefore, from Theorem 3.6 with $\beta = 0$, we see that we have $\mathscr{A}^{PRP} \leq \alpha_1\ldots\alpha_k$

3.5.3 A Composition Theorem to Eliminate a "Hole"

J denotes, as above, the set of all q pairwise distinct values of $\{0,1\}^N$.

Theorem 3.13 *Let G_1 and G_2 be two permutation generators with the same key space K. Let H_1 (respectively H_2) denotes the H-coefficients for G_1 (respectively G_2).*

If:

(1) For all sequences of pairwise distinct elements a_i, $1 \leq i \leq q$, and for all sequences of pairwise distinct $b_i \in E_1$, $1 \leq i \leq q$, we have

$$H_1 \geq \frac{|K|}{2^N(2^N-1)\ldots(2^N-q+1)}(1-\alpha_1)$$

with $|E_1| \geq |J|(1-\epsilon_1)$.

(2) *Similarly, for all sequences of pairwise distinct elements a_i, $1 \le i \le q$, and for all sequences of pairwise distinct $b_i \in E_2$, $1 \le i \le q$, we have*

$$H_2 \ge \frac{|K|}{2^N(2^N-1)\dots(2^N-q+1)}(1-\alpha_2)$$

with $|E_2| \ge |J|(1-\epsilon_2)$.

Then: for the composition generator $G_2^{-1} \circ G_1$, for all sequences of pairwise distinct elements a_i, and for all sequences of pairwise distinct b_i, we have

$$H' \ge \frac{|K|^2}{2^N(2^N-1)\dots(2^N-q+1)}(1-\epsilon_1-\epsilon_2)(1-\alpha_1)(1-\alpha_2)$$

where H' denotes the H-coefficients for $G_2^{-1} \circ G_1$ (wa have no hole). Moreover, if $E_1 = E_2$, then

$$H' \ge \frac{|K|^2}{2^N(2^N-1)\dots(2^N-q+1)}(1-\epsilon_1)(1-\alpha_1)(1-\alpha_2)$$

Proof For $G' = G_2^{-1} \circ G_1$, we have $H'(a,b) = \sum_{t\in J} H_1(a,t)H_2(t,b)$, with $\sum_{t\in J} H_1(a,t) = |K|$ and $\sum_{t\in J} H_2(t,b) = |K|$. Let $\tilde{H}_1 = \frac{|K|}{|J|}$, $\tilde{H}_2 = \frac{|K|}{|J|}$, and $\tilde{H}' = \frac{|K|^2}{|J|} = \tilde{H}_1\tilde{H}_2|J|$. We have $|J| = 2^N(2^N-1)\dots(2^N-q+1)$. Let $P_1 = J \setminus E_1$ and $P_2 = J \setminus E_2$. Then

$$\begin{aligned}
H'(a,b) &\ge \sum_{t\in J\setminus P_1\setminus P_2} H_1(a,t)H_2(t,b)\\
&\ge \sum_{t\in J\setminus P_1\setminus P_2} \tilde{H}_1(1-\alpha_1)\tilde{H}_2(1-\alpha_2)\\
&\ge |J\setminus P_1\setminus P_2|\tilde{H}_1(1-\alpha_1)\tilde{H}_2(1-\alpha_2)\\
&\ge |J|(1-\epsilon_1-\epsilon_2)\tilde{H}_1(1-\alpha_1)\tilde{H}_2(1-\alpha_2)\\
&\ge \frac{|K|^2}{|J|}(1-\epsilon_1-\epsilon_2)(1-\alpha_1)(1-\alpha_2)
\end{aligned}$$

as claimed. □

3.5.4 Comments About the Composition Theorems

This very simple theorem of composition is not very well known because the classical theorems of composition (with more difficult proofs) usually do not consider hypothesis in terms of the values on the H coefficients. For example, the famous "two weak make one strong" theorem of Maurer (2002), Maurer et al. (2007) says that if F and G are NCPA secure, then the composition $G^{-1} \circ F$ is CCA secure. This result only holds in the information-theoretic setting, not in the computational setting (cf. Myers 2004; Pietrzak 2005). Another example is this theorem (Cogliati et al. 2014):

Theorem 3.14 *Let E, F and G be 3 block ciphers with the same message space M. Denote $\epsilon_E = \mathscr{A}_E^{NCPA}(q)$, $\epsilon_F = \mathscr{A}_F^{NCPA}(q)$, $\epsilon_{F^{-1}} = \mathscr{A}_{F^{-1}}^{NCPA}(q)$ and $\epsilon_{G^{-1}} = \mathscr{A}_{G^{-1}}^{NCPA}(q)$, where q is the number of queries. We have*

$$\mathscr{A}_{G \circ F \circ E}^{CCA}(q) \leq \epsilon_E \epsilon_F + \epsilon_E \epsilon_{G^{-1}} + \epsilon_{F^{-1}} \epsilon_{G^{-1}} + \min \{\epsilon_E \epsilon_F, \epsilon_E \epsilon_{G^{-1}}, \epsilon_{F^{-1}} \epsilon_{G^{-1}}\}.$$

Why do we have 3 rounds in this theorem and only 2 rounds in Theorem 3.11 for the product of the advantages? (Moreover Theorem 3.14 was also proved by using the H-coefficient technique Cogliati et al. 2014). This is because in Theorem 3.11, we used the additional property that there are no "holes" in the hypothesis that H is greater than or equal to the mean value $H(1 - \epsilon)$, i.e., that this property was true for any q pairwise distinct inputs and q pairwise distinct outputs.

References

Chen, S., R. Lampe, Y. Lee, Y. Seurin, and J. P. Steinberger. 2014. Minimizing the two-round even-mansour cipher. In *Advances in cryptology—CRYPTO 2014*, vol. 8616, ed. J. A. Garay and R. Gennaro. Lecture notes in computer science, 39–56. Berlin: Springer.

Chen, S., and J. P. Steinberger. 2014. Tight security bounds for key-aleternating ciphers. In *Advances in cryptology—EUROCRYPT '14*, vol. 8441, ed. P. Q. Nguyen and E. Oswald. Lecture notes in computer science, 327–350. Berlin: Springer.

Cogliati, B., J. Patarin, and Y. Seurin. 2014. Security amplification for the composition of block cipher: simpler proofs and new results. In *Selected areas in cryptography—SAC '14*, vol. 8781, ed. A. Y. Joux. lecture notes in computer science, 129–146. Berlin: Springer.

Gilbert, H., and M. Minier. 2001. New results on the pseudorandomness of some blockcipher constructions. In *Fast software encrytion–FSE '01*, vol. 2355, ed. M. Matsui. Lecture notes in computer science, 248–266. Berlin: Springer.

Hoang, V. T., and S. Tessaro. 2016. Key-alternating ciphers and key-length extension: exact bounds and multi-user security. In *Advances in cryptology—CRYPTO 2016*, vol. 9814, ed. M. Robshaw and M. Katz. Lecture notes in computer science, 3–32. Berlin: Springer.

Luby, M., and C. Rackoff. 1988. How to construct pseudorandom permutations from pseudorandom functions. *SIAM Journal on Computing* 17: 373–386.

Maurer, U. 2002. Indistinguishability of random systemes. In *Advances in cryptology—EUROCRYPT '02*, vol. 2332, ed. L. R. Knudsen. Lecture notes in computer science, 110–132. Berlin: Springer.

Maurer, U., K. Pietrzak, and R. Renner. 2007. Indistinguishability amplification. In *Advances in cryptology—CRYPTO '07*, vol. 4622, ed. A. Menezes. Lecture notes in computer science, 130–149. Berlin: Springer.

Myers, S. 2004. Black-box composition does not imply adaptive security. In *Advances in cryptology—EUROCRYPT '04*, vol. 3027, ed. J. L. Camenisch. Lecture Notes in Computer Science, 189–206. Berlin: Springer.

Patarin, J. 2025. Pseudorandom Permutations based on the DES Scheme. In *Eurocode'90, LNCS*, vol. 514, 204th ed., 193. Berlin: Springer.

Patarin, J. 1991. *Étude des Générateurs de Permutations Pseudo-aléatoires basés sur le schéma du D.E.S.*, Ph.D., November 1991.

Patarin, J. 1997. Improved security bounds for pseudorandom permutations. In *4th ACM conference on computer and communications security*, 142–150. Switzerland: Zurich.

Patarin, J. 1998. About feistel schemes with six (or more) rounds. In *Fast software encrytion–FSE '98*, vol. 1372, ed. S. Vaudenay. Lecture notes in computer science, 103–121. Berlin: Springer.

Patarin, J. 2003. Luby-Rackoff: 7 rounds are enough for $2^{n(1-\epsilon)}$ security. In *Advances in cryptology—CRYPTO 2003*, vol. 2729, ed. D. Boneh. Lecture notes in computer science, 513–529. Berlin: Springer.

Patarin, J. 2009. The "coefficient H" technique. In *Selected areas in cryptography–SAC '08*, vol. 5381, ed. R. Avanzi, L. Keliher, and F. Sica. Lecture notes in computer science, 328–345. Berlin: Springer.

Pieprzyk, J.:How to Construct Pseudorandom Permutations from Single Pseudorandom Functions. Advances in Cryptology - EUROCRYPT '90, Ivan Bjerre Damgård (ed), 140-150, Springer-Verlag, Lecture Notes in Computer Science, 473, (1991)

Pietrzak, K. 2005. Composition does not imply adaptive security. In *Advances in cryptology—CRYPTO '05*, vol. 3621, ed. V. Shoup. Lecture notes in computer science, 55–65. Berlin: Springer.

Piret, G., and J. J. Quisquater. 2005. Security of the MISTY structure in the luby-rackoff model: improved results. In *Selected areas in cryptography—SAC'04*, vol. 3357, ed. H. Handschuh and A. Hasan. Lecture notes in computer science, 100–115. Berlin: Springer.

Yun, A., J. H. Park, and J. Lee. 2011. Lai-Massey scheme and quasi-feistel networks des. *Codes Cryptography* 58: 45–72.

Chapter 4
Luby–Rackoff Theorems

Abstract In this chapter we will give complete proofs of the two results of Michael Luby and Charles Rackoff published in their important paper of 1988 (Luby and Rackoff 1988). **These two results are**

1. A *three (or more) rounds Feistel scheme* with random round functions (or with pseudo-random round functions) will give us an invertible pseudo-random permutation generator. This means that we have a cryptosystem which is secure against chosen plaintext attacks.
2. A *four (or more) rounds Feistel scheme* with random round functions (or with pseudo-random round functions) will give us an invertible super pseudo-random permutation generator. This means that we have a cryptosystem which is secure against chosen plaintext and chosen ciphertext attacks.

Different kinds of proofs exist for these famous theorems. We will give here Jacques Patarin's proof of 1990 (Patarin 2025) based on "H-properties", i.e., counting arguments on the keys. This proof technique is called the "H-coefficient technique" and was presented in Chap. 3.

4.1 Pseudo-Randomness Notions

Definition 4.1 A *pseudo-random bit generator* is a sequence of functions $f_n : \{0, 1\}^n \to \{0, 1\}^{\ell(n)}$, where $\ell(n)$ is a polynomial in n such that

1. There exists a polynomial algorithm such that given n and x as input, it computes $f_n(x)$. (This is the *uniform case*. If we only have a polynomial $P(n)$ such that for all n there exists an algorithm which given x as input, computes $f_n(x)$ in less than $P(x)$ instructions, we call this is the *non-uniform case*. In this case, the algorithm depends on n).

J. Patarin et al., *Guide to Feistel Ciphers*, Information Security and Cryptography,
https://doi.org/10.1007/978-3-031-99850-8_4

2. For all polynomial $Q(n)$, for all polynomial distinguisher D using an element A of $\{0, 1\}^{\ell(n)}$ as input, and which gives 0 or 1 as output:

$$\mathscr{A}_n(\mathsf{D}) \stackrel{\text{def}}{=} \Big|\Pr\left[a \leftarrow_\$ \{0,1\}^n \,:\, \mathsf{D}(f_n(a)) \to 1\right] - \Pr\left[A \leftarrow_\$ \{0,1\}^{\ell(n)} \,:\, \mathsf{D}(A) \to 1\right]\Big|.$$

To summarize, a pseudo-random bit generator is a polynomial technique to extend a real random number of n bits to a pseudo-random number of $\ell(n)$ bits.

If a is randomly chosen in $\{0, 1\}^n$, then the pseudo-random $f_n(a)$ cannot be distinguished (in a time polynomial in n) from a real random of $\{0, 1\}^{\ell(n)}$.

Definition 4.2 Let $P(n)$ be a polynomial in n. We will call an element of $\{0, 1\}^{P(n)}$ a key. A *pseudo-random function generator* is a sequence of functions $f_n : \{0, 1\}^{P(n)} \to \mathscr{F}_n$ such that

1. For all keys $k \in \{0, 1\}^{P(n)}$, for all $\alpha \in \{0, 1\}^n$, $f_n(k)(\alpha)$ can be computed in polynomial time in n. (To be precise, we should say here whether the algorithm used depends on n or not: this gives the non-uniform or the uniform case, as in Definition 4.1).
2. For all polynomials $Q(n)$, and for all polynomials in n distinguisher D, D *using* a function $f : \{0, 1\}^n \to \{0, 1\}^n$ as input (that is to say that D can generate some numbers α and obtain the numbers $f(\alpha)$ in a time considered as unity), and D giving 0 or 1 as output:

$$\mathscr{A}_n(\mathsf{D}) \stackrel{\text{def}}{=} \Big|\Pr\left[f \leftarrow_\$ \mathscr{F}_n \,:\, \mathsf{D}(f) \to 1\right] - \Pr\left[k \leftarrow_\$ \{0,1\}^{P(n)} \,:\, \mathsf{D}(f_n(k)) \to 1\right]\Big|.$$

To summarize, a pseudo-random function generator is a polynomial technique to obtain a pseudo random function from a random key k.

If k is randomly chosen in $\{0, 1\}^{P(n)}$, then the pseudo-random function $f_n(k)$ cannot be distinguished (in a polynomial time in n) from a real random function by an algorithm which can generate, when needed some numbers x and obtaining $f_n(x)$ (in a time considered as unity).

Definition 4.3
- A *pseudo-random permutation generator* is a pseudo-random function generator for which all the $f_n(k)$ obtained are permutations (that is $f_n(k)$ is 1–1 onto).
- A *pseudo-random invertible permutation generator* is a pseudo-random permutation generator for which all the $f_n^{-1}(k)$ are obtained (as the $f_n(k)$) from a pseudo-random permutation generator.
- A *super pseudo-random invertible permutation generator* is a pseudo-random invertible permutation generator where:
if k is randomly chosen in $\{0, 1\}^{P(n)}$, then the pseudo-random permutation $f_n(k)$

cannot be distinguished (in polynomial time in n) from a real random permutation by an algorithm which can generate some numbers x and obtaining $f_n(x)$ and $f_n^{-1}(x)$.

With these definitions, the attacks on Feistel ciphers given in Chap. 2 show that Ψ and Ψ^2 are not pseudo-random and that Ψ^3 is not super pseudo-random.

4.2 Results on Ψ^3

4.2.1 The "H-Property of Ψ^3

Theorem 4.1 ("H-property" for Ψ^3) *Let q be an integer. Let $L_i, R_i, S_i, T_i, 1 \le i \le q$, be elements of $\{0, 1\}^n$ such that if $i \ne j$ then $[L_i, R_i] \ne [L_j, R_j]$ (i.e., $L_i \ne L_j$ or $R_i \ne R_j$), and such that $i \ne j \Rightarrow S_i \ne S_j$.*

Then the number H of 3-tuples of functions (f_1, f_2, f_3) such that $\forall\, i, 1 \le i \le q$, $\Psi^3(f_1, f_2, f_3)[L_i, R_i] = [S_i, T_i]$, satisfies:

$$H \ge \frac{|\mathscr{F}_n|^3}{2^{2nq}} \left(1 - \frac{q(q-1)}{2^{n+1}}\right),$$

with $|\mathscr{F}_n| = (2^n)^{2^n} = 2^{n \cdot 2^n}$.

Proof We want to evaluate the number of functions (f_1, f_2, f_3) such that

$$\forall\, i, 1 \le i \le q, \begin{cases} f_2(L_i \oplus f_1(R_i)) = S_i \oplus R_i \\ f_3(S_i) = T_i \oplus L_i \oplus f_1(R_i). \end{cases}$$

First we will show that when i and j are fixed, $1 \le i < j \le q$, there is at most $|\mathscr{F}_n|/2^n$ functions f_1 such that

$$L_i \oplus f_1(R_i) = L_j \oplus f_1(R_j). \tag{4.1}$$

First case $R_i = R_j$. Then $L_i \ne L_j$ by hypothesis. Therefore no function f_1 realizes (4.1).

Second case $R_i \ne R_j$. Whatever $f_1(R_i)$ is, the Eq. (4.1) just fixes the value of f_1 in (R_j). Therefore exactly $|\mathscr{F}_n|/2^n$ functions f_1 satisfy (4.1).

Thus, there are at most $\frac{|\mathscr{F}_n|}{2^n}\frac{q(q-1)}{2}$ functions f_1 such that $\exists\,(i, j), 1 \le i < j \le q$, such that $L_i \oplus f_1(R_i) = L_j \oplus f_1(R_j)$. (Because there are $q(q-1)/2$ pairs (i, j) such that $1 \le i < j \le q$).

Let f_1 be a function which is *not* like this. For f_1, we also have at least $|\mathscr{F}_n| - |\mathscr{F}_n|\frac{q(q-1)}{2^{n+1}}$ possibilities. For such an f_1, there are exactly $|\mathscr{F}_n|/2^{nq}$ functions f_2 such that $\forall\, i, 1 \le i \le q$, $f_2(L_i \oplus f_1(R_i)) = S_i \oplus R_i$. (Because this is the number of functions *possible* when their values are fixed at q *distinct* points).

For such f_1 and f_2, there are exactly $|\mathcal{F}_n|/2^{nq}$ functions f_3 such that $\forall i, 1 \le i \le q$, $f_3(S_i) = T_i \oplus L_i \oplus f_1(R_i)$ for the same reason.

In conclusion, we always have at least

$$\left(|\mathcal{F}_n| - |\mathcal{F}_n|\frac{q(q-1)}{2^{n+1}}\right)\left(\frac{|\mathcal{F}_n|}{2^{nq}}\right)\left(\frac{|\mathcal{F}_n|}{2^{nq}}\right) \text{ solutions.}$$

This is what was claimed. □

4.2.2 "Main Lemma" of Luby and Rackoff for Ψ^3 from the "H-Property"

Theorem 4.2 ("Main Lemma for Ψ^3")

$$\mathscr{A}^{CPA}(\Psi^3) \le \frac{q(q-1)}{2^n}$$

where Adv^{CPA} denotes the advantage to distinguish Ψ^3 when (f_1, f_2, f_3) are randomly and independently chosen on $\mathcal{F}_n$ from a random function of $\{0,1\}^{2n}$ with an adaptive chosen plaintext attack.

Proof This comes immediately from Theorem 4.1 ("H-property" for Ψ^3) and Theorem 3.4 ("H-coefficient technique" in CPA). □

4.3 Results on Ψ^4

4.3.1 The "H-Property" for Ψ^4

Theorem 4.3 (H-Property for Ψ^4) *Let $[L_i, R_i]$, $1 \le i \le m$, be distinct inputs, and let $[S_i, T_i]$, $1 \le i \le q$ be distinct outputs. (Distinct means that $i \ne j \Rightarrow S_i \ne S_j$ or $T_i \ne T_j$)*

Then the number H of 4-tuples of functions (f_1, f_2, f_3, f_4) such that $\forall\, i$, $1 \le i \le q$, $\Psi^4(f_1, f_2, f_3, f_4)[L_i, R_i] = [S_i, T_i]$, satisfies:
$H \ge \frac{|\mathcal{F}_n|^4}{2^{2nm}}\left(1 - \frac{q(q-1)}{2^{n+1}}\right)^2$ *where* $|\mathcal{F}_n| = 2^{n \cdot 2^n}$.

Remark 4.1 For Ψ^4 we do not need the hypothesis $S_i \ne S_j$ if $i \ne j$. (And this is why Ψ^4 will be a *super* pseudo-random permutation generator and not Ψ^3).

Proof We require that, $\forall\, i$, $1 \le i \le q$,

$$f_3(R_i \oplus f_2(L_i \oplus f_1(R_i)) = S_i \oplus L_i \oplus f_1(R_i)$$
$$f_4(S_i) = T_i \oplus R_i \oplus f_2(L_i \oplus f_1(R_i)).$$

Let k be the number of independent equalities of the form $S_i = S_j$ (k may be zero).

First we will take a function f_1 such that $\forall i,\ j,\ i \neq j,\ L_i \oplus f_1(R_i) \neq L_j \oplus f_1(R_j)$. As we did for Ψ^3 we have at least $|\mathcal{F}_n|\left(1 - q(q-1)/2^{n+1}\right)$ possibilities for f_1.

We take f_2 such that

1. All the $R_i \oplus f_2(L_i \oplus f_1(R_i))$ are distinct $(\forall i, \forall j, 1 \leq i < j \leq q)$.
2. There are k equations of the form $T_i \oplus R_i \oplus f_2(L_i \oplus f_1(R_i)) = T_j \oplus R_j \oplus f_2(L_j \oplus f_1(R_j))$ which are consequences of the k equalities $S_i = S_j$.

We have $|\mathcal{F}_n|/2^{kn}$ functions f_2 which satisfy item 2.

First case. (i, j) is such that $S_i = S_j$. Then $T_i \neq T_j$. Therefore from item 2 we have: $R_i \oplus f_2(L_i \oplus f_1(R_i)) \neq R_j \oplus f_2(L_j \oplus f_1(R_j))$. Therefore item 1 is always true for these (i, j).

Second case. (i, j) is such that $S_i \neq S_j$. (We have at most $q(q-1)/2$ such (i, j), $1 \leq i < j \leq q$).
Let S be the system comprising the k equations of item 2 and the equation $R_i \oplus f_2(L_i \oplus f_1(R_i)) = R_j \oplus f_2(L_j \oplus f_1(R_j))$. S is a system of $k + 1$ independent equations. Therefore there are at most $|\mathcal{F}_n|/2^{(k+1)n}$ solutions for S.

Conclusion: There are at least $\frac{|\mathcal{F}_n|}{2^{kn}}\left(1 - \frac{q(q-1)}{2^{n+1}}\right)$ functions f_2 which satisfy items 1 and 2.

When f_1 and f_2 are chosen in this way, we take f_3 such that

$$\forall i,\ 1 \leq i \leq q,\ f_3(R_i \oplus f_2(L_i \oplus f_1(R_i))) = S_i \oplus L_i \oplus f_1(R_i).$$

(This is possible because the $R_i \oplus f_2(L_i \oplus f_1(R_i))$ are all pairwise distinct). There are exactly $|\mathcal{F}_n|/2^{nq}$ possibilities for f_3.

Then, we write the $q - k$ independent equalities of the form $f_4(S_i) = T_i \oplus R_i \oplus f_2(L_i \oplus f_1(R_i))$. (The S_i are all different in these equalities). These are $|\mathcal{F}_n|/2^{n(q-k)}$ such functions f_4.

Finally, we have at least:

$$\underbrace{|\mathcal{F}_n|\left(1 - \frac{m(m-1)}{2^{n+1}}\right)}_{\text{choice of } f_1} \quad \underbrace{\frac{|\mathcal{F}_n|}{2^{kn}}\left(1 - \frac{m(m-1)}{2^{n+1}}\right)}_{\text{choice of } f_2 \text{ when } f_1 \text{ is fixed}} \quad \underbrace{\frac{|\mathcal{F}_n|}{2^{nm}}}_{\text{choice of } f_3} \quad \underbrace{\frac{|\mathcal{F}_n|}{2^{n(m-k)}}}_{\text{choice of } f_4}$$

4-tuples (f_1, f_2, f_3, f_4) solutions.

Therefore there are at least $\frac{|\mathcal{F}_n|^4}{2^{2nq}}\left(1 - \frac{q(q-1)}{2^{n+1}}\right)^2$ solutions. □

4.3.2 "Main Lemma" of Luby and Rackoff for Ψ^4 from the "H-Property" of Ψ^4

Theorem 4.4

$$\mathscr{A}^{CCA}(\Psi^4) \le \frac{q^2}{2^n}$$

where Adv^{CCA} denotes the advantage to distinguish Ψ^4 when (f_1, f_2, f_3, f_4) are randomly and independently chosen on $\mathscr{F}_n$ from a random function of $\{0, 1\}^{2n}$ with an adaptive chosen plaintext and ciphertex attack.

Proof This comes immediately from Theorem 4.3 ("H-property for ψ^4") and Theorem 3.5 ("H-coefficient technique" in CCA).

4.4 Conclusion : Ψ^3 Is Pseudo-Random, Ψ^4 is Super Pseudo-Random

We can now precisely formulate the results of Luby and Rackoff, and the new results on the subject of pseudo-random permutation based on the Feistel scheme.

Convention. Let F be a pseudo-random function generator (the existence of a pseudo-random function generator can be obtained from any one-way function, however this is not the topic of this book) with key of length $P(n)$. We will denote by Ψ^r the sequence of permutations obtained as follows :

1. The keys K used are integers of length $r\ P(n)$.
2. f_1 is the function generated by F with the first $P(n)$ bits of K, f_2 is the function generated by F with the next $P(n)$ bits of K, ..., f_r is the function generated by F with the last $P(n)$ bits of K.
3. Ψ^r is then the permutation $\Psi^k(f_1, \ldots, f_r)$.

Theorem 4.5 (Luby and Rackoff) *If F is a pseudo-random function generator, then Ψ^3 is a pseudo-random invertible permutation generator.*

Proof Let us consider the following permutations:

1. $\Psi^3(f_1, f_2, f_3)$, where the f_i functions are generated by F.
2. $\Psi^3(f_1, f_2, f_3)$, where the f_i functions are truly random.
3. A permutation g which is randomly chosen in $\mathscr{P}_{2n}$.

From the "main lemma" for Ψ^3 we know that for items 2 and 3, we have: $|P_1 - P_1^*| \le \frac{q(q-1)}{2^n}$. Similarly, if F is pseudo-random; it is not possible to distinguish permutations from items 1 and 2 in polynomial time. Thus (triangle inequality) it is not possible to distinguish permutations from items 1 and 3 in polynomial time and therefore Ψ^3, is a pseudo-random invertible permutation generator. □

Theorem 4.6 *$\forall r \geq 3$, Ψ^r is a pseudo-random permutation generator.*

Proof This comes immediately from Theorem 4.5 and the fact that the pseudo-random security of $f \circ g$ where f and g are two permutations with independent and random keys is always greater than the pseudo-random security of g, since an attacker is always able to add to g a permutation f for which he has chosen randomly the keys. This shows that if he is able to distinguish $f \circ g$ from a random permutation, by choosing the inputs, then he is also able to distinguish g from a random permutation. □

Theorem 4.7 *If F is a pseudo-random function generator, then Ψ^4 is a super pseudo-random invertible permutation generator. (And all the Ψ^r, $r \geq 4$ are also super pseudo-random permutations generators).*

Proof Exactly the same as for Theorem 4.5, but using the "Main Lemma for Ψ^4" of Sect. 4.3.2. □

4.4.1 Comments About Luby–Rackoff Theorems

We can express Theorem 4.5 in the following way:

Theorem 4.8 *If $f_1, \ldots, f_r$ are random functions from $\{0,1\}^n \to \{0,1\}^n$, or if $f_1, \ldots, f_r$ are pseudo-random functions from $\{0,1\}^n \to \{0,1\}^n$ obtained by a pseudo-random function generator, then $\Psi^k(f_1, \ldots, f_r)$ $(r \geq 3)$ is a permutation which cannot be distinguished (in polynomial time) from a truly random permutation from $\{0,1\}^{2n} \to \{0,1\}^{2n}$ by an adaptive chosen plaintext attack (CPA) and similarly for $\Psi^k(f_1, \ldots, f_r)$ $(r \geq 4)$ by an adaptive chosen plaintext and ciphertext attack (CCA).*

In a way, Luby–Rackoff Theorems show how to solve these two problems:

1. Create pseudo-random permutations from pseudo-random functions.
2. How to double the lengths: from function of $\{0,1\}^n \to \{0,1\}^n$, we generate permutations $\{0,1\}^{2n} \to \{0,1\}^{2n}$.

What was particularly original in Luby–Rackoff result was to evaluate the best possible attacks from the number of queries: the number of computations is not bounded, but we know that it is **always** at least as large as the number of queries: this was a clever way to avoid difficult problems in the evaluation of the number of computations (for example $P \neq NP$). This was original in 1988, but thanks to Luby–Rackoff, this is not very classical since a huge number of papers have been now published with these ideas.

4.5 Other Results

It is possible to use only two, or even one function to obtain a pseudo-random permutation generator, with the following theorems.

Theorem 4.9 *If f and g are independently generated by a pseudo-random function generator, then $\Psi^3(f, f, g)$ is a pseudo-random permutation.*

Proof The demonstration is exactly the same as for Theorem 4.1. □

Theorem 4.10 *For all integers i, j, k, $\Psi^3(f^k, f^j, f^i)$ is not pseudo-random (where f^ℓ is the composition of f, ℓ times).*

Proof See Pieprzyk (1991) □

Theorem 4.11 *$\Psi^4(f, f, f, f^2)$ is pseudo-random.*

Proof See Patarin (1997).

In part IV, we will see how to obtain stronger security results, when the number of rounds is greater than or equal to 5.

Problems

4.1 For Ψ^3 (in CPA) and Ψ^4 in CCA, Luby–Rackoff Theorems show that $\mathcal{A} \leq \frac{q^2}{2^n}$, therefore we have security when $q \ll \sqrt{2^n}$. Prove that for any Ψ^r, we always have an attack when $q \geq \frac{r \cdot 2^n}{2}$. (Remember that we are limited in the number of queries but not in the number of computations). Therefore, when we study Ψ^r, $\sqrt{2^n}$ is called the "birthday bound" and 2^n is called the "information bound". Notice that in the "switching lemma" of Sect. 3.2, in another context, the birthday bound was $\sqrt{2^N} = \sqrt{2^{2n}}$ unlike $\sqrt{2^n}$ here.

4.2 Let G^r be the composition of a permutation generator r times, with r independent keys. When r increases, the security of G^r increases (or stay the same, but cannot decrease). This is because an adversary can always add more rounds of G with random key, at the beginning, or at the end. Show that this property is not always true anymore if the keys are not independent.

4.3 Similarly show that the security of G^r can decrease when r increases if G is a function generator, but not a permutation generator, with r independent keys.

4.4 In the proofs of this chapter we have used a "H-property" for Ψ^4 that says that H is always greater than or almost equal to its mean value $\tilde{H}$ when the number of queries $q \leq \sqrt{2^n}$. Do we have a similar result for H less than or almost equal instead of H greater than or equal? We will see in the exercises below that the answer is no. Even with a small number of queries it is possible to have $H \geq \tilde{H}$.

- Let G be a permutation generator, such that G involves ℓ different pseudorandom functions of $\mathscr{F}_n$ to compute a permutation of $\mathscr{P}_{2n}$. We denote by K the set of all ℓ-uples of functions $(f_1, ..., f_\ell)$ of $\mathscr{F}_n$ (i.e., $K = \mathscr{F}_n^\ell$). Thus G associates to each $k \in K$ a permutation $G(k)$ of $\mathscr{P}_{2n}$. K can be seen as the set of the keys of G, and $k \in K$ as a secret key.
- Let $\alpha_1, ..., \alpha_q$ be q distinct elements of $\{0, 1\}^{2n}$, and let $\beta_1, ..., \beta_q$ be also q distinct elements of $\{0, 1\}^{2n}$. We denote by $H(\alpha_1, ..., \alpha_q, \beta_1, ..., \beta_q)$ the number of keys k of K such that

$$\forall i,\ 1 \leq i \leq q,\ G(k)(\alpha_i) = \beta_i.$$

Definition 4.4 We say that G is a "homogeneous" permutation generator if there exist a function $\varepsilon(q, n) : \mathbf{N}^2 \to \mathbf{R}$ such that, for any integer q:

1. For all $\alpha_1, ..., \alpha_q$ being q distinct elements of $\{0, 1\}^{2n}$, and for all $\beta_1, ..., \beta_q$ being q distinct elements of $\{0, 1\}^{2n}$, we have:

$$\left| H(\alpha_1, ..., \alpha_q, \beta_1, ..., \beta_q) - \frac{|K|}{2^{2nq}} \right| \leq \varepsilon(q, n) \frac{|K|}{2^{2nq}}.$$

2. For any polynomial $P(n)$ and any $\alpha > 0$, an integer n_0 exists such that

$$\forall n \geq n_0,\ \forall q \leq P(n),\ \varepsilon(q, n) \leq \alpha.$$

Remark 4.2 This definition might look a bit complex but in fact this notion of "homogeneous" permutations is a very natural notion: roughly speaking, a permutation generator is homogeneous when for all set of q plaintext/ciphertext pairs, there are always **about the same number** of possible keys that send all the plaintexts on the ciphertexts.

Prove that Ψ^4 is not homogeneous.

4.5 Prove that Ψ^5 is not homogeneous.

Remark 4.3 In exercise, at the end of Chap. 17, we will prove more generally that $\forall r \geq 1$, Ψ^r is not homogeneous.

References

Luby, M., and C. Rackoff. 1988. How to construct pseudorandom permutations from pseudorandom functions. *SIAM Journal on Computing* 17: 373–386.

Patarin, J. 2025. Pseudorandom Permutations based on the DES Scheme. In *Eurocode'90, LNCS*, vol. 514, 193–204. Berlin: Springer.

Patarin, J. 1997. Improved security bounds for pseudorandom permutations. In *4th ACM conference on computer and communications security*, 142–150. Switzerland: Zurich.
Pieprzyk, J. 1991. How to construct pseudorandom permutations from single pseudorandom functions. In *Advances in cryptology—EUROCRYPT '90*, vol. 473, ed. I. B. Damgård. Lecture notes in computer science, 140–150. Berlin: Springer.

Part II
Generic Attacks

Chapter 5
Introduction to Cryptanalysis and Generic Attacks

Abstract In this chapter, we describe the method that we will use in most of attacks used in this book. We call it the variance method. For the attacks, we determine conditions that have to be satisfied by inputs and outputs. The conditions appear at random but with a cipher, well chosen differential characteristics can lead to the conditions on the outputs. This is due to the structure of the cipher. Then one has to compare the number of plaintext/ciphertext verifying the conditions. The variance method is a tool that allow to measure efficiently if the difference between the number obtained with a random permutation and the number obtained with a cipher is significant.

5.1 Generic Attacks—Distinguishers

The attacks described in this part of the book are distinguishers that allow to distinguish a permutation produced by a cipher from a permutation chosen randomly in the set of permutations. The round functions are chosen at random and are not known to the adversary. Moreover, we assume that the round functions are independent of each other. The attacks described in this book exploit differential characteristics that appear with a high probability, unlike impossible differential cryptanalysis (Biham et al. 1999) which exploits differentials that cannot happen, i.e., have probability of zero.

Differential cryptanalysis on encryption schemes was invented in the early 1990s by Biham and Shamir who applied it against DES (Biham and Shamir 1990, 1992). Differential cryptanalysis exploits the high probability of certain occurrences of plaintext differences and differences into the last round of the cipher. Consider two inputs M' and M'' with corresponding outputs Y' and Y''. The input difference is ΔM and the output difference is ΔY. In an ideally randomizing cipher, the probability that a particular output difference ΔY occurs given a particular difference input ΔM is almost $\frac{1}{2^n}$ where n is the number of bits of M. Differential cryptanalysis seeks for a scenario where a particular ΔY occurs given a particular input difference ΔM with a very high probability. This method uses pairs of plaintext/ciphertexts. They allowed

J. Patarin et al., *Guide to Feistel Ciphers*, Information Security and Cryptography,
https://doi.org/10.1007/978-3-031-99850-8_5

to produce attacks on classical Feistel schemes with random functions (Luby and Rackoff 1988; Lucks 1996; Patarin 2001) or random permutations (Knudsen 1998; Treger and Patarin 2009), unbalanced Feistel schemes with expanding (Jutla 1998; Schneier and Kelsey 1996; Patarin et al. 2007; Volte et al. 2010) or contracting functions (Naor and Reingold 1999; Patarin et al. 2006), Misty schemes (Gilbert and Minier 2001; Nachef et al. 2010), generalized Feistel schemes of types 1, 2, and 3 (Nachef et al. 2013). We will also present attacks that use differentials characteristics on t-uples of points (Jutla 1998; Schneier and Kelsey 1996; Patarin et al. 2007; Volte et al. 2010). This is explained in the next section.

Most of the time, we consider KPA and NCPA. For each kind of cipher, we will give the maximal number of rounds that we can attack in KPA and NCPA and we will describe the best attacks known so far up to the maximal number of rounds. However, for classical Feistel, we will also give CPA and CPCA.

5.2 2-Point Attacks and φ-Point Attacks and the Variance Method

5.2.1 *General Description of the Attacks*

We will investigated several attacks allowing to distinguish a permutation produced by a cipher from a random permutation. Depending on the number of rounds and the structure of the cipher, some attacks are more efficient than others. All attacks are using sets of plaintext/ciphertext pairs. We will study two kinds of attacks. First, we will use couples of plaintext/ciphertext pairs: these attacks are called 2-point attacks. Secondly, we will have φ tuples of plaintext/ciphertext pairs with φ even and $\varphi \geq 4$. This will be represented by a rectangle structure as shown in Fig. 5.1. Here we have $\varphi = 6$, the input is denoted by $I = [I_1, I_2, I_3]$ and the output by $S = [S_1, S_2, S_3]$. The corresponding attacks are called φ-point attacks. As shown in Fig. 5.1, we see that there exist several possibilities for equalities on the inputs and outputs. Depending to these type of equalities, we will define 4 types of φ-point attacks called R1, R2, R3, and R4. This will be defined precisely in Chap. 9.

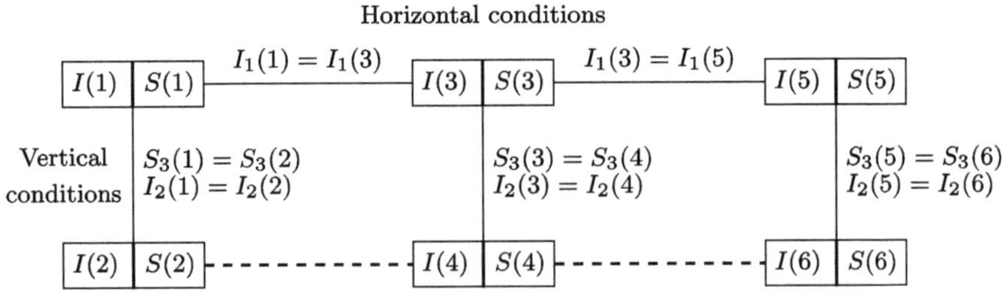

Fig. 5.1 Rectangle structure for $\varphi = 6$

In the attacks, we set conditions on the input and output variables. According to the structure of the cipher and the number of rounds, the condition will appear more frequently for a cipher than for a random permutation. Then we have to estimate how this allow to distinguish a random permutation from a permutation produced by a cipher.

5.2.2 *Distinguishing Attacks*

The attacks will make use of a distinguisher D whose aim is to decide whether the studied permutation is a random permutation or is produced by the cipher. Depending on the number of rounds and the structure of the cipher, for a given attack, conditions are set on the input and output variables. In KPA, D will make q queries to the oracle and receive the corresponding ciphertexts. In CPA, D will choose q inputs with given conditions and will receive again the corresponding ciphertexts. Then it will use a random variable $\mathcal{N}$ that counts the number of couples of input/output pairs in the case of a 2-point attack or the number of φ tuples of input/output pairs in the case of a φ-point attack. This number is denoted by $\mathcal{N}_{perm}$ for a random permutation and by $\mathcal{N}_{cipher}$ when the permutation is produced by the cipher. Then the distinguisher compares $\mathcal{N}_{cipher}$ with $\mathcal{N}_{perm}$. The attacks are successful, i.e., it is possible to distinguish a permutation generated by a Feistel-type cipher from a random permutation with q messages and we do not need a number of messages greater than the whole codebook, in three cases. The first case occurs when we can choose q such that $\mathcal{N}_{cipher}$ is significantly greater than $\mathcal{N}_{perm}$. For expanding Feistel schemes, the attacks used this fact (Patarin et al. 2007; Patarin 2007; Volte et al. 2010). The second case happens the q is such that $\mathcal{N}_{cipher}$ is significantly smaller than $\mathcal{N}_{perm}$ (this is the case for impossible attacks or improbable attacks). For the third case, the choice of q will imply that $\mathcal{N}_{cipher}$ and $\mathcal{N}_{perm}$ have the same order but the difference $|\mathbb{E}(\mathcal{N}_{cipher}) - \mathbb{E}(\mathcal{N}_{perm})|$ is much larger than both standard deviations $\sigma(\mathcal{N}_{perm})$ and $\sigma(\mathcal{N}_{cipher})$, where $\mathbb{E}$ denotes the expectation. In that case, the attacks work thanks to the Chebychev formula, which states that for any random variable X, and any $\epsilon > 0$, we have $\mathbb{P}(|X - \mathbb{E}(X)| \geq \epsilon\sigma(x)) \leq \frac{1}{\epsilon^2}$. Using this formula, it is then possible to construct a prediction interval for $\mathcal{N}_{cipher}$, in which future computations will fall, with a good probability. In order to compute these values, we need to take into account the fact that the structures obtained from the plaintext/ciphertext couples or tuples are not independent. However their mutual dependence is very small. To compute $\sigma(\mathcal{N}_{perm})$ and $\sigma(\mathcal{N}_{cipher})$, we will use this well-known formula as in Hoel et al. (1971) that we will call the "Covariance Formula":

$$V\Big(\sum X_i\Big) = \sum_i V(X_i) + \sum_{i<j}\big[E(X_i, X_j) - E(X_i)E(X_j)\big]$$

where the X_i are random variables. This last technique has been used for classical Feistel schemes in Patarin (2001), for contracting Feistel schemes in Patarin et al. (2006) and for generalized Feistel schemes of types 1, 2, and 3 (Nachef et al. 2013).

5.3 Attacks with More Than 2^{kn} Computations

5.3.1 Attacks on Generators

In practice, for example in designing block ciphers we need to consider generators of pseudo-random permutations. We now describe attacks against a generator of permutations (and not only against a single permutation randomly generated by a generator of permutations), i.e., we will be able to study several permutations generated by the generator. This allows more than 2^{kn} computations.

Let G be a Feistel type cipher, i.e., from a binary string K, G generates a r round Feistel type permutation. Let G' be a truly random permutation generator, i.e., from a string K, G' generates a truly random permutation G'_K of $\mathscr{P}_{kn}$. Let G'' be a truly random even permutation generator, i.e., from a string K, G'' generates a truly random permutation G''_K of $\mathscr{A}_{kn}$, with $\mathscr{A}_{kn}$ being the group of all the permutations of $\{0, 1\}^{kn} \to \{0, 1\}^{kn}$ with even signature. We are looking for attacks that distinguish G from G', and also for attacks that will distinguish G from G''.

Remark 5.1 Most Feistel type ciphers will have an even signature, see Sect.5.3.3.

Suppose that, for a 2-point attack, we want to test μ permutations and for one permutation we have the following results:

$$\begin{aligned}
&\mathbb{E}(\mathcal{N}_{perm}) \simeq \tfrac{q(q-1)}{2\cdot 2^{xn}}\\
&\mathbb{E}(\mathcal{N}_{cipher}) \simeq \tfrac{q(q-1)}{2\cdot 2^{xn}} + O(\tfrac{q(q-1)}{2\cdot 2^{yn}}),\ y > x\\
&\sigma(\mathcal{N}_{perm}) \simeq \sqrt{\mathbb{E}(\mathcal{N}_{perm})} \simeq \tfrac{q}{2^{xn/2}}\\
&\sigma(\mathcal{N}_{cipher}) \simeq \sqrt{\mathbb{E}(\mathcal{N}_{cipher})} \simeq \tfrac{q}{2^{xn/2}}.
\end{aligned}$$

With μ permutations, we obtain

$$\begin{aligned}
&\mathbb{E}(\mathcal{N}_{perm}) \simeq \mu\tfrac{q(q-1)}{2\cdot 2^{xn}}\\
&\mathbb{E}(\mathcal{N}_{cipher}) \simeq \mu\tfrac{q(q-1)}{2\cdot 2^{xn}} + O(\mu\tfrac{q(q-1)}{2\cdot 2^{yn}}),\ y > x\\
&\sigma(\mathcal{N}_{perm}) \simeq \sqrt{\mathbb{E}(\mathcal{N}_{perm})} \simeq \sqrt{\mu}\tfrac{q}{2^{xn/2}}\\
&\sigma(\mathcal{N}_{cipher}) \simeq \sqrt{\mathbb{E}(\mathcal{N}_{cipher})} \simeq \sqrt{\mu}\tfrac{q}{2^{xn/2}}.
\end{aligned}$$

In order to have the difference between the expectations greater than both standard deviations, we must have $\mu\frac{q^2}{2^{yn}} \geq \sqrt{\mu}\frac{q}{2^{xn/2}}$. This implies $\mu q^2 \geq 2^{(2y-x)n}$. If $q = 2^{kn}$, then we have $\mu \geq 2^{(2y-x-2k)n}$ and the complexity of the attack is $\lambda = \mu q = 2^{(2y-x-k)n}$.

5.3.2 *Brute Force Attacks*

A possible attack is an exhaustive search for the r round functions $f_1, \ldots, f_r$. Let us give an example with balanced Feistel construction. The round functions are from $\{0, 1\}^n$ to $\{0, 1\}^n$. This attack always exists, but since we have $2^{r \cdot n \cdot 2^n}$ possibilities for $f^1, \ldots, f^r$, this attack requires about $2^{r \cdot n \cdot 2^n}$ computations and about $\frac{r}{2} \cdot 2^n$ random queries but only for one permutation of the generator. This attacks means that an adversary with infinite computing power will be able to distinguish a Feistel type cipher from a random permutation (or from a truly random permutation with even signature) when $q \geq \frac{r}{2} \cdot 2^n$. This attack can be performed against other Feistel structures as well.

5.3.3 *Attack by the Signature*

Definition 5.1 Let f be a function from $\{0, 1\}^\beta$ to $\{0, 1\}^\alpha$. Then one round of an unbalanced Feistel network denoted by $\Phi(f)$ is a permutation from $\{0, 1\}^{\alpha+\beta}$ to $\{0, 1\}^{\alpha+\beta}$ defined by $\Phi(f)[A, B] = [B, A \oplus f(B)]$. Let $f_1, \ldots, f_r$ be r functions from $\{0, 1\}^\beta$ to $\{0, 1\}^\alpha$. Then $\Phi^r(f_1, \ldots, f_r) = \Phi(f_r) \circ \Phi(f_{r-1}) \circ \ldots \circ \Phi(f_1)$ is an r-unbalanced Feistel network.

The following theorem (Patarin 2007) is a generalization of Theorem 2.2.

Theorem 5.1 *Let* Φ^r *be an unbalanced Feistel permutation from* $\{0, 1\}^{\alpha+\beta}$ *to* $\{0, 1\}^{\alpha+\beta}$ *with round functions of* $\{0, 1\}^\beta \to \{0, 1\}^\alpha$. *Then if* $\alpha \geq 2$ *and* $\beta \geq 1$, Φ *has an even signature.*

Proof It is enough to prove Theorem 5.1 for one round since the composition of even permutations is an even permutation. Let $A \in \{0, 1\}^\alpha$ and $B \in \{0, 1\}^\beta$ and let f_1 be a function of $\{0, 1\}^\beta \to \{0, 1\}^\alpha$. Then for one round, we have

$$\Phi(f_1)[A, B] = [B, A \oplus f_1(B)].$$

So $\Phi(f_1) = \sigma \circ \Phi'(f_1)$, where $\Phi'(f_1)[A, B] = [A \oplus f_1(B), B]$ and σ is a rotation of α bits.

Signature of $\Phi'(f_1)$. We have $\Phi'(f_1) \circ \Phi'(f_1) = \text{Identity}$. In $\Phi'(f_1)$ we have only cycles with 1 or 2 elements, so the signature is the number of cycles with 2 elements modulo 2. The number of cycles with 2 elements is exactly $\frac{2^\alpha . k}{2}$, where k is the number of values B such that $f_1(B) \neq 0$. So when $\alpha \geq 2$, the signature of $\Phi'(f_1)$ is even.

Signature of σ. σ is a rotation of α bits. It is enough to show that a rotation of one bit have an even signature. Let us suppose that σ is a rotation of one bit. Let $N = \alpha + \beta$. Then signature $(\sigma) = (-1)^{k(\sigma)}$ where $k(\sigma)$ is the number of inversions of σ, i.e., the number of (x, y) with $x < y$ and $\sigma(x) > \sigma(y)$. Let us write $x = 0x'$ to say that the first bit of x is 0 and the last $N - 1$ bits of x is x'. Similarly for y. The only way to get an inversion is when $x = 0x'$ and $y = 1y'$ and $x' > y'$. So $k(\sigma)$ is equal to the number of (x', y') in $\{0, 1\}^{N-1}$ such that $x' > y'$, and this number is exactly $\frac{2^{N-1}(2^{N-1}-1)}{2}$. So when $N \geq 3$, the rotation of one bit have an even signature and by composition, the rotation of α bits has an even signature. □

Remark 5.2 If $\alpha = 1$, then the permutations obtained do not always have an even signature. For example, if the number of B such that $f_1(B) \neq 0$ is odd, if there is only one round and if we change only one bit per round, then the signature will be odd.

Attack by the Signature.

Let f be a permutation from kn bits to kn bits. Then using $O(2^{kn})$ computations on the 2^{kn} input/output values of f, we can compute the signature of f. To achieve this we just compute all the cycles c_i of f, $f = \prod_{i=1}^{\alpha} c_i$ and use the formula:

$$\text{signature}(f) = \prod_{i=1}^{\alpha} (-1)^{length(c_i)+1}.$$

The consequence is that it is possible to distinguish G a generator of unbalanced Feistel permutations from a generator of truly random permutations from kn bits to kn bits after $O(2^{kn})$ computations on $O(2^{kn})$ input/output values.

Remark 5.3 To compute the signature of a permutation g we need however to know all the input/outputs of g (or all of them minus one, since the last one can be found from the others if g is a permutation).

5.4 Further Readings

In this chapter, we have described the variance method used in the case of differential cryptanalysis. However, many kinds of attacks have been developed against block ciphers: truncated differential cryptanalysis (Knudsen 1995), impossible differential crypanalysis (Biham et al. 1999; Borst et al. 1999), multiple differential crypanalysis (Blondeau and Gérard 2011; Wang et al. 2012), boomerang attacks (Wagner 1999), differential-algebraic cryptanalysis, linear cryptanalysis, multiple linear cryptanalysis, and zero-correlation linear cryptanalysis. Differential and linear cryptanalysis provide most of the time the best attacks on block ciphers.

References

Albrecht, M., and C. Cid. 2009. Algebraic techniques in differential cryptanalysis. In *Fast software encrytion—FSE '09*, vol. 5665, ed. O. Dunkelman. Lecture notes in computer science, 193–208. Berlin: Springer.

Biham, E., and A. Shamir. 1990. Differential cryptanalysis of DES-like cryptosystems. In *Advances in cryptology—CRYPTO '90*, vol. 537, ed. A. Menezes and P. S. A. Vanstone. Lecture notes in computer science, 2–21. Berlin: Springer.

Biham, E., and A. Shamir. 1992. Differential cryptanalysis of the full 16-round DES. In *Advances in cryptology—CRYPTO '92*, vol. 740, ed. E. F. Brickell. Lecture notes in computer science, 487–496. Berlin: Springer.

Biham, E., A. Biryukov, and A. Shamir. 1999. Cryptanalysis of skipjack reduced to 31 rounds using impossible differentials. In *Advances in cryptology—EUROCRYPT '99*, vol. 1592. Lecture notes in computer science, 12–23.

Biryukov, A., C. De Cannière, and M. Quisquater. 2004. On multiple linear approximations. In *Advances in cryptology—CRYPTO '04*, vol. 3152, ed. M. Franklin. Lecture notes in computer science, 1–22. Berlin: Springer.

Blondeau, C., and B. Gérard. 2011. .: Multiple differential cryptanalysis: theory and practice. In *Fast software encrytion—FSE '11*, vol. 6733, ed. A. Joux. Lecture notes in computer science, 35–54. Berlin: Springer.

Borst, J., L. Knudsen, and V. Rijmen. 1999. Two attacks on reduced IDEA. In *Advances in cryptology—EUROCRYPT '99*, vol. 1233. Lecture notes in computer science, 1–13.

Gilbert, H., and M. Minier. 2001. New results on the pseudorandomness of some blockcipher constructions. In *ast software encrytion–FSE '01*, vol. 2355, ed. M. Matsui. Lecture notes in computer science, 248–266. Berlin: Springer.

Hoel, P. G., S. C. Port, and C. J. Stone. 1971. *Introduction to probability theory*. Houghton Mifflin Company.

Jutla, C. S. 1998. Generalized birthday attacks on unbalanced feistel networks. In *Advances in cryptology—CRYPTO 1998*, vol. 1462, ed. H. Krawczyk. Lecture notes in computer science, 186–199. bERLIN: Springer.

Kaliski, B., and M. Robshaw. 1994. Linear cryptanalysis using multiple approximations. In *Advances in cryptology—CRYPTO '94*, vol. 839, ed. Y. G. Desmedt. Lecture notes in computer science, 26–39. Berlin: Springer.

Knudsen, L. R. 1995. Truncated and higher order differentials. In *Fast software encrytion—FSE '94*, vol. 1008, ed. B. Preneel. Lecture notes in computer science, 196–211. Berlin: Springer.

Knudsen, L. R. 1998. *DEAL—A 128-bit block cipher*, vol. 1551. Norway: University of Bergen, Department of Informatics.

Luby, M., and C. Rackoff. 2014. Linear hulls with correlation zero and linear cryptanalysis of block ciphers designs. *Codes and Cryptography* 70 (3): 369–383.

Luby, M., and C. Rackoff. 1988. How to construct pseudorandom permutations from pseudorandom functions. *SIAM Journal on Computing* 17: 373–386.

Lucks, S. 1996. Faster luby-rackoff ciphers. In *Fast software encrytion—FSE '96*, vol. 1039, ed. D. Gollmann. Lecture notes in computer science, 189–203. Berlin: Springer.

Matsui, M. 1993. Linear crytpanalysis method for DES cipher. In *Advances in cryptology—EUROCRYPT '93*, vol. 765, ed. T. Hellseth. Lecture notes in computer science, 389–397.

Nachef, V., J. Patarin, and J. Treger. 2010. Generic attacks on misty schemes. In *Progress in cryptology–LATINCRYPT 2010*, vol. 6612, ed. M. Abdalla and P. S. L. M. Barreto. Lecture notes in computer science, 222–240. Berlin: Springer.

Nachef, V., E. Volte, and J. Patarin. 2013. Differential attacks on generalized feistel schemes. In *CANS 2013*, vol. 8257, ed. M. Abdalla, C. Nita-Rotaru, and R. Dahab. Lecture notes in computer science, 1–19. Berlin: Springer.

Naor, M., and O. Reingold. 1999. On the construction of pseudorandom permutations: Luby-Rackoff revisited. *Journal of Cryptology* 12: 29–66.

Patarin, J., V. Nachef, and C. Berbain. 2006. Generic attacks on unbalanced feistel schemes with contracting functions. In *Advances in cryptology—ASIACRYPT 2006*, vol. 4284, ed. X. Lai and K. Chen. Lecture notes in computer science, 396–411. Berlin: Springer.

Patarin, J., V. Nachef, and C. Berbain. 2007. Generic Attacks on unbalanced feistel schemes with expanding functions. In *Advances in cryptology—ASIACRYPT 2007*, vol. 4833, ed. K. Kurosawa. Lecture notes in computer science, 325–341. Berlin: Springer.

Patarin, J. 2001. Generic attacks on feistel schemes. In *Advances in cryptology—ASIACRYPT 2001*, vol. 2248, ed. C. Boyd. Lecture notes in computer science, 222–238. BERLIN: Springer.

Patarin, J. 2007. Generic attacks on unbalanced feistel schemes with expanding functions. In *Cryptology ePrint Archive: Report 2007/449.*

Patarin, J. 2008. Generic attacks on feistel schemes. In *Cryptology ePrint Archive: Report 2008/036.*

Schneier, B., and J. Kelsey. 1996. Unbalanced feistel networks and block cipher design. In *Fast software encrytion—FSE '96*, vol. 1039, ed. D. Gollmann. Lecture notes in computer science, 121–144. Berlin: Springer.

Treger, J., and J. Patarin. 2009. Generic attacks on feistel networks with internal permutations. In *Progress in cryptology—AFRICACRYPT 2009*, vol. 5080, ed. B. Preneel. Lecture notes in computer science, 41–59. Berlin: Springer.

Volte, E., V. Nachef, and J. Patarin. 2010. Improved generic attacks on unbalanced feistel schemes with expanding functions. In *Advances in cryptology–ASIACRYPT 2010*, vol. 6477, ed. M. Abe. Lecture notes in computer science, 94–111. Berlin: Springer.

Wagner, D. 1999. The boomerang attack. In *Fast software encryption—FSE '99*, vol. 1636, ed. L. Knudsen. Lecture notes in computer science, 156–170. Berlin: Springer.

Wang, M., Y. Sun, E. Mouha, and B. Preneel. 2011. Algebraic techniques in differential cryptanalysis revisited. In *ACISP '11*, vol. 6812, ed. P. H. Parampalli. Lecture notes in computer science, 120–141. Berlin: Springer.

Wang, M. Q., Y. Sun, E. Tischhauser, and B. Preneel. 2012. A model for structure attacks, with aplications to PRESENT and SERPENT. In *Fast software encrytion—FSE '12*, vol. 7549, ed. A. Canteaut. Lecture notes in computer science, 49–68. Berlin: Springer.

Chapter 6
Generic Attacks on Classical Feistel Ciphers

Abstract In this chapter, we will give a complete description of best known attacks on classical Feistel ciphers.

6.1 Introduction

Many secret key algorithms used in cryptography are Feistel ciphers, for example DES, 3DES, many AES candidates, etc. In order to be as fast as possible, it is interesting to have not too many rounds. However, for security reasons it is important to have a sufficient number of rounds. Generally, when a Feistel cipher is designed for cryptography, the designer either uses many (say ≥ 16 as in DES) very simple rounds, or uses very few (for example 8 as in DFC) more complex rounds. A natural question is: what is the minimum number of rounds required in a Feistel cipher to avoid all the "generic attacks", i.e., all the attacks effective against most of the schemes, and with a complexity negligible compared with a search on all the possible inputs of the permutation.

Let assume that we have a permutation from $2n$ bits to $2n$ bits. Then a generic attack will be an attack with a complexity negligible compared to $O(2^{2n})$, since there are 2^{2n} possible inputs on $2n$ bits.

In Chap. 2, we have seen for a Feistel cipher with only one round there is a generic attack with only 1 query of the permutation and $O(1)$ computations: just check if the first half (n bits) of the output are equal to the second half of the input. We also described a generic NCPA on two rounds with with a complexity of $O(1)$ chosen inputs. We will see in Sect. 6.2 that there exists a generic KPA with $(2^{\frac{n}{2}})$ random inputs.

Also in Luby and Rackoff (1988), M. Luby and C. Rackoff have shown their famous result: for more than 3 rounds all generic attacks on Feistel ciphers require at least $O(2^{\frac{n}{2}})$ inputs, even for chosen inputs.

Moreover for 4 rounds all the generic attacks on Feistel ciphers require at least $O(2^{\frac{n}{2}})$ inputs, even for a stronger attack that combines chosen inputs and chosen outputs (see Luby and Rackoff 1988 and a proof in Patarin 2025, that shows that

J. Patarin et al., *Guide to Feistel Ciphers*, Information Security and Cryptography,
https://doi.org/10.1007/978-3-031-99850-8_6

the Luby–Rackoff construction with 4 rounds is super pseudo-random, a.k.a strong pseudo-random). This was explained in Chap. 4.

However it was discovered in Patarin (1992) (and independently in Aiollo and Venkatesan 1996) that these lower bounds on 3 and 4 rounds are tight, i.e., there exist a generic attack on all Feistel ciphers with 3 or 4 rounds with $O(2^{\frac{n}{2}})$ chosen inputs with $O(2^{\frac{n}{2}})$ computations.

For 5 rounds or more the question is difficult. In Patarin (1992) it was proved that for 5 rounds (or more) the number of queries must be at least $O(2^{\frac{2n}{3}})$ (even with unbounded computation complexity), and in Patarin (1998) it was shown that for 6 rounds (or more) the number of queries must be at least $O(2^{\frac{3n}{4}})$ (even with unbounded computations). Finally in Patarin (2004, 2005), it was proved that for 5 rounds (or more) the number of queries must be at least $O(2^n)$.

It can be noticed (see Patarin 1992) that if we have access to unbounded computations, then we can make an exhaustive search on all the possible round functions of the Feistel cipher, and this will give an attack with only $O(2^n)$ queries (see Patarin 1992) so the bound $O(2^n)$ of the number of queries is optimal. However here we have a gigantic complexity $\geq O(2^{n2^n})$. This "exhaustive search" attack always exists, but since the complexity is far much larger than the exhaustive search on plaintexts in $O(2^{2n})$, it was still an open problem to know if generic attacks, with a complexity $\ll O(2^{2n})$, exist on 5 rounds (or more) of Feistel cipher.

Finally, in Patarin (2004), it was shown that there exist generic attacks on 5 rounds of the Feistel cipher, with a complexity $\ll O(2^{2n})$. Indeed, there are two attacks on 5 round Feistel ciphers:

1. An attack with $O(2^{\frac{3n}{2}})$ computations on $O(2^{\frac{3n}{2}})$ **random** input/output pairs.
2. An attack with $O(2^n)$ computations on $O(2^n)$ **chosen** inputs.

For 6 rounds (or more) there exist attacks with a complexity much smaller than $O(2^{n2^n})$ of exhaustive search, but still $\geq O(2^{2n})$. So these attacks on 6 rounds and more are generally not interesting against a single permutation. However they may be useful when several permutations are used, i.e., they will be able to distinguish some permutation generators. These attacks show for example that when several small permutations must be generated (for example in the Graph Isomorphism scheme, or as in the Permuted Kernel scheme) then we must not use a 6-round Feistel construction.

6.2 Generic Attacks on 1, 2, 3 and 4 Rounds

We describe the best known generic attacks.

Let f be a permutation of $\mathcal{P}_{2n}$. For a value $[L_i, R_i] \in \{0, 1\}^{2n}$ we will denote by $[S_i, T_i] = f[L_i, R_i]$.

6.2.1 1 Round

This attack was described in Chap. 2. With one message, the distinguisher tests if $S_1 = R_1$.

6.2.2 2 Rounds

6.2.2.1 NCPA with $q = 2$ Messages

This attack was described in Chap. 2. With two messages $[L_1, R]$ and $[L_2, R]$ such that $L_1 \neq L_2$, the distinguisher tests if $S_1 \oplus S_2 = L_1 \oplus L_2$

6.2.2.2 KPA with $m \simeq 2^{n/2}$

It is possible to transform the previous NCPA into a KPA as we will see now. Consider the following distinguisher D that makes $q \simeq 2^{n/2}$ queries to the oracle $\mathscr{O}$ and receives q random plaintext $[L_i, R_i]$, $1 \leq i \leq q$ together with $[S_i, T_i] = \mathscr{O}((L_i, R_i])$. By the birthday paradox, with a good probability, let us say greater than $\frac{1}{2}$, D obtains two indices i and j such that $i \neq j$ and $R_i = R_j$. Then for this messages i and j, D outputs 1 if $S_i \oplus S_j = L_i \oplus L_j$ and 0 otherwise. Then, when $\mathscr{O}$ is a random permutation, D outputs 1 with probability $\frac{2^n}{2^{2n}-1}$ and when $\mathscr{O} = \Psi^2(f_1, f_2)$ it outputs 1 with probability 1. This we obtain

$$\mathscr{A}_{\Psi^2}^{\text{KPA}}(\mathsf{D}) \geq \frac{1}{2}\left(1 - \frac{2^n}{2^{2n}-1}\right).$$

This KPA requires $O(2^{\frac{n}{2}})$ random queries and $O(2^{\frac{n}{2}})$ computations.

6.2.3 3 Rounds

6.2.3.1 KPA with $q \simeq 2^{n/2}$

We use the following KPA distinguisher D.

1. D chooses q random distinct $[L_i, R_i]$, $1 \leq i \leq q$.
2. D asks for the values $[S_i, T_i] = \mathscr{O}[L_i, R_i]$, $1 \leq i \leq m$.
3. D counts the number $\mathscr{N}$ of equalities of the form $R_i \oplus S_i = R_j \oplus S_j, i < j$.
4. Let $\mathscr{N}_{perm}$ be the expected value of $\mathscr{N}$ when $\mathscr{O}$ is a random permutation, and $\mathscr{N}_{scheme}$ be the expected value of $\mathscr{N}$ when $\mathscr{O} = \Psi^3(f_1, f_2, f_3)$, with randomly chosen f_1, f_2, f_3.

Then $\mathcal{N}_{scheme} \simeq 2\mathcal{N}_{perm}$, because when $\mathcal{O} = \Psi^3(f_1, f_2, f_3)$, $R_i \oplus S_i = f_2(L_i \oplus f_1(R_i))$ so $f_2(L_i \oplus f_1(R_i)) = f_2(L_j \oplus f_1(R_j)), i < j$, if $L_i \oplus f_1(R_i) \neq L_j \oplus f_1(R_j)$ and $f_2(L_i \oplus f_1(R_i)) = f_2(L_j \oplus f_1(R_j))$ <u>or</u> if $L_i \oplus f_1(R_i) = L_j \oplus f_1(R_j)$.

So by counting $\mathcal{N}$ we will obtain a way to distinguish 3 round Feistel permutations from random permutations. This generic KPA requires $0(2^{\frac{n}{2}})$ random queries and $O(2^{\frac{n}{2}})$ computations (just store the values $R_i \oplus S_i$ and count the collisions).

Remark 6.1 Here $\mathcal{N}_{scheme} \simeq 2 \cdot \mathcal{N}_{perm}$ when f_1, f_2, f_3 are randomly chosen. Therefore this attack is effective on most of 3 round Feistel ciphers but not necessarily on all 3 round Feistel ciphers (however very special f_1, f_2, f_3 may create other attacks, as we will see for example with the Knudsen attack in Sect. 6.3).

6.2.3.2 CCA with $q = 3$

This attacks was described in Chap. 2

6.2.4 4 Rounds

6.2.4.1 NCPA with $m \simeq 2^{n/2}$

Here the attacker chooses $R_i = 0$ (or R_i constant), and counts the number $\mathcal{N}$ of equalities of the form $S_i \oplus L_i = S_j \oplus L_j$, $i < j$. In fact, when $f = \Psi^4(f_1, f_2, f_3, f_4)$, then $S_i \oplus L_i = f_3(f_2(L_i \oplus f_1(0))) \oplus f_1(0)$. So the probability of such an equality is about the double in this case (as long as f_1, f_2, f_3 are randomly chosen) than in the case where f is a random permutation (because if $f_2(L_i \oplus f_1(0)) = f_2(L_j \oplus f_1(0))$ this equality holds, and if $\beta_i = f_2(L_i \oplus f_1(0)) \neq f_2(L_j \oplus f_1(0)) = \beta_j$ but $f_3(\beta_i) = f_3(\beta_j)$, this equality also holds).

So by counting $\mathcal{N}$ we obtain a way to distinguish 4 round Feistel permutations from random permutations. This generic attack requires $O(2^{\frac{n}{2}})$ non-adaptive chosen queries and $O(2^{\frac{n}{2}})$ computations (just store the values $S_i \oplus L_i$ and count the collisions).

Remark 6.2 These attacks for 3 and 4 rounds have been first published in Patarin (1992), and independently re-discovered in Aiollo and Venkatesan (1996).

6.2.4.2 KPA with $q \simeq 2^n$

When $q \geq O(2^n)$, it is possible to transform this attack in a known plaintext attack. We will count the number $\mathcal{N}$ of (i, j), $1 \leq i < j \leq q$ such that $R_i = R_j$ and $S_i \oplus$

$L_i = S_j \oplus L_j$. For a random permutation $\mathcal{N}_{perm} \simeq \frac{q^2}{2 \cdot 2^{2n}}$, and for a Ψ^4 we have $\mathcal{N}_{scheme} \simeq \frac{q^2}{2^{2n}}$ (i.e., about double).

Remark 6.3 Here the number of computations to be done is $O(q)$ if we have $O(q)$ in memory (for all i compute $S_i \oplus L_i$ and store $+1$ at the address $R_i || S_i \oplus L_i$).

6.3 Generic Attacks on Ψ^5

6.3.1 NCPA on Ψ^5

Let us assume that $R_i =$ constant, $\forall i, 1 \leq i \leq m, m \simeq 2^n$. We will simply count the number $\mathcal{N}$ of $(i, j), i < j$ such that $S_i = S_j$ and $L_i \oplus T_i = L_j \oplus T_j$. This number N will be about double for Ψ^5 compared with a truly random permutation.

Proof We recall the notations : $(L, R) \to (R, X) \to (X, Y) \to (Y, Z) \to (Z, S) \to (S, T)$.

If $S_i = S_j$:

$$\begin{aligned} L_i \oplus T_i = L_j \oplus T_j &\Leftrightarrow L_i \oplus Z_i = L_j \oplus Z_j \\ &\Leftrightarrow f_1(R_1) \oplus f_3(Y_i) = f_1(R_1) \oplus f_3(Y_j) \\ &\Leftrightarrow f_3(R_1 \oplus f_2(L_i \oplus f_1(R_1))) = f_3(R_1 \oplus f_2(L_j \oplus f_1(R_1))) \end{aligned}$$

This will occur if $f_2(L_i \oplus f_1(R_1)) = f_2(L_j \oplus f_1(R_1))$, or if these values are distinct but when Xored with R_1, they have the same images by f_3, so the probability is about two times larger. □

Remark 6.4 1. By storing the $S_i || L_i \oplus T_i$ values and looking for collisions, the complexity is in $\lambda \simeq O(2^n)$.

2. With a single value for R_i, we will get very few collisions. However this attack becomes significant if we have a few values R_i and for all these values about 2^n values L_i.

6.3.2 KPA on Ψ^5

This NCPA attack can immediately be transformed in a KPA: for random $[L_i, R_i]$, we will simply count the number $\mathcal{N}$ of $(i, j), i < j$ such that $R_i = R_j$, $S_i = S_j$, and $L_i \oplus T_i = L_j \oplus T_j$. Then we have $\mathcal{N}_{perm} \simeq \frac{q(q-1)}{2.2^{3n}}$ and $\mathcal{N}_{scheme} \simeq \frac{q(q-1)}{2^{3n}}$. This KPA is efficient when q^2 becomes not negligible compared with 2^{3n}, i.e., when $q \geq 2^{3n/2}$.

Remark 6.5 If we count the number $\mathcal{N}$ of $(i, j), i < j$ such that $R_i \oplus R_j = S_i \oplus S_j$, we get another KPA attack with a similar complexity.

Remark 6.6 Similar attacks on 5-round Feistel ciphers are described by Knudsen (1998) when f_2 and f_3 are permutations (in this chapter, f_2 and f_3 are random functions). As we will see in Chap. 7, the attacks provided by Knudsen are based on impossible differentials.

Remark 6.7 The generic attacks presented so far for 3, 4, and 5 rounds are effective against most Feistel ciphers, or when the round functions are randomly chosen. However it can occur that for specific choices of the round function, the attacks, performed exactly as described, may fail. Most of the times, we can modify this attack by taking account of these specific round functions.

6.4 Attacks on Ψ^r Generators, $r \geq 6$

As we have seen in Chap. 5, since Feisel permutation are even permutations, this provides an attack by the signature. Then, it is however much more difficult to distinguish generators of Feistel permutations from random permutations of $\mathcal{A}_{2n}$, with $\mathcal{A}_{2n}$ being the group of all the permutations of $\mathcal{P}_{2n}$ with even signature. In this section we present the best known attacks for this problem. The attacks are 2-point attacks and we will make use of the H-coefficients computed in Sect. 3.4.

6.4.1 *KPA with r Even*

Here we suppose that we are attacking μ permutations. The attacker counts the number of $(i, j), i < j$ such that

$$R_i = R_j, \text{ and } S_i \oplus S_j = L_i \oplus L_k.$$

Then it is easy to see that

$$\mathbb{E}(\mathcal{N}_{perm}) = \mu \frac{q(q-1)}{2} \frac{2^{5n}(2^n-1)}{(2^{2n}(2^{2n}-1))^2} = \mu \frac{q(q-1)}{2} \frac{(1-\frac{1}{2^n})}{2^{2n}(1-\frac{1}{2^{2n}})^2} \simeq \mu \frac{q^2}{2.2^{2n}}.$$

With a r-round balanced Feistel cipher, we see that for this attack, the conditions are those of case 3 of Theorem 3.8. Then we have

Proposition 6.1 *With the conditions*

$$R_i = R_j, \text{ and } S_i \oplus S_j = L_i \oplus L_k$$

we have

$$\mathbb{E}(\mathcal{N}_{\Psi^r}) = \mu \frac{q(q-1)}{2} \frac{2^{5n}(2^n-1)}{2^{2n}(2^{2n}-1)} \frac{H}{|\mathcal{F}_n|^r}$$

where H is the number of functions $f_1, \ldots, f_r$ such that $\Psi^r(f_1, \ldots, f_r)$ transforms each $[L_i, R_i]$ into $[S_i, T_i]$ for $i \in 1, \ldots, q$.

Proof The probability to have a couple with the given relations on their blocks multiplied by the number of acceptable outputs is $\frac{2^{2n}(2^n-1)2^{3n}}{2^{2n}(2^{2n}-1)}$ and the probability for Ψ^r to output a specific couple is $\frac{H}{|\mathcal{F}_n|^r}$ Since we have μ permutations and $\frac{q(q-1)}{2}$ possibilities for the indices i, j with $i < j$ with a 2-point attack, we obtain $\mathbb{E}(\mathcal{N}_{\Psi^r}) = \mu \frac{q(q-1)}{2} \frac{2^{5n}(2^n-1)}{2^{2n}(2^{2n}-1)} \frac{H}{|\mathcal{F}_n|^r}$ as claimed. □

The computation of the standard deviation $\sigma(\mathcal{N}_{perm})$ and $\sigma(\mathcal{N}_{\Psi^r})$ shows that $\sigma(\mathcal{N}_{perm}) \simeq \sqrt{\mu}\frac{q}{2^n}$ and $\sigma(\mathcal{N}_{\Psi^r}) \simeq \sqrt{\mu}\frac{q}{2^n}$. Now, we now that from Case 3 of Theorem 3.8, we have

$$H = \frac{1}{1-\frac{1}{2^{2n}}} \frac{|\mathcal{F}_n|^r}{2^{4n}} \left(1 + \frac{1}{2^{(\frac{r}{2}-2)n}} - \frac{1}{2^{(\frac{r}{2}-1)n}} - \frac{2}{2^{\frac{rn}{2}}} + \frac{1}{2^{(r-1)n}}\right).$$

This gives

$$\mathbb{E}(\mathcal{N}_{\Psi^r}) = \mu \frac{q(q-1)}{2} \frac{(1-2^n)}{2^{2n}(1-\frac{1}{2^{2n}})^2} \left(1 + \frac{1}{2^{(\frac{r}{2}-2)n}} - \frac{1}{2^{(\frac{r}{2}-1)n}} - \frac{2}{2^{\frac{rn}{2}}} + \frac{1}{2^{(r-1)n}}\right).$$

As explained in Chap. 5, we can distinguish when the difference of the expectations is greater than the standard deviations. The condition is $\mu\frac{q^2}{2^{\frac{rn}{2}}} \geq \sqrt{\mu}\frac{q}{2^n}$. Then we have

$$\mu \frac{q^2}{2^{\frac{rn}{2}}} \geq \sqrt{\mu}\frac{q}{2^n} \Leftrightarrow \mu q^2 \geq 2^{(r-2)n}.$$

When $q = 2^{2n}$, we can choose $\mu = 2^{(r-6)n}$ and the complexity of the attack is $\lambda = \mu 2^{2n} = 2^{(r-4)n}$.

Remark 6.8 In Sect. 11.6.2, another generic attack against ψ^6, with the same complexity, will be presented and translated into a quantum attack with a quantum query complexity of $2^{8n/5}$.

6.4.2 KPA with r Odd

When r is odd, the attacker will count the number of $(i, j), i < j$ such that

$$R_i = R_j,\ S_i = S_j,\ \text{and } T_i \oplus T_j = L_i \oplus L_j.$$

Here, we are in the Case 4 of Theorem 3.9. Then, we have

$$H = \frac{1}{1-\frac{1}{2^{2n}}}\frac{|\mathcal{F}_n|^r}{2^{4n}}\left(1+\frac{1}{2^{(\frac{r}{2}-\frac{5}{2})n}}-\frac{1}{2^{(\frac{r}{2}-\frac{3}{2})n}}-\frac{2}{2^{(\frac{r}{2}-\frac{1}{2})n}}+\frac{1}{2^{(r-2)n}}\right).$$

With computations similar to those done for r even, we obtain $\mathbb{E}(\mathcal{N}_{perm}) \simeq \mu\frac{q^2}{2.2^{3n/2}}$ and $\mathbb{E}(\mathcal{N}_{scheme}) \simeq \mu\frac{q^2}{2.2^{3n/2}} + O(\mu\frac{q^2}{2^{\frac{(r+1)n}{2}}})$. As in previous subsection, the standard deviations $\sigma(\mathcal{N}_{perm})$ and $\sigma(\mathcal{N}_{\Psi^r})$ behave like the square root of the expectations and this attacks succeeds as soon as the difference of the expectations is greater than the standard deviations. This gives the condition: $\mu\frac{q^2}{2^{\frac{(r+1)n}{2}}} \geq \sqrt{\mu}\frac{q}{2^{3n/2}}$. Again, with $q = 2^{2n}$, we can choose $\mu = 2^{(r-6)n}$ and the complexity of the attack is $\lambda = \mu 2^{2n} = 2^{(r-4)n}$.

Remark 6.9 If the attacker counts the number (i, j), $i < j$ such that $R_i \oplus R_j = S_i \oplus S_j$, this gives another KPA with the same complexity.

6.5 Summary of the Best Known Results on Random Feistel Ciphers

The results are summarized in Table 6.1. For CPA or CCA, there are no better attacks KPA when $r \geq 6$.

Table 6.1 Results on Ψ^r. For $r \geq 6$, more than one permutation is needed or more than 2^{2n} computations are needed in the best known attacks to distinguish Ψ^r from a random permutation with an even signature

	KPA	NCPA	CCA
Ψ^1	1	1	1
Ψ^2	$2^{\frac{n}{2}}$	2	2
Ψ^3	$2^{\frac{n}{2}}$	$2^{\frac{n}{2}}$	3
Ψ^4	2^n	$2^{n/2}$	$2^{\frac{n}{2}}$
Ψ^5	$2^{3n/2}$	2^n	2^n
Ψ^6	2^{2n}	2^{2n}	2^{2n}
Ψ^7	2^{3n}	2^{3n}	2^{3n}
Ψ^8	2^{4n}	2^{4n}	2^{4n}
$\Psi^r, r \geq 8$	$2^{(r-4)n}$	$2^{(r-4)n}$	$2^{(r-4)n}$

6.6 Conclusion

Attacks on balanced Feistel ciphers show that when we have to generate several small pseudo-random permutations, it is not recommended to use a Feistel cipher generator with only 6 rounds (whatever the length of the secret key may be). As an example, it is possible to distinguish most generators of 6 round Feistel permutations from truly random permutations on 32 bits, within approximately 2^{32} computations and 2^{32} chosen plaintexts (and this whatever the length of the secret key may be).

Similar attacks can be generalized for any number of rounds r, but they require to analyze many more permutations and they have a larger complexity when r increases.

Problems

6.1 Let us assume that we want to generate a pseudo-random permutation from 100 bits to 100 bits with a balanced Feistel cipher with k rounds. What minimal value for k would be recommended?

6.2 Same question to generate a pseudo-random permutation from 32 bits to 32 bits.

References

Aiollo, W., and R. Venkatesan. 1996. Foiling birthday attacks in length-doubling transformations—benes: a non-reversible alternative to feistel. In *Advances in cryptology–EUROCRYPT 1996*, vol. 1070, ed. U. Maurer. Lecture notes in computer science, 307–320. Berlin: Springer.

Knudsen, L. R., and V. Rijmen. 1999. On the decorrelated fast cipher (DFC) and its theory. In *Fast Software Encrytion—FSE '99*, vol. 1636, ed. L. Knudsen. Lecture notes in computer science, 81–94. Berlin: Springer.

Knudsen, L. R. 1998. DEAL—A 128-bit Block Cipher. Technical report #151, University of Bergen, Department of Informatics, Norway. Submitted as a candidate for the Advanced Encryption Standard. Available at http://www.ii.uib.no/~larsr/newblock.html

Luby, M., and C. Rackoff. 1988. How to construct pseudorandom permutations from pseudorandom functions. *SIAM Journal on Computing* 17 (2): 373–386 April.

Nachef, V. 2025. Random Feistel schemes for $m = 3$. Available from the author at: Valerie.nachef@math.u-cergy.fr.

Naor, M., and O. Reingold. 1999. On the construction of pseudo-random permutations: Luby-Rackoff revisited. *Journal of Cryptology* 12: 29–66. Extended abstract in: Proceedings 29th Annual ACM Symposium on Theory of Computing, 1997, 189–199

Patarin, J. 1992. New results on pseudorandom permutation generators based on the DES scheme. In *Advances in Cryptology—CRYPTO 1991*, vol. 576, ed. J. Feigenbaum. Lecture notes in computer science, 301–312. Berlin: Springer.

Patarin, J. 1998. About feistel schemes with six (or More) rounds. In *Fast software encryption—FSE '98*, vol. 1372, ed. S. Vaudenay. Lecture notes in computer science, 103–121. Berlin: Springer.

Patarin, J. 2004. Generic attacks on feistel schemes. In *Advances in cryptology—ASIACRYPT 2001*, vol. 2248, ed. C. Boyd. Lecture notes in computer science, 222–238. Berlin: Springer.

Patarin, J. 2004. Security of random feistel schemes with 5 or more rounds. In *Advances in cryptology—CRYPTO 2004*, vol. 3152, ed. M. Franklin. Lecture notes in computer science, 106–122. Berlin: Springer.
Patarin, J. 2005. On linear systems of equations with distinct variables and small block size. In *Information security and cryptology—ICISC 2005*, vol. 3935, ed. D. Won and K. Seungioo. Lecture notes in computer science, 299–321. Berlin: Springer.
Patarin, J. 2025. Pseudorandom permutations based on the DES scheme. In *Eurocode'90, LNCS 514*, 193–204. Berlin: Springer.
Patarin; J. 1991. Etude des générateurs de permutations basés sur le schéma du DES. Ph.D. Thesis, INRIA, Domaine de Voluceau, Le Chesnay, France.

Chapter 7
Generic Attacks on Classical Feistel Ciphers with Internal Permutations

Abstract In this chapter, generic attacks on Feistel networks with internal permutations instead of Feistel networks with internal functions as designed originally are studied. As always in generic attacks, the internal permutations are supposed to be random. However, as we will see, Feistel networks with internal permutations do not always behave like the original Feistel networks with round functions.

7.1 Introduction

This chapter is devoted to the study of r round balanced Feistel networks with round permutations, instead of round functions as in balanced Feistel networks. Some symmetric ciphers based on Feistel networks using internal permutations have been developed. Nyberg (1995), Aoki et al. (2000), and DEAL (Knudsen 1998) belong to that category. The fact that the internal functions are permutations do have an influence on the study of the security. Rijmen et al. (1997) exposed an attack which worked for Feistel networks with internal functions presenting bad surjectivity properties. Biham (1997) gives an example where the bijectiveness of the internal functions leads to an attack. Another instance is given by Knudsen's attack on a 5-round Feistel network with round permutations (Knudsen 2002).

As for balanced Feistel networks with internal function, 2-point attacks are the most efficient attacks against Feistel networks with round permutations too. Indeed, when the number of rounds is small, the security bounds found by Piret (2006) are the same as the complexities of the best 2-point attacks (except for the adaptive chosen plaintext and ciphertext on 3 rounds, where the best attack is a 3-point attack). For $r \leq 5$ rounds, the complexities are the same as for balanced Feistel networks with internal functions, except for 3 rounds, where there exists a known plaintext attack in $O(2^n)$ messages instead of $O(2^{n/2})$ for Feistel networks with round functions. Then, a difference appears on all $3i$ rounds, $i \geq 1$. This fact is illustrated in Sect. 7.5, Table 7.5. When the number r of rounds is greater, attacks enable to distinguish a r-round Feistel network generator from a random permutation generator.

J. Patarin et al., *Guide to Feistel Ciphers*, Information Security and Cryptography,
https://doi.org/10.1007/978-3-031-99850-8_7

The technique used to find these attacks is based on the computation of the so called *H-coefficients*.[1] From these H-coefficients, attacks using correlations between pairs of messages (2-point attacks) are deduced. General formulas enable to compute the H-coefficient values for all possible pairs of input/output couples, and to deduce the best possible generic 2-point attacks.

7.2 Generic Attacks for a Small Numbers of Rounds ($r \le 5$)

Balanced Feistel schemes are described in Chap. 2. The notation and definition are identical to those described in that chapter, except the fact that the round functions are permutations. In order to avoid confusion in the notation, $\widetilde{\Psi}^r$ is used when the round functions are permutations.

In this Section, we present the best known generic attacks on r-round Feistel networks with internal permutations, for $r \le 5$, are exposed. The generic attacks on the first two rounds are identical to the ones on Feistel networks with rounds functions of Chap. 2. The properties of the internal permutations first appear on attacks after 3 rounds.

7.2.1 *Generic Attacks on 3-Rounds Feistel Networks with Internal Permutations*

The following 2-point attacks are different from those described in Chap. 6 for Feistel networks with internal functions. After 3 rounds, the attacks used for balanced Feistel networks with round functions do not work anymore. One has to take into account the fact that the internal functions are permutations. This allows to produce attacks that are based on impossible differentials.

7.2.1.1 $NCPA$ with $2^{n/2}$ Messages

Consider the following distinguisher D:

1. D chooses L and about $q \simeq 2^{n/2}$ distinct values for R. Then he queries $[S_i, T_i] := \mathcal{O}([L, R_i])$ for $1 \le i \le q$.
2. D waits until he gets two different outputs $[S_i, T_i]$ and $[S_j, T_j]$ verifying $R_i \oplus R_j = S_i \oplus S_j$.
3. If D obtains two distinct indices i and j such that $R_i \oplus R_j = S_i \oplus S_j$ then D outputs 1, otherwise it outputs 0.

[1] Here, a H-coefficient, for specific input/output couples, is the number of r-tuples of permutations $(f_1, ..., f_r)$, such that the r-round Feistel network using $f_1, ... f_r$ as internal permutations applied to the inputs gives the corresponding outputs.

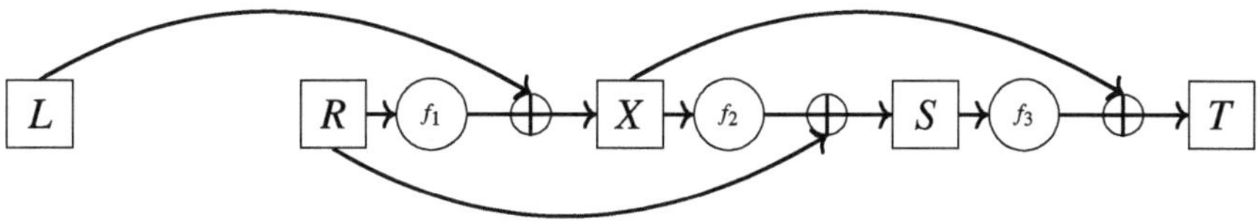

Fig. 7.1 $\widetilde{\Psi}^3(f_1, f_2, f_3)$

By the birthday paradox, if D computes $q \simeq 2^{n/2}$ messages and if $\mathcal{O}$ is a random permutation, the probability to obtain two distinct indices i and j such that $R_i \oplus R_j = S_i \oplus S_j$, is greater than or equal to $1/2$. This means that the probability for D to output 1 is greater than or equal to $1/2$. Suppose now that $\mathcal{O}$ is a 3-round Feistel network with internal permutations and that there exists such i and j, $i \neq j$ such that $R_i \oplus R_j = S_i \oplus S_j$.

A look at Fig. 7.1 shows that $S_i \oplus S_j = R_i \oplus R_j \Leftrightarrow f_2(X_i) = f_2(X_j)$ with $X_i = L_i \oplus f_1(R_i)$ and $X_j = L_j \oplus f_1(R_j)$. Since f_2 is a permutation, $f_2(X_i) = f_2(X_j) \Leftrightarrow X_i = X_j$. As $L_i = L_j$ and f_1 is a permutation, we get $R_i = R_j$. The assumption leads to a contradiction, therefore this case cannot happen with a 3-round Feistel network with internal permutations. Here, the probability that D outputs 1 is equal to 0. This gives

$$\mathcal{A}^{\mathrm{NCPA}}_{\widetilde{\Psi}^3}(1) \geq 1/2.$$

This gives a non-adaptive chosen plaintext attack ($NCPA$) and thus a CPA, a $NCCA$ and a CCA working with $O(2^{n/2})$ messages and $O(2^{n/2})$ computations.

7.2.1.2 KPA with 2^n Messages

The previous NCPA can be transformed to get a KPA with 2^n messages. The attacker waits until he gets $O(2^{n/2})$ input pairs verifying: $L_i = L_j$, $R_i \neq R_j$ and then he computes the preceding attack. By the birthday paradox, with 2^n messages, the probability to obtain such input pairs is greater than $1/2$. This provides a KPA with $O(2^n)$ messages and we have

$$\mathcal{A}^{\mathrm{KPA}}_{\widetilde{\Psi}^3}(1) \geq 1/2.$$

7.2.1.3 CCA

The adaptive chosen plaintext and cipher text attack, described in Luby and Rackoff (1988, p. 385), that needs 3 messages, still works here. It is a three points attack with only three messages. The input/output messages are $[L_1, R_1]$ and $[S_1, T_1]$, $[L_2, R_1]$ and $[S_2, T_2]$, $[L_3, R_3]$ and $[S_1, T_1 \oplus L_1 \oplus L_2]$. One just can check whether $R_3 =$

$S_2 \oplus S_3 \oplus R_2$ or not. This last equality happens with probability 1 for a 3-rounds Feistel network and probability $1/2^n$ for a random permutation. This shows that

$$\mathscr{A}_{\widetilde{\Psi}^3}^{\mathrm{CCA}}(1) \geq 1 - \frac{1}{2^n}.$$

Remark 7.1 With a number q of messages small compared to $O(2^{n/2})$, it is not possible to distinguish random internal permutations from random internal functions. Therefore, the attack here is the same as for a 3-round Feistel network with internal functions.

7.2.2 Generic Attacks on 4-Rounds Feistel Networks with Internal Permutations

For 4 rounds, the desired equations on the input/output pairs leading to the best attack is the same as for Feistel networks with internal functions (see Patarin 2001 or Aiollo and Venkatesan 1996). Still, the attacks are different as they are based in this case on impossible differentials.

7.2.2.1 $NCPA$ with $2^{n/2}$ Messages

This NCPA is similar to the one performed of 3 rounds except that we require that $L_i \oplus L_j = S_i \oplus S_j$ and the attacker generate inputs of the form $[L_i, R]$, $1 \leq i \leq q$, where the L_i values are pairwise distinct. This is justified by Fig. 7.1. Again, we have

$$\mathscr{A}_{\widetilde{\Psi}^4}^{\mathrm{NCPA}}(1) \geq 1/2.$$

Thus, there is a $NCPA$ with $O(2^{n/2})$ messages based on impossible differentials.

7.2.2.2 KPA with 2^n Messages

As previously, it is possible to adapt the above NCPA into a KPA with 2^n messages.

7.2.3 Generic Attacks on 5 Rounds Feistel Networks with Internal Permutations

For 5 rounds, the attacks are those by Knudsen given in Knudsen (2002). Again, they are based on impossible differentials and Theorem 7.1 (see Knudsen 1998).

Theorem 7.1 *Let $[L_1, R_1]$ and $[L_2, R_2]$ be two inputs of a 5-round Feistel cipher, and let $[S_1, T_1]$ and $[S_2, T_2]$ be the outputs. Let us assume that the round functions f_2 and f_3 are permutations (therefore they are not random functions of $\mathscr{F}_n$). Then, if $R_1 = R_2$ and $L_1 \neq L_2$, it is impossible to have simultaneously $S_1 = S_2$ and $L_1 \oplus L_2 = T_1 \oplus T_2$.*

Proof This comes immediately from 6.1 above. □

More precisely, it is not possible to obtain two input/output pairs ($[L_1, R_1]$, $[S_1, T_1]$) and ($[L_2, R_2]$, $[S_2, T_2]$), such that $L_1 \neq L_2$, $R_1 = R_2$, $S_1 = S_2$, $T_1 \oplus T_2 = L_1 \oplus L_2$ with a 5-round Feistel network with internal permutation. In other words, there is no $(f_1, \ldots, f_5) \in \mathscr{P}_n^5$, for which there exists such a pair of input/output couples.[2]

7.2.3.1 $NCPA$ with 2^n Messages

The attacker chooses inputs whose left and right blocks verify the above first two conditions and waits for outputs verifying the two last conditions. If he computes about 2^n messages, by the birthday paradox, the equations will hold with a probability $\geq 1/2$ for a random permutation. Thus there is a non-adaptive chosen plaintext attack with $O(2^n)$ messages.

7.2.3.2 KPA with $2^{3n/2}$ Messages

As for the previous attacks, this $NCPA$ can be turned into a known plaintext attack by waiting for enough wanted pairs of inputs to be generated. This leads to an attack with $O(2^{3n/2})$ messages.

Remark 7.2 Beyond 5 rounds, the attacks are not based on impossible differentials. This can be deduced from the H-coefficient values as explained in the following sections.

7.3 Generic Attacks for Any Number of Rounds. General Method

This section is devoted to present a method that allows to systematically analyze all possible 2-point attacks. That is, for each relation on the input and output blocks, the complexity of the corresponding attack is computed. This provides the best known generic attacks on Feistel networks with internal permutations. The previous attacks

[2] This can also be verified by the H-values as explained in the next sections.

were based on impossible differentials. The method exposed here does not use impossible differentials any more. As usual, the attacker is confronted to a permutation black box which is either a random permutation of $\{0,1\}^{2n}$ or a Feistel network with round permutations. Her goal is to determine with a high probability which of these permutations is used. For this, she will have to compute expectations of random variables defined by the relations imposed on input and output blocks as explained in Chap. 5.

7.3.1 Computation of the Probabilities

In order to get the wanted expectations, the attacker needs to compute the probability (for both a random permutation and a Feistel network with round permutations) to have input/output pairs satisfying some relations. These probabilities depend on the type of attacks that are performed: KPA, CPA.

Definition 7.1 Let us consider some equalities between the input and output blocks of two different pairs of messages the attacker is looking at $[L_1, R_1]/[S_1, T_1]$ and $[L_2, R_2]/[S_2, T_2]$: for example, he considers Equalities of the type $L_1 = L_2$, $L_1 \oplus L_2 = T_1 \oplus T_2$. Then, depending the type of attack which is considered, the number n_e, is defined by

- in KPA, n_e is the total number of equalities,
- in CPA, n_e denotes the total number of equalities minus the number of equalities involving the input blocks only,
- in CCA, n_e denotes the total number of equalities minus the number of equalities involving the blocks which have been chosen by the attacker only.

n_e will be called the number of *non-imposed equalities.*

Remark 7.3 Among given equalities between input and output blocks that the attacker wants to test on a pair of messages, there are some that can always be verified, due to the type of attack. For example, in CPA, the equality $R_1 = R_2$ happens with probability 1, because the attacker can impose it. This is why the remaining ones are called "non-imposed".

Proposition 7.1 *For a random permutation, the probability to get two input/output pairs* $[L_1, R_1]/[S_1, T_1] \neq [L_2, R_2]/[S_2, T_2]$ *verifying some relations on their blocks can be approximated by*

$$p_{perm} = \frac{2^{(4-n_e)\cdot n}}{2^{2n}(2^{2n}-1)}.$$

Proof The probability to have a pair of inputs with the right relations on their blocks multiplied by the number of acceptable output couples is approximated by $\frac{1}{2^{n_e \cdot n}} \cdot 2^{4n}$. The probability for the permutation to output one specific couple is $\frac{1}{2^{2n}(2^{2n}-1)}$. The result is obtained by multiplying these two values. □

Remark 7.4 The exact number of "acceptable" output pairs (in the sense that the wanted relations between the blocks are verified) would be difficult to express, the proposition gives only an approximation. However, the approximation is very close to the exact value and sufficient for the use made of it.

Next, we give the definition of the H-coefficients for $\widetilde{\Psi}^r$. It is the same as for Ψ^k. Thus, we keep the same notation.

Definition 7.2 Let $[L_1, R_1] \neq [L_2, R_2]$ and $[S_1, T_1] \neq [S_2, T_2]$ be four elements of $\{0, 1\}^{2n}$. The H-coefficient for $\widetilde{\Psi}^r$ (denoted by $H(L_1, R_1, L_2, R_2, S_1, T_1, S_2, T_2)$ or simply H) is the number of r-tuples $(f_1, \ldots, f_r) \in \mathscr{P}_n^r$, such that

$$\begin{cases} \widetilde{\Psi}^r(f_1, \ldots, f_r)([L_1, R_1]) = [S_1, T_1] \\ \widetilde{\Psi}^r(f_1, \ldots, f_r)([L_2, R_2]) = [S_2, T_2] \end{cases}.$$

Proposition 7.2 *For a r-round Feistel network with random internal permutations* $\widetilde{\Psi}^r$, *the probability to get two input/output couples* $[L_1, R_1]/[S_1, T_1] \neq [L_2, R_2]/[S_2, T_2]$ *verifying some relations on their blocks is approximated by*

$$p_{scheme} = \frac{2^{(4-n_e)\cdot n} \cdot H}{|\mathscr{P}_n|^r}.$$

Proof As before, the probability to have a couple of inputs with the right relations on their blocks multiplied by the number of acceptable output pairs is approximated by $2^{(4-n_e)\cdot n}$. The probability for $\widetilde{\Psi}^r$ to output one specific couple is $\frac{H}{|\mathscr{P}_n|^r}$. The result is obtained by multiplying these two values. □

Remark 7.5 Same remark as for Proposition 7.1.

7.3.2 *All Possible 2-Point Attacks*

The value of H depends on the following relations:

$$\begin{cases} L_1 = L_2, \text{ or not} \\ R_1 = R_2, \text{ or not} \\ S_1 = S_2, \text{ or not} \\ T_1 = T_2, \text{ or not} \end{cases} \qquad \begin{cases} L_1 \oplus L_2 = S_1 \oplus S_2, \text{ or not, when } k \text{ is even} \\ R_1 \oplus R_2 = T_1 \oplus T_2, \text{ or not, when } k \text{ is even} \\ L_1 \oplus L_2 = T_1 \oplus T_2, \text{ or not, when } k \text{ is odd} \\ R_1 \oplus R_2 = S_1 \oplus S_2, \text{ or not, when } k \text{ is odd} \end{cases}$$

This leads to 13 different cases when k is odd and 11 when k is even.[3] The value of H in case j will be denoted by H_j, or simply by H when it is clear that the case j is considered.

The significant cases are given below

When the number k of rounds is *odd*:

$$L_1 \neq L_2, R_1 \neq R_2, S_1 \neq S_2, T_1 \neq T_2, L_1 \oplus L_2 \neq T_1 \oplus T_2, R_1 \oplus R_2 \neq S_1 \oplus S_2 \quad (7.1)$$
$$L_1 \neq L_2, R_1 \neq R_2, S_1 \neq S_2, T_1 \neq T_2, L_1 \oplus L_2 = T_1 \oplus T_2, R_1 \oplus R_2 \neq S_1 \oplus S_2 \quad (7.2)$$
$$L_1 \neq L_2, R_1 \neq R_2, S_1 \neq S_2, T_1 \neq T_2, L_1 \oplus L_2 \neq T_1 \oplus T_2, R_1 \oplus R_2 = S_1 \oplus S_2 \quad (7.3)$$
$$L_1 \neq L_2, R_1 \neq R_2, S_1 = S_2, T_1 \neq T_2, L_1 \oplus L_2 \neq T_1 \oplus T_2, R_1 \oplus R_2 \neq S_1 \oplus S_2 \quad (7.4)$$
$$L_1 = L_2, R_1 \neq R_2, S_1 \neq S_2, T_1 \neq T_2, L_1 \oplus L_2 \neq T_1 \oplus T_2, R_1 \oplus R_2 \neq S_1 \oplus S_2 \quad (7.5)$$
$$L_1 = L_2, R_1 \neq R_2, S_1 = S_2, T_1 \neq T_2, L_1 \oplus L_2 \neq T_1 \oplus T_2, R_1 \oplus R_2 \neq S_1 \oplus S_2 \quad (7.6)$$
$$L_1 \neq L_2, R_1 \neq R_2, S_1 = S_2, T_1 \neq T_2, L_1 \oplus L_2 = T_1 \oplus T_2, R_1 \oplus R_2 \neq S_1 \oplus S_2 \quad (7.7)$$
$$L_1 \neq L_2, R_1 = R_2, S_1 = S_2, T_1 \neq T_2, L_1 \oplus L_2 \neq T_1 \oplus T_2, R_1 \oplus R_2 = S_1 \oplus S_2 \quad (7.8)$$
$$L_1 = L_2, R_1 \neq R_2, S_1 \neq S_2, T_1 \neq T_2, L_1 \oplus L_2 \neq T_1 \oplus T_2, R_1 \oplus R_2 = S_1 \oplus S_2 \quad (7.9)$$
$$L_1 = L_2, R_1 \neq R_2, S_1 \neq S_2, T_1 = T_2, L_1 \oplus L_2 = T_1 \oplus T_2, R_1 \oplus R_2 \neq S_1 \oplus S_2 \quad (7.10)$$
$$L_1 \neq L_2, R_1 \neq R_2, S_1 \neq S_2, T_1 \neq T_2, L_1 \oplus L_2 = T_1 \oplus T_2, R_1 \oplus R_2 = S_1 \oplus S_2 \quad (7.11)$$
$$L_1 \neq L_2, R_1 = R_2, S_1 = S_2, T_1 \neq T_2, L_1 \oplus L_2 = T_1 \oplus T_2, R_1 \oplus R_2 = S_1 \oplus S_2 \quad (7.12)$$
$$L_1 = L_2, R_1 \neq R_2, S_1 \neq S_2, T_1 = T_2, L_1 \oplus L_2 = T_1 \oplus T_2, R_1 \oplus R_2 = S_1 \oplus S_2. \quad (7.13)$$

When the number k of rounds is *even* :

$$L_1 \neq L_2, R_1 \neq R_2, S_1 \neq S_2, T_1 \neq T_2, L_1 \oplus L_2 \neq S_1 \oplus S_2, R_1 \oplus R_2 \neq T_1 \oplus T_2 \quad (7.1)$$
$$L_1 \neq L_2, R_1 = R_2, S_1 \neq S_2, T_1 \neq T_2, L_1 \oplus L_2 \neq S_1 \oplus S_2, R_1 \oplus R_2 \neq T_1 \oplus T_2 \quad (7.2)$$
$$L_1 = L_2, R_1 \neq R_2, S_1 \neq S_2, T_1 \neq T_2, L_1 \oplus L_2 \neq S_1 \oplus S_2, R_1 \oplus R_2 \neq T_1 \oplus T_2 \quad (7.3)$$
$$L_1 \neq L_2, R_1 \neq R_2, S_1 \neq S_2, T_1 \neq T_2, L_1 \oplus L_2 = S_1 \oplus S_2, R_1 \oplus R_2 \neq T_1 \oplus T_2 \quad (7.4)$$
$$L_1 = L_2, R_1 \neq R_2, S_1 \neq S_2, T_1 = T_2, L_1 \oplus L_2 \neq S_1 \oplus S_2, R_1 \oplus R_2 \neq T_1 \oplus T_2 \quad (7.5)$$
$$L_1 \neq L_2, R_1 = R_2, S_1 = S_2, T_1 \neq T_2, L_1 \oplus L_2 \neq S_1 \oplus S_2, R_1 \oplus R_2 \neq T_1 \oplus T_2 \quad (7.6)$$
$$L_1 \neq L_2, R_1 = R_2, S_1 \neq S_2, T_1 = T_2, L_1 \oplus L_2 \neq S_1 \oplus S_2, R_1 \oplus R_2 = T_1 \oplus T_2 \quad (7.7)$$
$$L_1 \neq L_2, R_1 = R_2, S_1 \neq S_2, T_1 \neq T_2, L_1 \oplus L_2 = S_1 \oplus S_2, R_1 \oplus R_2 \neq T_1 \oplus T_2 \quad (7.8)$$
$$L_1 \neq L_2, R_1 \neq R_2, S_1 \neq S_2, T_1 = T_2, L_1 \oplus L_2 = S_1 \oplus S_2, R_1 \oplus R_2 \neq T_1 \oplus T_2 \quad (7.9)$$
$$L_1 \neq L_2, R_1 \neq R_2, S_1 \neq S_2, T_1 \neq T_2, L_1 \oplus L_2 = S_1 \oplus S_2, R_1 \oplus R_2 = T_1 \oplus T_2 \quad (7.10)$$
$$L_1 \neq L_2, R_1 = R_2, S_1 \neq S_2, T_1 = T_2, L_1 \oplus L_2 = S_1 \oplus S_2, R_1 \oplus R_2 = T_1 \oplus T_2. \quad (7.11)$$

[3] There were more in fact, but it is possible to avoid some of them because of the following relation: $\widetilde{\Psi}^k(f_1, \ldots, f_k)([L, R]) = [S, T] \iff \widetilde{\Psi}^k(f_k, \ldots, f_1)([T, S]) = [R, L]$.

7.3.3 The Attacks

Suppose that we have q random plaintext/ciphertext couples. Let $\mathcal{N}_{\widetilde{\Psi}^r,j}$ be the random variable that counts the number of pairs of these couples belonging to the j-th case among the cases previously defined ($1 \leq j \leq 13$, when r is odd, and $1 \leq j \leq 11$, when r is even), for a r-round Feistel network with internal permutations. Similarly, the random variable $\mathcal{N}_{perm,j}$ gives the number of pairs of these couples belonging to the k-th case for a random permutation. We have

$$\mathbb{E}(\mathcal{N}_{\widetilde{\Psi}^r,j}) \simeq \frac{q(q-1)}{1} p_{\widetilde{\Psi}^r,j} \text{ and } \mathbb{E}(\mathcal{N}_{perm,j}) \simeq \frac{q(q-1)}{2} p_{perm}.$$

As explained in Chap. 5, the attacks are successful when the difference of the expectations is greater than the standard deviations. Then Propositions 7.1 and 7.2 show that

$$|\mathbb{E}(\mathcal{N}_{\widetilde{\Psi}^r,j}) - \mathbb{E}(\mathcal{N}_{perm,j})| = \frac{q(q-1)}{2 \cdot 2^{n_e n}} \left(\frac{2^{4n} H}{|\mathcal{P}_n|^r} - \frac{1}{1 - \frac{1}{2^{2n}}} \right).$$

This shows that we need to know the leading term of $\frac{2^{4n} H}{|\mathcal{P}_n|^r} - \frac{1}{1-\frac{1}{2^{2n}}}$. This is the aim of the next section.

7.4 Computation of the H-Coefficients

Once we know the H-coefficients, we can obtain the probabilities as shown by Proposition 7.2. Then it is possible to compute the expectation involved in the attacks. This section is devoted to the computation of the H-coefficients.

7.4.1 General Ideas for the Computation of the H-Coefficients

The computation of the formulas for the H-coefficients is rigorously done in this section. First, let us give an overview of the basic ideas. Let $[L_1, R_1]$, $[L_2, R_2]$ and $[S_1, T_1]$, $[S_2, T_2]$ be four elements of $\{0, 1\}^{2n}$. By definition of the H-coefficients, one has to determine the number of $(f_1, \ldots, f_r) \in \mathcal{P}_n^r$ such that

$$\begin{cases} \widetilde{\Psi}^r(f_1, \ldots, f_r)([L_1, R_1]) = [S_1, T_1] \\ \widetilde{\Psi}^r(f_1, \ldots, f_r)([L_2, R_2]) = [S_2, T_2] \end{cases}$$

Let us introduce the following notation for the internal variables X^t that are defined at round t, $1 \leq t \leq r$:

$$\begin{cases} X^1 = L \oplus f_1(R) \\ X^2 = R \oplus f_2(X^1) \\ X^3 = X^1 \oplus f_3(X^2) \\ \forall t,\ 3 \leq t \leq r,\ X^s = X^{t-2} \oplus f_t(X^{t-1}) \end{cases}$$

More generally, by setting $X^{-1} = L$ and $X^0 = R$, one obtains: $\forall 1 \leq t \leq r,\ X^t = X^{t-2} \oplus f_t(X^{r-1})$. Notice that $X^{r-1} = S$ and $X^r = T$.

The following result holds:

Theorem 7.2 *A formula for H is given by*

$$H = \sum_{\text{possible } s} (2^n - 1)!^{e(s)} (2^n - 2)!^{d(s)} \cdot (2^n)^{r-2} \cdot N(d_1) \cdots N(d_{r-2}),$$

where

- *s denotes a sequence of relations verified by the internal variables, namely $X_1^i = X_2^i$ or $X_1^i \neq X_2^i$, for $i = -1, \ldots, r$,*
- *$e(s)$ is for the number of equalities in s,*
- *$d(s)$ is for the number of differences in s,*
- *$N(d_i)$, for $i = 1, \ldots, r-2$, is the number of possible values for $X_1^i \oplus X_2^i$, for a fixed sequence s.*

Proof In a first step, let s be a fixed sequence as in Theorem 7.2, and let us evaluate the number of possibilities $H(s)$ for $(f_1, \ldots, f_r)$. The second step then consists in summing up over all possible sequences s.

Let us detail the first step. In the following, $f_1, \ldots, f_r$ will always denote the permutations that satisfy the condition in the definition of H. A sequence s as in Theorem 7.2 is fixed.

- For $i = 1 \ldots r-2$, $N(d_i) \cdot 2^n$ is the number of possibilities for the pair (X_1^i, X_2^i).
- For $i = 1, \ldots, r$, $f_i(X^{i-1}) = X^{i-2} \oplus X^i$. Thus, $N(d_1) \cdot 2^n$ is the number of possibilities for the pair $(f_1(X_1^0), f_1(X_2^0))$, and for $i = 2, \ldots, r-2$, $N(d_i) \cdot 2^n$ is the number of possibilities for the pair $(f_i(X_1^{i-1}), f_i(X_2^{i-1}))$, when $f_1, \ldots, f_{i-1}$ are fixed.
- Let us denote by $F_1(s)$ the number of possibilities for f_1, and $F_i(s)$ for $i = 2, \ldots, r$, the number of possibilities for f_i, when $f_1, \ldots, f_{i-1}$ are fixed. Then:

$$\begin{cases} \text{if } X_1^{i-1} \neq X_2^{i-1} : F_i(s) := N(d_i) 2^n (2^n - 2)!, \text{ for } i = 1, \ldots r-2, \\ \qquad\qquad\qquad\quad F_{r-1}(s) = F_r(s) = (2^n - 2)! \\ \text{if } X_1^{i-1} = X_2^{i-1} : F_i(s) := N(d_i) 2^n (2^n - 1)!, \text{ for } i = 1, \ldots r-2, \\ \qquad\qquad\qquad\quad F_{r-1}(s) = F_r(s) = (2^n - 1)! \end{cases}$$

This results from the preceding point and from noticing that $(2^n-2)!$ (respectively $(2^n-1)!$) is the number of permutations, for which the image of two elements (respectively one element) is imposed. For $i=r-1$, all values in the equation $f_i(X^{i-1})=X^{i-2}\oplus X^i$ are fixed, thus the number of possibilities for the pair $(f_i(X_1^{i-1}), f_i(X_2^{i-1}))$ is 1.

- Finally, for a fixed sequence s, the number of possibilities for $(f_1,\ldots,f_r)$ is

$$H(s)=\prod_{i=1}^{r}F_i(s)=(2^n-1)!^{e(s)}(2^n-2)!^{d(s)}\cdot(2^n)^{r-2}\cdot N(d_1)\cdots N(d_{r-2}).$$

The final formula for the H-coefficients is then:

$$H=\sum_{\text{possible } s}(2^n-1)!^{e(s)}(2^n-2)!^{d(s)}\cdot(2^n)^{r-2}\cdot N(d_1)\cdots N(d_{r-2}),$$

as claimed. □

7.4.2 *Exact Formulas for H-Coefficients*

7.4.2.1 Notation

Here are some more notations.

- For $-1\le i\le r$, d_i denotes the value $X_1^i\oplus X_2^i$.
- Let s be a sequence of relations of length $r+2$, such that s_i is the symbol $=$ if $X_1^i=X_2^i$, s_i is the symbol $\neq$ if $X_1^i\neq X_2^i$. The number of possible d_i's, $-1\le i\le r$ is denoted by $N(s,d_i)$ or simply $N(d_i)$,
- Let $s=(s_{-1},\ldots,s_r)$ be a fixed sequence of $\{=,\neq\}^{r+2}$. The number of $=$'s intervening in the sequence $(s_{i_1},\ldots s_{i_\ell})$ is denoted by $e(s_{i_1},\ldots,s_{i_\ell})$ $(i_j\in\{-1,\ldots,r\}$, $\ell\le r+2)$ and the number of $\neq$'s intervening in that sequence is denoted by $d(s_{i_1},\ldots s_{i_\ell})$. For more convenience, $e(s)$ stands for $e(s_{-1},\ldots,s_r)$ and $d(s)$ stands for $d(s_{-1},\ldots,s_r)$.
- $$\alpha_{odd}=1 \text{ if } L_1\oplus L_2=\begin{cases} S_1\oplus S_2 \text{ when } r \text{ is even}\\ T_1\oplus T_2 \text{ when } r \text{ is odd}\end{cases},\ 0 \text{ elsewhere}$$

$$\alpha_{ev}=1 \text{ if } R_1\oplus R_2=\begin{cases} T_1\oplus T_2 \text{ when } r \text{ is even}\\ S_1\oplus S_2 \text{ when } r \text{ is odd}\end{cases},\ 0 \text{ elsewhere}$$

- ℓ_{odd} (respectively ℓ_{ev}) will stand for the number of odd (respectively even) intermediate values (i.e., different from L, R, S, T, in fact, the internal value X^i's), appearing during the computation of $\widetilde{\Psi}^r$.

From Theorem 7.2, we know that the expression

$$F(s) := (2^n - 1)!^{e(s)} \cdot (2^n - 2)!^{d(s)} \cdot (2^n)^{r-2} \cdot N(d_1) \ldots N(d_{r-2})$$

counts the wanted number of r-tuples $(f_1, \ldots, f_r)$, when the particular sequence s is fixed. All possible sequences s and the product $N(d_1) \ldots N(d_{r-2})$ have to be determined for each case.

7.4.2.2 Possible Sequences s

By using the definition of the internal variables and taking into account the fact that the f_i's are permutations, one obtains some conditions on the s_i's and d_i's:

- For $-1 \leq i \leq r$, if s_i is $=$, then s_{i+1}, s_{i+2}, s_{i-1} and s_{i-2} are $\neq$ (when those are well-defined).
- For $-1 \leq i \leq r-2$, if s_i, s_{i+1} and s_{i+2} are $\neq$, then $d_i \neq d_{i+2}$.
- For $-1 \leq i \leq r$, if $d_i = 0$, then $d_{i-1} = d_{i+1}$ and d_{i+2} can take all values but 0 (when those are well-defined).

These conditions imply that we have different valid sequences s, depending on the initial values s_{-1}, s_0, s_{k-1} and s_k.

7.4.2.3 Exact Computation of the Product $N(d_1) \cdots N(d_{k-2})$, for a Fixed Sequence s

Let $s = (s_{-1}, \ldots, s_r)$ be a fixed sequence of $\{=, \neq\}^{r+2}$. Finding the value of the product $\Pi := N(d_1) \cdots N(d_{r-2})$, leads to considering different situations.

1. When at least one s_i is $=$ among the even i's and at least one s_i is $=$ among the odd i's:
$$\Pi_1 = (2^n - 1)^{e(s)-2}(2^n - 2)^{r-3e(s)+e(s_0,s_{r-1})+2e(s_{-1},s_r)}.$$

2. When at least one s_i is $=$ among the even i's and none among the odd i's :

$$\Pi_2 = \left(\sum_{j=0}^{M_{odd}} C_{j,odd} \cdot (2^n - 2)^{j+\alpha_{odd}} (2^n - 3)^{P_{j,odd}} \right) \cdot (2^n - 1)^{e(s)-1} (2^n - 2)^{Q_{odd}}.$$

With

$$\begin{aligned} M_{odd} &= \frac{\ell_{odd} - \alpha_{odd} - e(s) + e(s_{-1}, s_r)}{2} \\ C_{j,odd} &= \binom{\ell_{odd} - e(s) + e(s_{-1}, s_r) - \alpha_{odd} - j}{j} \\ P_{j,odd} &= \ell_{odd} - e(s) + e(s_{-1}, s_r) - \alpha_{odd} - 2j \\ Q_{odd} &= \ell_{ev} - 2e(s) + e(s_{-1}, s_0, s_{r-1}, s_r) + 1 \end{aligned}$$

3. When at least one s_i is $=$ among the odd i's and none among the even i's :

$$\Pi_3 = \left(\sum_{j=0}^{M_{ev}} C_{j,ev}\cdot(2^n-2)^{j+\alpha_{ev}}(2^n-3)^{P_{j,ev}}\right)\cdot(2^n-1)^{e(s)-1}(2^n-2)^{Q_{ev}}\,.$$

With

$$\begin{aligned} M_{ev} &= \frac{\ell_{ev}-\alpha_{ev}-e(s)+e(s_{-1},s_r)}{2} \\ C_{j,ev} &= \binom{\ell_{ev}-e(s)+e(s_{-1},s_r)-\alpha_{ev}-j}{j} \\ P_{j,ev} &= \ell_{ev}-e(s)+e(s_{-1},s_r)-\alpha_{ev}-2j \\ Q_{ev} &= \ell_{odd}-2e(s)+e(s_{-1},s_0,s_{r-1},s_r)+1 \end{aligned}$$

4. When $e(s)=0$:

$$\Pi_4 = \left(\sum_{j=0}^{M_{ev}} C_{j,ev}\cdot(2^n-2)^{j+\alpha_{ev}}\cdot(2^n-3)^{P_{j,ev}}\right)\cdot\left(\sum_{j=0}^{M_{odd}} C_{j,odd}\cdot(2^n-2)^{j+\alpha_{odd}}\cdot(2^n-3)^{P_{j,odd}}\right).$$

In fact, these formulas hold with a special convention sometimes, when the number of rounds is 1, 2, or 3. This will be specified in Sect. 7.4.2.4.

7.4.2.4 General Formulas for the H-Coefficients

Here, given two different couples $([L_1, R_1], [S_1, T_1])$ and $([L_2, R_2], [S_2, T_2])$, the H-coefficient formulas are provided. The formulas are obtained from Theorem 7.2 of Sect. 7.4.4.1, where the right expression for $\prod_{i=1}^{r-2} d_i$, is applied, depending on the sequence s considered. All possibilities are dispatched in four situations.[4]

The formula $F(s)$ for a fixed sequence s only uses the number of $=$'s and $\neq$'s in this sequence (plus the initial equalities on the blocks, see Sect. 7.4.2.3). Thus, instead of summing up $F(s)$ over all possible sequences s to get H, we sum up over the number A of possible $=$'s in the sequences and multiply the formula $F(s)$ for a fixed s by the number of possible sequences s with A $=$'s.

Remark 7.6 Note that $r = 1$ is a very particular case. The H-values for $k = 1$ are easily obtained directly (see Sect. 7.4.3).

We now give the expression of H for the 4 situations

[4] The values $\Pi_1, \Pi_2, \Pi_3, \Pi_4$ below are the products of Sect. 7.4.2.3, with $e(s) - e(s_{-1}, s_0, s_{k-1}, s_r) = A$. Also the bound M is $\frac{r-2e(s_0,s_{r-1})-e(s_{-1},s_r)}{3}$.

1. First situation:

$$\begin{array}{l} r \text{ even and } L_1 = L_2,\ R_1 \neq R_2,\ S_1 \neq S_2,\ T_1 = T_2 \\ \text{or } r \text{ even and } L_1 \neq L_2,\ R_1 = R_2,\ S_1 = S_2,\ T_1 \neq T_2 \\ \text{or } r \text{ odd and } L_1 = L_2,\ R_1 \neq R_2,\ S_1 = S_2,\ T_1 \neq T_2 \\ \text{or } r \text{ odd and } L_1 \neq L_2,\ R_1 = R_2,\ S_1 \neq S_2,\ T_1 = T_2 \end{array}$$

(cases 5 and 6 for r even, and case 6 for r odd).

$$H = \sum_{A=0}^{M} (2^n - 1)!^{A+e(s_0,s_{k-1})} \cdot (2^n - 2)!^{r-A-e(s_0,s_{r-1})} \cdot 2^{n(r-2)} \cdot \binom{r-2e(s_0,s_{r-1})-e(s_{\;1},s_r)-2A}{A} \cdot \Pi_1.$$

This formula is always true for $r > 1$.

2. Second situation:

$$\begin{array}{l} r \text{ even and } L_1 \neq L_2,\ R_1 = R_2,\ S_1 \neq S_2,\ T_1 = T_2 \\ \text{or } r \text{ even and } L_1 \neq L_2,\ R_1 = R_2,\ S_1 \neq S_2,\ T_1 \neq T_2 \\ \text{or } r \text{ even and } L_1 \neq L_2,\ R_1 \neq R_2,\ S_1 \neq S_2,\ T_1 = T_2 \\ \text{or } r \text{ odd and } L_1 \neq L_2,\ R_1 \neq R_2,\ S_1 = S_1,\ T_1 \neq T_2 \\ \text{or } r \text{ odd and } L_1 \neq L_2,\ R_1 = R_2,\ S_1 \neq S_2,\ T_1 \neq T_2 \\ \text{or } r \text{ odd and } L_1 \neq L_2,\ R_1 = R_2,\ S_1 = S_2,\ T_1 \neq T_2 \end{array}$$

(cases 2, 7, 8, 9, and 11 for r even, and cases 4, 7, 8, and 12 for r odd).

$$H = \sum_{A=0}^{M} (2^n - 1)!^{A+e(s_0,s_{r-1})} \cdot (2^n - 2)!^{r-A-e(s_0,s_{r-1})} \cdot 2^{n(r-2)} \cdot \Big[\binom{\ell_{ev}+1-A-e(s_{-1},s_0,s_{r-1},s_k)}{A} \cdot \Pi_2 + \Big[\binom{r-2e(s_0,s_{r-1})-e(s_{-1},s_r)-2A}{A} - \binom{\ell_{ev}+1-A-e(s_{-1},s_0,s_{r-1},s_r)}{A}\Big] \cdot \Pi_1 \Big].$$

This formula is always true for $r > 2$, and also for $r = 2$ if the sum in Π_2 equal to 1 when $\alpha_{odd} = e(s_0)$.

3. Third situation:

$$\begin{array}{l} r \text{ even and } L_1 \neq L_2,\ R_1 \neq R_2,\ S_1 = S_2,\ T_1 \neq T_2 \\ \text{or } r \text{ even and } L_1 = L_2,\ R_1 \neq R_2,\ S_1 \neq S_2,\ T_1 \neq T_2 \\ \text{or } r \text{ even and } L_1 = L_2,\ R_1 \neq R_2,\ S_1 = S_2,\ T_1 \neq T_2 \\ \text{or } r \text{ odd and } L_1 = L_2,\ R_1 \neq R_2,\ S_1 \neq S_2,\ T_1 \neq T_2 \\ \text{or } r \text{ odd and } L_1 \neq L_2,\ R_1 \neq R_2,\ S_1 \neq S_2,\ T_1 = T_2 \\ \text{or } r \text{ odd and } L_1 = L_2,\ R_1 \neq R_2,\ S_1 \neq S_2,\ T_1 = T_2 \end{array}$$

(case 3 for r even, and cases 5, 9, 10, and 13 for r odd).

$$H = \sum_{A=0}^{M} (2^n - 1)!^{A+e(s_0,s_{r-1})} \cdot (2^n - 2)!^{k-A-e(s_{r-1})} \cdot 2^{n(r-2)} \cdot$$

$$\left[\binom{\ell_{odd}+1-A-e(s_{-1},s_0,s_{r-1},s_k)}{A} \cdot \Pi_3 + \left[\binom{r-2e(s_0,s_{r-1})-e(s_{-1},s_r)-2A}{A}\right.\right.$$

$$\left.\left. -\binom{\ell_{odd}+1-A-e(s_{-1},s_0,s_{r-1},s_r)}{A}\right] \cdot \Pi_1 \right].$$

This formula is always true for $r > 2$, and also for $r = 2$ if the sum in Π_3 equal to 1 when $e(s_{-1}, s_{r-1}) = 1$, and $\alpha_{ev} = e(s_{r-1})$.

4. Fourth situation:

$$r \in \mathbb{N}^*,\ L_1 \neq L_2,\ R_1 \neq R_2,\ S_1 \neq S_2,\ T_1 \neq T_2$$

(cases 1, 4 and 10 for r even, and cases 1, 2, 3, and 11 for r odd).

$$H = (2^n - 2)!^r \cdot (2^n)^{r-2} \cdot \Pi_4 + \sum_{A=1}^{r/3} (2^n - 1)!^A \cdot (2^n - 2)!^{r-A} \cdot 2^{n(r-2)} \cdot$$

$$\left[\binom{\ell_{ev}+1-A}{A}\Pi_2 + \binom{\ell_{odd}+1-A}{A}\Pi_3 + \left[\binom{r-2A}{A} - \binom{\ell_{ev}+1-A}{A} - \binom{\ell_{odd}+1-A}{A}\right] \cdot \Pi_1\right].$$

This formula is always true for $r > 3$ and also for $r = 2, 3$, if:

- the sum in Π_4, whose upperbound is M_{ev} is equal to 1, when $r = 2$ and $\alpha_{ev} = 0$,
- the sum in Π_4, whose upperbound is M_{odd} is equal to 1, when $r = 2$ and $\alpha_{ev} = 0$,
- the sum in Π_3 is equal to 1, when $r = 3$ and $\alpha_{ev} = 1 = e(s)$.

7.4.3 Exact H-Coefficient Values for $r \leq 5$

In Tables 7.1, 7.2, and 7.3 below, are given some exact values of the H-coefficients, computed from the formulas of Sect. 7.4.2.4.

Remark 7.7 For 5 rounds, H in case 12 is 0 (this is linked with Knudsen's attack Knudsen 2002). However, as soon as $r > 5$, $\forall i\ H_i \neq 0$. The attacks get then more complex.

Table 7.1 H-coefficient values in the different cases of Sect. 7.3.2, for 1, 2, and 3 rounds. Here $N = 2^n$

Case	1 Round	2 Rounds	3 Rounds
1	0	$((N-2)!)^2$	$((N-2)!)^3\,N\,(N-3)$
2	0	0	$((N-2)!)^3\,N\,(N-2)$
3	$(N-2)!$	$((N-2)!)^2$	$(N-1)!\,((N-2)!)^2\,N$
4	0	0	$(N-1)!\,((N-2)!)^2\,N$
5	0	$((N-2)!)^2$	$((N-2)!)^3\,N\,(N-2)$
6	0	0	$(N-1)!\,((N-2)!)^2\,N$
7	0	0	0
8	0	$(N-1)!\,(N-2)!$	0
9	0	0	0
10	0	0	$((N-2)!)^3\,N\,(N-1)$
11	0	0	$(N-1)!\,((N-2)!)^2\,N$
12	$(N-1)!$		0
13	0		0

Table 7.2 H-coefficient values in the different cases of Sect. 7.3.2, for 4 rounds. Here $N = 2^n$

Case	4 Rounds
1	$((N-2)!)^4\,N^2\,(N-3)^2 + 2\,(N-1)!\,((N-2)!)^3\,N^2$
2	$(N-1)!\,((N-2)!)^3\,N^2\,(N-2)$
3	$((N-2)!)^4\,N^2\,(N-2)\,(N-3) + (N-1)!\,((N-2)!)^3\,N^2$
4	$((N-2)!)^4\,N^2\,(N-2)\,(N-3) + (N-1)!\,((N-2)!)^3\,N^2$
5	$((N-2)!)^4\,N^2\,(N-2)^2$
6	$((N-1)!)^2\,((N-2)!)^2\,N^2$
7	$(N-1)!\,((N-2)!)^3\,N^2\,(N-1)$
8	0
9	$((N-2)!)^4\,N^2\,(N-2)^2 + (N-1)!\,((N-2)!)^3\,N^2$
10	$((N-2)!)^4\,N^2\,(N-2)^2 + (N-1)!\,((N-2)!)^3\,N^2$
11	0
12	
13	

Table 7.3 H-coefficient values in the different cases of Sect. 7.3.2, for 5 rounds. Here $N = 2^n$

Case	5 Rounds
1	$((N-2)!)^5\, N^3\left((N-3)^2+N-2\right)(N-3)+$ $(N-1)!\,((N-2)!)^4\, N^3\,(3N-7)$
2	$((N-2)!)^5\, N^3\,(N-2)\,(N-3)^2+(N-1)!\,((N-2)!)^4\, N^3\,(3N-6)$
3	$((N-2)!)^5\, N^3\left((N-3)^2+N-2\right)(N-2)+$ $(N-1)!\,((N-2)!)^4\, N^3\,(N-3)$
4	$(N-1)!\,((N-2)!)^4\, N^3\,(N-2)\,(N-3)+((N-1)!)^2\,((N-2)!)^3\, N^3$
5	$((N-2)!)^5\, N^3\,(N-2)^2\,(N-3)+(N-1)!\,((N-2)!)^4\, N^3\,(2N-3)$
6	$(N-1)!\,((N-2)!)^4\, N^3\,(N-2)^2$
7	$(N-1)!\,((N-2)!)^4\, N^3\,(N-2)^2+((N-1)!)^2\,((N-2)!)^3\, N^3$
8	$((N-1)!)^2\,((N-2)!)^3\, N^3\,(N-1)$
9	$((N-2)!)^5\, N^3\,(N-2)^3+(N-1)!\,((N-2)!)^4\, N^3\,(N-2)$
10	$((N-2)!)^5\, N^3\,(N-3)\,(N-1)\,(N-2)+(N-1)!\,((N-2)!)^4\, N^3\,(N-1)$
11	$((N-2)!)^5\, N^3\,(N-2)^2\,(N-3)+(N-1)!\,((N-2)!)^4\, N^3\,(N-2)$
12	0
13	$((N-2)!)^5\, N^3\,(N-2)^2\,(N-1)+(N-1)!\,((N-2)!)^4\, N^3\,(N-1)$

7.4.4 *Table of Leading Terms of* $\frac{H\cdot 2^{4n}}{|\mathcal{P}_n|^k}-\frac{1}{1-1/2^{2n}}$ *and Example of Attack*

7.4.4.1 Table of Leading Terms of $\frac{H\cdot 2^{4n}}{|\mathcal{P}_n|^k}-\frac{1}{1-1/2^{2n}}$

At Sect. 7.3, it was specified that the coefficient $\frac{H\cdot 2^{4n}}{|\mathcal{P}_n|^k}-\frac{1}{1-1/2^{2n}}$ allows to easily find the best 2-point attacks. Table 7.4 below gives the leading term of $\frac{H\cdot 2^{4n}}{|\mathcal{P}_n|^k}-\frac{1}{1-1/2^{2n}}$, for each case exposed in Sect. 7.3.2.[5]

7.4.4.2 Example of Attack Given by the General Method: 3 Rounds, KPA

First, the number of equalities required in each case of Sect. 7.3.2 is examined. For an odd number of rounds and in the case of a KPA, the case 1 does not require any equations on the inputs and outputs, cases 2 to 5 require 1 equation, cases 6 to 11 require 2 equations, and cases 12 and 13 require 3 equations. Then, from Table 7.4,

[5] The reader has to be careful because the case i is not the same depending on the parity of k (see Sect. 7.3.2).

Table 7.4 Order of the leading term of $\frac{H \cdot 2^{4n}}{|\mathcal{P}_n|^k} - \frac{1}{1-1/2^{2n}}$. From these values, we can easily get the best attacks using correlations between pairs of messages. Here $N = 2^n$

Number of rounds	Case												
	1	2	3	4	5	6	7	8	9	10	11	12	13
1	1	1	N^2	1	1	1	1	1	1	1	1	N^3	1
2	N^{-1}	1	N^{-1}	1	N^{-1}	1	1	N	1	1	1		
3	N^{-2}	N^{-1}	N^{-1}	N^{-1}	N^{-1}	N^{-1}	1	1	1	N^{-1}	N^{-1}	1	1
4	N^{-3}	N^{-1}	N^{-2}	N^{-2}	N^{-2}	N^{-1}	N^{-1}	1	N^{-1}	N^{-2}	1		
5	N^{-2}	N^{-2}	N^{-1}	N^{-2}	N^{-3}	N^{-2}	N^{-1}	N^{-1}	N^{-2}	N^{-2}	N^{-1}	1	N^{-1}
6	N^{-3}	N^{-3}	N^{-3}	N^{-2}	N^{-3}	N^{-2}	N^{-2}	N^{-2}	N^{-3}	N^{-2}	N^{-1}		
7	N^{-3}	N^{-3}	N^{-2}	N^{-3}	N^{-3}	N^{-3}	N^{-3}	N^{-2}	N^{-2}	N^{-3}	N^{-2}	N^{-1}	N^{-3}
8	N^{-4}	N^{-3}	N^{-5}	N^{-3}	N^{-4}	N^{-3}	N^{-3}	N^{-2}	N^{-3}	N^{-3}	N^{-3}		
9	N^{-5}	N^{-4}	N^{-4}	N^{-5}	N^{-4}	N^{-4}	N^{-3}	N^{-3}	N^{-3}	N^{-5}	N^{-4}	N^{-3}	N^{-3}
10	N^{-6}	N^{-4}	N^{-6}	N^{-5}	N^{-5}	N^{-4}	N^{-5}	N^{-3}	N^{-4}	N^{-5}	N^{-3}		
11	N^{-5}	N^{-5}	N^{-4}	N^{-6}	N^{-6}	N^{-5}	N^{-4}	N^{-5}	N^{-5}	N^{-6}	N^{-4}	N^{-3}	N^{-4}
12	N^{-6}	N^{-6}	N^{-6}	N^{-5}	N^{-6}	N^{-5}	N^{-6}	N^{-5}	N^{-6}	N^{-5}	N^{-4}		

it appears that the cases leading to the best generic attacks should be cases 7, 8 and 9. For $i = 7, 8, 9$, the following values (with the notations of Sect. 7.3) are obtained: $|\mathbb{E}(\mathcal{N}_{\widetilde{\Psi}^r,j}) - \mathbb{E}(\mathcal{N}_{perm,j})| \simeq \frac{q(q-1)}{2 \cdot 2^{2n}}$, $\sigma(\mathcal{N}_{\widetilde{\Psi}^r,j}) \simeq \frac{q}{2^n}$ and $\sigma(\mathcal{N}_{perm,j}) \simeq \frac{q}{2^n}$. Solving the equation $\frac{q^2}{2^{2n}} \geq \frac{\sqrt{q}}{2^n}$, gives $q \geq 2^n$. We can take $q = O(2^n)$.

Therefore, the best generic attack, exploiting only correlations between pairs of input/output couples works with a complexity $O(2^n)$, in the case of a KPA.

Remark 7.8 It can be noticed that it is possible to find this way the attack given in Sect. 7.2.1. In fact, the case 9 is the one corresponding to the attack on 3 rounds described in Sect. 7.2.1. This also shows that the attacks are not unique.

7.5 Table of Results for Any Number of Rounds

In this section is given a table of results (Table 7.5), showing the complexities of the attacks we obtained. These results are obtained by computer. All values, except for the special case of Sect. 7.2.1.3, are obtained by computing the formulas for H and applying the reasoning of Sect. 7.3 (an example is given in Sect. 7.4.4.2).

Remark 7.9 As long as the number of queries to the round functions is small compared to $O(2^{n/2})$, it is not possible to distinguish a function of $\{0, 1\}^n$ from a permutation of $\{0, 1\}^n$. When the complexities are small compared to $2^{n/2}$, it is then normal to find the same attack complexities for Feistel networks with round permutations than for those with round functions (as for 10 values on the first 3 rounds, in Table 7.5).

Remark 7.10 In Sect. 11.6.2, another generic attack against ψ^6, with the same complexity, will be presented and translated into a quantum attack with a quantum query complexity of $2^{8n/5}$.

Table 7.5 gives the best known generic attacks on Feistel networks with internal permutations. For $r \leq 5$, the final results are similar to the ones on classical Feistel networks, except for the known plaintext attack on 3 rounds. Table 7.5 gives also the number of computations needed to distinguish with high probability a r-round Feistel permutation generator (with round permutations), from a random permutation generator. The computations are quite similar to those performed for attacks against one permutation and we omitted them. Here, things are a little different than for Feistel networks with round functions. For instance for 6 rounds, the attacks can a priori no longer be done with $O(2^{2n})$ computations. In fact, more generally for $3i$ rounds ($i \geq 2$), the attacks seem always harder to perform.

Table 7.5 Maximum number of computations needed to get an attack on a r-round Feistel network with internal *permutations*. We write (+) when the complexity is worse than for classical Feistel networks

Number r of rounds	KPA	NCPA	CPA	CCA-1	CCA
1	1	1	1	1	1
2	$2^{n/2}$	2	2	2	2
3	2^n(+)	$2^{n/2}$	$2^{n/2}$	$2^{n/2}$	3
4	2^n	$2^{n/2}$	$2^{n/2}$	$2^{n/2}$	$2^{n/2}$
5	$2^{3n/2}$	2^n	2^n	2^n	2^n
6	2^{2n}	2^{2n}	2^{2n}	2^{2n}	2^{2n}
7	2^{3n}	2^{3n}	2^{3n}	2^{3n}	2^{3n}
8	2^{4n}	2^{4n}	2^{4n}	2^{4n}	2^{4n}
9	2^{6n}(+)	2^{6n}(+)	2^{6n}(+)	2^{6n}(+)	2^{6n}(+)
10	2^{6n}	2^{6n}	2^{6n}	2^{6n}	2^{6n}
11	2^{7n}	2^{7n}	2^{7n}	2^{7n}	2^{7n}
12	2^{9n}(+)	2^{9n}(+)	2^{9n}(+)	2^{9n}(+)	2^{9n}(+)
$r \geq 6$, $r = 0$ *mod* 3	$2^{(r-3)n}$(+)	$2^{(r-3)n}$(+)	$2^{(r-3)n}$(+)	$2^{(r-3)n}$(+)	$2^{(r-3)n}$(+)
$r \geq 6$, $r =$ 1 or 2 *mod* 3	$2^{(r-4)n}$	$2^{(r-4)n}$	$2^{(r-4)n}$	$2^{(r-4)n}$	$2^{(r-4)n}$

References

Aiollo, W., and R. Venkatesan. 1996. Foiling birthday attacks in length-doubling transformations—benes: a non-reversible alternative to feistel. In *Advances in cryptology—EUROCRYPT 1996*, vol. 1070, ed. U. Maurer. Lecture notes in computer science, 307–320. Berlin: Springer.

Aoki, K., T. Itchikawa, M. Kanda, M. Matsui, S. Moriai, J. Nakajima, and T. Tokita. 2000. CamelliaÃƒÂ¢Ã¢âŁšÂ¬Ã‚Â¯: A 128-bit block cipher suitable for multiple platforms—design and analysis. In *Selected areas in cryptography—SAC '00*, ed. D. R. Stinson and S. Tavares. Lecture notes in computer science, 39–56. Berlin: Springer.

Biham, E. 1997. Cryptanalysis of ladder-DES. In *Fast software encryption—FSE '97*, vol. 1267, ed. E. Biham. Lecture notes in computer science, 134–138. Berlin: Springer.

Knudsen, L. R. 1998. DEAL—A 128-bit block cipher. Technical report #151, University of Bergen, Department of Informatics, Norway. Submitted as a candidate for the Advanced Encryption Standard. Available at http://www.ii.uib.no/~larsr/newblock.html

Knudsen, L. R. 2002. The security of feistel ciphers with six rounds or less. *Journal of Cryptology* 15: 207–222.

Luby, M., and C. Rackoff. 1988. How to construct pseudorandom permutations from pseudorandom functions. *SIAM Journal on Computing* 17 (2): 373–386 April.

Nyberg, K. 1995. Linear approximation of block ciphers. In *Advances in cryptology—EUROCRYPT 1994*, vol. 950, ed. A. de Santis. Lecture notes in computer science, 439–444. Berlin: Springer.

Patarin, J. 2004. Generic attacks on feistel schemes. In *Advances in cryptology- ASIACRYPT 2001*, vol. 2248, ed. C. Boyd. Lecture notes in computer science, 222–238. Berlin: Springer.

Piret, G. 2006. Luby-rackoff revisited: on the use of permutations as inner functions of a feistel scheme. *Designs, Codes and Criptography* 39 (2): 233–245.

Rijmen, R., B. Preneel, and E. De Win. 1997. On weakness of non-surjective round functions. *Designs, Codes and Criptography* 12 (3): 253–266.

Treger, J., and J. Patarin. 2009. Generic attacks on feistel networks with internal permutations. In *Progress in cryptology—AFRICACRYPT 2009. AFRICACRYPT 2009*, vol. 5580, ed. B. Preneel. Lecture notes in computer science. Berlin: Springer.

Chapter 8
Generic Attacks on Contracting Feistel Ciphers

Abstract This chapter deals with generic attacks on unbalanced Feistel ciphers with contracting functions. These ciphers are used to construct pseudo-random permutations from kn bits to kn bits by using d pseudo-random functions from $(k-1)n$ bits to n bits. The study concerns KPA and NCPA against these schemes with less than 2^{kn} plaintext/ciphertext pairs and complexity strictly less than $O(2^{kn})$ for a number of rounds $d \leq 2k-1$. Consequently at least $2k$ rounds are necessary to avoid generic attacks. For $k=3$, there exists attacks up to 6 rounds, so 7 rounds are required. When $r \geq 2k$, it is possible to attack permutation generators instead of one permutation. Some results on contracting Feistel schemes or on small transformations of these schemes can be found in Lucks (1996), Naor and Reingold (1999). In Naor and Reingold (1999), Naor and Reingold studied the security of contracting Feistel schemes that begin and end with pairwise independent permutations. They provide lower bounds for the security of such schemes. Lucks (1996) gives some security results on contracting Feistel schemes built with hash functions. Birthday bound security results are given in Yun et al. (2011), first results above the birthday bound are proved in Patarin (2010). Security results based on the coupling method are given in Hoang and Rogaway (2010). Generic attacks on contracting Feistel ciphers are studied in Patarin et al. (2006). A large number of attacks use the variance method described in Chap. 5.

8.1 Definition Notation

Let us start with the definition of the 1-round unbalanced Feistel transformation with one contracting function.

Definition 8.1 Let $k \in \mathbb{N}$ and f a function from $\{0,1\}^{(k-1)n}$ to $\{0,1\}^n$. The *1-round unbalanced Feistel network with contracting functions*, denoted $G_k(f)$, is a function from $\{0,1\}^{kn}$ to $\{0,1\}^{kn}$ defined by (see also Fig. 8.1):

$$\forall (I_1, \ldots, I_k) \in (\{0,1\}^n)^k,\ G_k(f)([I_1, \ldots, I_k]) = [I_2, \ldots, I_{k-1}, I_1 \oplus f([I_2, \ldots, I_k])]$$

J. Patarin et al., *Guide to Feistel Ciphers*, Information Security and Cryptography,
https://doi.org/10.1007/978-3-031-99850-8_8

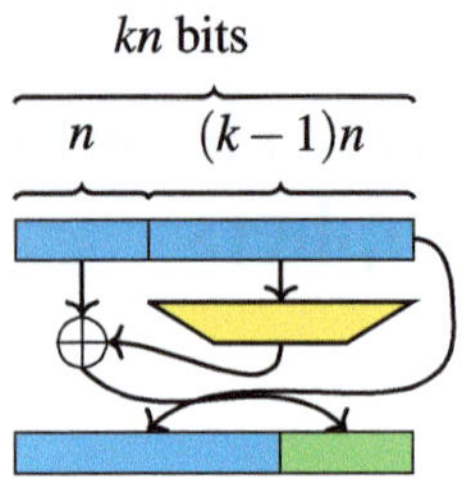

Fig. 8.1 One round for an unbalanced Feistel network with contracting functions

So, if $G_k(f)([I_1, \dots, I_k]) = [S_1, \dots, S_k]$ we have

$$\begin{cases} \forall j,\ 1 \le j \le k-1,\ S_j = I_{j+1} \\ S_k = I_1 \oplus f([I_2, \dots, I_k]). \end{cases}$$

Proposition 8.1 *For any function f from $\{0, 1\}^{(k-1)n}$ to $\{0, 1\}^n$, $G_k(f)$ is a permutation of $\{0, 1\}^{kn}$.*

Proof We have $\forall j,\ 2 \le j \le k,\ I_j = S_{j+1}$ and $I_1 = S_k \oplus f_1[S_1, \dots, S_{k-1}]$.

Definition 8.2 Let $k \in \mathbb{N}$, $r \ge 1$ and let $f_1, \dots, f_r$ be r function from $\{0, 1\}^{(k-1)n}$ to $\{0, 1\}^n$. The *r-round unbalanced Feistel network with contracting functions* associated with $f_1, \dots, f_r$, denoted $G_k^r(f_1, \dots, f_r)$ is the function from $\{0, 1\}^{kn}$ to $\{0, 1\}^{kn}$ defined by

$$G_k^r(f_1, \dots, f_r) = G_k(f_r) \circ \dots \circ G_k(f_2) \circ G_k(f_1).$$

Theorem 8.1 *For any functions $f_1, \dots, f_r$ from $\{0, 1\}^{(k-1)n}$ to $\{0, 1\}^n$, the construction $G_k^r(f_1, \dots, f_r)$ is a permutation of $\{0, 1\}^{kn}$.*

Proof $G_k^r(f_1, \dots, f_r)$ is a permutation of $\{0, 1\}^{kn}$ since it is the composition of r permutations of $\{0, 1\}^{kn}$. □

Definition 8.3 Let $r \ge 1$. The *r-round unbalanced Feistel transformation with contracting functions*, denoted G_k^r, maps a tuple of functions $(f_1, \dots, f_r) \in \left((\{0, 1\}^n)^{\{0,1\}^{(k-1)n}}\right)^r$ to the permutation $G_k^r(f_1, \dots, f_r)$ of $\{0, 1\}^{kn}$ as defined by Definition 2.2.

From unbalanced Feistel networks with contracting functions, we can finally define unbalanced Feistel ciphers with contracting functions, by letting round functions depend on secret keys.

Definition 8.4 Let $r \geq 1$ and let $F = (f_K)$ be a family of functions in $(\mathscr{F}_n^{\mathscr{F}(k-1)n})$ indexed by a set $\mathscr{K}$. The *r-round unbalanced Feistel cipher with contracting functions* associated with F is the block cipher with key space $\mathscr{K}^r$ and message space $\{0, 1\}^{kn}$ which maps a key $(K_1, \ldots, K_r) \in \mathscr{K}^r$ and a plaintext $[I_1, \ldots, I_k] \in \{0, 1\}^{kn}$ to the ciphertext

$$G_k^r(f_{K_1}, \ldots, f_{K_r})([I_1, \ldots, I_k]).$$

In other words, the permutation of $\{0, 1\}^{kn}$ associated with key $(K_1, \ldots, K_r)$ is the unbalanced Feistel network with contracting functions $G_k^r(f_{K_1}, \ldots, f_{K_r})$.

The definition of G_k^r shows that at each intermediate round t, $1 \leq t \leq r$, there is only one n-bit value which is modified and becomes the k-th coordinates of the output. This value is called an internal variable and it is denoted by X^t. For example, we have

$$\begin{aligned}
X^1 &= I_1 \oplus f_1([I_2, \ldots, I_k]),\\
X^2 &= I_2 \oplus f_2([I_3, \ldots, I_k, X^1]),\\
X^3 &= I_3 \oplus f_3([I_4, \ldots, I_k, X^1, X^2]),\\
&\ldots\\
X^k &= I_k \oplus f_k([X^1, X^2, \ldots, X^{k-1}]),\\
\forall t,\ k \leq t \leq r, \quad X^t &= X^{t-k} \oplus f_t([X^{t-k+1}, \ldots, X^{t-1}]).
\end{aligned}$$

After r rounds, the output is $[S_1, \ldots, S_k] = [X^{r-k+1}, X^{r-k+2}, \ldots, X^{r-1}, X^r]$.

The attacks described in this chapter against unbalanced Feistel ciphers with contracting functions are 2-point attacks and they use pairs of points and partial differential on these pairs of points. Suppose that the attack needs q messages, then $\forall i, 1 \leq i \leq q, [I_1(i), \ldots, I_k(i)]$ represents the input numbered i and the corresponding output is denoted by $[S_1(i), \ldots, S_k(i)]$.

We end this section by giving the signature of contracting Feistel ciphers. It is a corollary of Theorem 5.1

Proposition 8.2 *G_k^r has an even signature as soon as $n \geq 2$.*

Proof For G_k^r, the round functions are from $\{0, 1\}^{(k-1)n}$ to $\{0, 1\}^n$. Thus, Theorem 5.1 shows that G_k^r has an even signature when $n \geq 2$. □

This allows attack by the signature.

8.2 Simple Attacks on the First k Rounds

This section is devoted to present simple attacks on the first k rounds.

8.2.1 Attacks on G_k^r for $1 \leq r \leq k-1$

It is easy to see that after rounds r, $1 \leq r \leq k-1$, we have $S_1 = I_{r+1}$. This leads to a KPA on G_k^r.

Let r such that $1 \leq r \leq k-1$. Consider the following KPA-distinguisher D:

1. D makes a query to the oracle, and receives a random plaintext $[I_1, \ldots, I_k]$ together with $[S_1, \ldots, S_k] = \mathcal{O}([I_1, \ldots, I_k])$;
2. if $S_1 = I_{r+1}$, D outputs 1, otherwise it outputs 0.

Clearly, when $\mathcal{O} = G_k^1(f_1, \ldots, f_r)$, D *always* outputs 1 since

$$G_k^1(f_1)([I_1, \ldots, I_k]) = [S_1, \ldots, S_k] \iff \begin{cases} S_1 = I_{r+1} \\ \forall t,\ 2 \leq t \leq k-r,\ S_t = I_{r+t} \\ \forall t,\ k-r+1 \leq t \leq r,\ S_t = X^{t-k+r}. \end{cases}$$

On the other hand, when $\mathcal{O}$ is a random permutation of $\{0,1\}^{kn}$, then $[S_1, \ldots, S_k]$ is uniformly random in $\{0,1\}^{kn}$, and the probability that $S_1 = I_{r+1}$ (and hence that D outputs 1) is exactly 2^{-n}. Therefore, by definition of the advantage (cf. Definition 1.3) we have

$$\mathcal{A}_{G_k^1}(\mathsf{D}) = 1 - \frac{1}{2^n}.$$

Since D makes exactly one query, it follows that

$$\mathcal{A}_{G_k^1}^{\text{KPA}}(1) \geq 1 - \frac{1}{2^n}.$$

Hence, there is a very efficient known-plaintext attack against G_k^r, making only one query and distinguishing G_k^r from a random permutation with probability negligibly close to one.

8.2.2 NCPA and KPA on G_k^k

Let r such that $1 \leq r \leq k-1$.

Consider the following NCPA-distinguisher D:

1. D chooses $I_1(1), I_1(2) \in \{0,1\}^n$ with $I_1(1) \neq I_1(2)$, and queries $[S_1(1), \ldots, S_k(1)] = \mathcal{O}([I_1(1), 0, \ldots, 0])$ and $[S_1(2), \ldots, S_k(2)] = \mathcal{O}([I_1(2), 0, \ldots, 0])$

2. Whether $S_1(1) \oplus S_1(2) = I_1(1) \oplus I_1(2)$; if this holds, D outputs 1, otherwise D outputs 0.

By definition of G_k^k, we have $S_1 = X^1 = I_1 \oplus f_1([0, 0, \ldots 0])$. Hence, when $\mathcal{O} = G_k^k(f_1, \ldots, f_k)$, D always outputs 1.

When $\mathcal{O}$ is a random permutation of $\{0, 1\}^{kn}$, then $[S_1(2), \ldots, S_k(2)]$ is uniformly random in $\{0, 1\}^{kn} \setminus \{[S_1(1), \ldots, S_k(1)]\}$. Since there are exactly $2^{(k-1)n}$ possible values of $\{0, 1\}^{kn}$, then $[S_1(2), \ldots, S_k(2)]$ in $\{0, 1\}^{kn} \setminus \{[S_1(1), \ldots, S_k(1)]\}$ such that $S_1(2) = S_1(1) \oplus I_1(1) \oplus I_1(2)$, D outputs 1 with probability

$$\frac{2^{(k-1)n}}{2^{kn} - 1}.$$

Hence, we have, by definition of the advantage,

$$\mathcal{A}_{G_k^k}(\mathsf{D}) = 1 - \frac{2^{(k-1)n}}{2^{kn} - 1}.$$

Since D makes exactly two queries, this implies

$$\mathcal{A}_{G_k^k}^{\mathrm{NCPA}}(\mathsf{D}) \geq 1 - \frac{2^{(k-1)n}}{2^{kn} - 1}.$$

Hence, there is a very efficient non-adaptive chosen-plaintext attack against G_k^k, making only two queries and distinguishing G_k^k from a random permutation with probability negligibly close to one.

This attack can be transformed into a KPA with $O(2^{\frac{(k-1)n}{2}})$ random queries as follows.

Consider the following KPA-distinguisher D:

1. D makes about $q = 2^{\frac{(k-1)n}{2}}$ to the oracle and receives q random plaintexts $[I_1(i), \ldots I_k(i)]$ together with $[S_1(i), \ldots S_k(i)] = \mathcal{O}([I_1(i), \ldots I_k(i)])$ for $1 \leq i \leq q$.
2. If D obtains two distinct indices i and j such that

$$\forall t,\ 2 \leq t \leq k,\ I_t(i) = I_t(j) \text{ and } S_1(i) \oplus S_1(j) = I_1(i) \oplus I_1(j)$$

D outputs 1, otherwise it outputs 0.

The probability that D outputs 1 is equal to the following probability:

$$\begin{aligned}
&\Pr[\exists i, j,\ i \neq j,\ \forall t,\ 2 \leq t \leq k,\ I_t(i) = I_t(j) \text{ and } S_1(i) \oplus S_1(j) = I_1(i) \oplus I_1(j)] = \\
&\Pr[S_1(i) \oplus S_1(j) = I_1(i) \oplus I_1(j) \mid \exists i, j,\ i \neq j,\ \forall t,\ 2 \leq t \leq k,\ I_t(i) = i_t(j)] \\
&\times \Pr[\exists i, j,\ i \neq j,\ \forall t,\ 2 \leq t \leq k,\ I_t(i) = I_t(j)].
\end{aligned}$$

When $\mathcal{O} = G_k^k(f_1, \ldots, f_k)$, we have

$$\Pr\left[S_1(i) \oplus S_1(j) = I_1(i) \oplus I_1(j) \mid \exists i, j,\ i \neq j,\ \forall t,\ 2 \leq t \leq k,\ I_t(i) = I_t(j)\right] = 1$$

and when $\mathcal{O}$ is a random permutation, we have

$$\Pr\left[S_1(i) \oplus S_1(j) = I_1(i) \oplus I_1(j) \mid \exists i, j,\ i \neq j,\ \forall t,\ 2 \leq t \leq k,\ I_t(i) = i_t(j)\right] = \frac{2^{(k-1)n}}{2^{kn} - 1}.$$

Moreover, when q is about $2^{\frac{(k-1)n}{2}}$, by the birthday paradox,

$$\Pr\left[\exists i, j,\ i \neq j,\ \forall t,\ 2 \leq t \leq k,\ I_t(i) = I_t(j)\right] \geq 1/2.$$

This shows that when $q = 2^{\frac{(k-1)n}{2}}$

$$\mathcal{A}_{G_k^k}^{\mathrm{KPA}}(\mathsf{D}) \geq 1/2\left(1 - \frac{2^{(k-1)n}}{2^{kn} - 1}\right).$$

8.3 Generic Attacks When $k = 3$

First, we study G_3^r ciphers since this case is slightly different from the general case $k \geq 4$. We have $[S_1(i), S_2(i), S_3(i)] = G_3^r([I_1(i), I_2(i), I_3(i)])$. Attacks are given for $r \geq 4$ since for $r = 1, 2, 3$ attacks are given in the previous section. For $r \geq 4$, the methods presented in Chap. 5 will be used.

8.3.1 Attacks on 4 Rounds: G_3^4

After 4 rounds, the output is $[S_1, S_2, S_3]$ with

$$\begin{aligned} S_1 &= X^2 = I_2 \oplus f_2([I_3, X^1]) \\ S_2 &= X^3 = I_3 \oplus f_3([X^1, X^2]) \\ S_3 &= X^4 = X^1 \oplus f_4([X^2, X^3]). \end{aligned}$$

8.3.1.1 NCPA with $O(2^{\frac{n}{2}})$ Queries and KPA with $0(2^n)$ Queries

The attacker chooses q messages $[I_1(i), I_2(i), I_3(i)]$ such that $\forall i,\ 1 \leq i \leq q, I_3(i) = 0$, and $\forall i, j,\ i \neq j,\ I_2(i) \neq I_2(j)$. Then the attacker sends them to the oracle $\mathcal{O}$ and receives the output $[S_1(i), S_2(i), S_3(i)]$. Then she will count the number of pairs

(i, j) with $i < j$ such that $I_2(i) \oplus I_2(j) = S_1(i) \oplus S_1(j)$. This defines a random variable. If $\mathcal{O}$ is a random permutation, this random variable is denoted by $\mathcal{N}_{perm}$. If $\mathcal{O} = G_3^4(f_1, f_2, f_3, f_4)$, it is denoted by $\mathcal{N}_{G_3^4}$.

Let us introduce the following random variables:

$$\begin{aligned} \delta_{i,j} &= 1 \text{ if } I_2(i) \oplus I_2(j) = S_1(i) \oplus S_1(j) \\ &= 0 \text{ otherwise.} \end{aligned}$$

Then $\mathcal{N}_{perm} = \sum_{i<j} \delta_{i,j}$ and $\mathcal{N}_{G_3^4} = \sum_{i<j} \delta_{i,j}$ as well.

Suppose that $\mathcal{O}$ is a random permutation. Let i, j such that $i < j$. Let us compute $\mathbb{E}(\delta_{i,j})$. For $I(i)$, there are 2^{2n} possibilities since $I_3(i) = 0$. When $I(i)$ is fixed, there are $(2^n - 1)2^n$ possibilities for $I(j)$ since $I_2(j) \neq I_2(i)$. Now for $S(i)$, there are 2^{3n} possibilities and when $S(i)$ is fixed, there are 2^{2n} possibilities for $S(j)$. The total number of possibilities for $I(i)$, $I(j)$ pairwise distinct and such that for all i, $I_3(i) = 0$ is $2^{2n}(2^{2n} - 1)$ and the total number of $S(i)$ and $S(j)$ pairwise distinct is $2^{3n}(2^{3n} - 1)$. This gives

$$\mathbb{E}(\delta_{i,j}) = \frac{2^{8n}(2^n - 1)}{2^{5n}(2^{2n} - 1)(2^{3n} - 1)}.$$

This implies that $\mathbb{E}(\mathcal{N}_{perm}) \simeq \frac{q(q-1)}{2.2^n}$.

For G_3^4, we have

$$I_2(i) \oplus I_2(j) = S_1(i) \oplus S_1(j) \Leftrightarrow f_2([0, X^1(i)]) = f_2([0, X^1(j)])$$

$$I_2(i) \oplus I_2(j) = S_1(i) \oplus S_1(j) \Leftrightarrow \begin{cases} X^1(i) = X^1(j) \\ \text{or} \\ X^1(i) \neq X^1(j) \\ \quad \text{and } f_2([0, X^1(i)]) = f_2([0, X^1(j)]). \end{cases}$$

Thus we obtain $\mathbb{E}(\delta_{i,j}) = \frac{2^n}{2^n(2^n-1)} + \left(1 - \frac{2^n}{2^n(2^n-1)}\right)\frac{1}{2^n}$ and $\mathbb{E}(\mathcal{N}_{G_3^4}) \simeq \frac{q(q-1)}{2^n}$. This shows that the expectation obtained for a G_3^4 network is twice the expectation obtained for a random permutation. Thus implies that when q is about $2^{n/2}$ it is possible to distinguish a random permutation from a G_3^4 network. This generic attack requires $O(2^{\frac{n}{2}})$ random queries and $O(2^{\frac{n}{2}})$ computations.

This attack can be transformed into KPA with $m \simeq 2^n$, as follows. The attacker receives q random plaintexts $[I_1(i), I_2(i), I_3(i)]$, $1 \leq i \leq q$, together with $[S_1(i), S_2(i), S_3(i)] = \mathcal{O}([I_1(i), I_2(i), I_3(i)])$. Then she will count the number of pairs (i, j) with $i < j$ such that

$$\begin{cases} I_3(i) = I_3(j) \\ I_2(i) \oplus I_2(j) = S_1(i) \oplus S_1(j). \end{cases}$$

By using similar computations to those done for the previous NCPA, it is possible to show that $\mathbb{E}(\mathcal{N}_{perm}) \simeq \frac{q(q-1)}{2.2^{2n}}$ and $\mathbb{E}(\mathcal{N}_{G_3^4}) \simeq \frac{q(q-1)}{2^{2n}}$. This gives a generic KPA with $O(2^n)$ random queries and $O(2^n)$ computations.

8.3.2 *Attacks on 5 Rounds: G_3^5*

After 5 rounds, the output is $[S_1, S_2, S_3]$ with

$$\begin{aligned} S_1 &= X^3 = I_3 \oplus f_3([X^1, X^2]) \\ S_2 &= X^4 = X^1 \oplus f_4([X^2, X^3]) \\ S_3 &= X^5 = X^2 \oplus f_4([X^3, X^4]). \end{aligned}$$

8.3.2.1 NCPA with 2^n Messages

The attacker chooses q messages $[I_1(i), I_2(i), I_3(i)]$ such that $\forall i, I_2(i) = 0, I_3(i) = 0$ and the $I_1(i)$ values are pairwise distinct. Notice that this directly implies $X^1(i) \oplus X^1(j) = I_1(i) \oplus I_1(j)$, so the $X^1(i)$ values are pairwise distinct. Then the attacker sends them to the oracle $\mathcal{O}$ and receives the output $[S_1(i), S_2(i), S_3(i)]$. Then she will count the number of pairs $(i, j), i < j$ such that $S_1(i) = S_1(j)$ and $I_1(i) \oplus I_1(j) = S_2(i) \oplus S_2(j)$. Suppose that $\mathcal{O}$ is a random permutation. Then it is easy to see that $\mathbb{E}(\mathcal{N}_{perm}) = \frac{q(q-1)}{2} \cdot \frac{2^n}{2^{3n}-1}$. Thus

$$\mathbb{E}(\mathcal{N}_{perm}) \simeq \frac{q(q-1)}{2.2^{2n}}.$$

Now if $\mathcal{O}$ is a G_3^5 network, since for all i, $I_2(i) = I_3(i) = 0$, we have

$$\begin{cases} S_1(i) = S_1(j) \\ \text{and} \\ I_1(i) \oplus I_1(j) = S_2(i) \oplus S_2(j) \end{cases} \Leftrightarrow \begin{cases} f_1([0, X^1(i)]) = f_1([0, X^1(j)]) \\ \text{and} \\ f_4([X^2(i), S_1(i)]) = f_4([X^2(j), S_1(j)]). \end{cases}$$

Since $X^1(i)$ and $X^1(j)$ are pairwise distinct, the probability to have

$$f_1([0, X^1(i)]) = f_1([0, X^1(j)])$$

is $\frac{1}{2^n}$. Once we know that $S_1(i) = S_1(j)$, then

$$f_4([X^2(i), S_1(i)]) = f_4([X^2(j), S_1(j)]) \Leftrightarrow \begin{cases} X^2(i) = X^2(j) \\ \text{or} \\ X^2(i) \neq X^2(j) \text{ and } f_4([X^2(i), S_1(i)]) = \\ \qquad f_4([X^2(j), S_1(j)]). \end{cases}$$

This implies that $\mathbb{E}(\mathcal{N}_{G_3^5}) = \frac{1}{2^n} \times \frac{1}{2^n} + \frac{1}{2^n}(1 - \frac{1}{2^n}) \times \frac{1}{2^n}$. Thus

$$\mathbb{E}(\mathcal{N}_{G_3^5}) \simeq \frac{q(q-1)}{2^{2n}}.$$

Again the expectation for a cipher is twice the expectation for a random permutation. This provides a NCPA with $O(2^n)$ random queries and $O(2^n)$ computations.

As before this attack leads to a KPA attack with 2^{2n} messages. But there is a better attack described below.

8.3.2.2 KPA with $m = 2^{\frac{3n}{2}}$ Messages

The attacker receives q random plaintexts $[I_1(i), I_2(i), I_3(i)]$, $1 \leq i \leq q$, together with $[S_1(i), S_2(i), S_3(i)] = \mathcal{O}([I_1(i), I_2(i), I_3(i)])$. The inputs and the outputs are pairwise distinct. Then she will count the number of pairs (i, j) with $i < j$ such that

$$\left\{ I_3(i) \oplus I_3(j) = S_1(i) \oplus S_1(j). \right.$$

Let us introduce the following random variables:

$$\begin{aligned} \delta_{i,j} &= 1 \text{ if } I_3(i) \oplus I_3(j) = S_1(i) \oplus S_1(j) \\ &= 0 \text{ otherwise.} \end{aligned}$$

Suppose that $\mathcal{O}$ is a random permutation. Let i, j such that $i < j$. Since the $\delta_{i,j}$ are Bernoulli variables, we have $\mathbb{E}(\delta_{i,j}) = \Pr[I_3(i) \oplus I_3(j) = S_1(i) \oplus S_1(j)]$. Then we have

$$\begin{aligned} &\Pr[I_3(i) \oplus I_3(j) = S_1(i) \oplus S_1(j)] = \\ &\Pr[I_3(i) \oplus I_3(j) = S_1(i) \oplus S_1(j)/(I_3(i) = I_3(j))] \Pr[I_3(i) = I_3(j)] \\ &+ \Pr[I_3(i) \oplus I_3(j) = S_1(i) \oplus S_1(j)/(I_3(i) \neq I_3(j)] \Pr[I_3(i) \neq I_3(j)]. \end{aligned}$$

First we study the case where $I_3(i) = I_3(j)$. The probability is given by

$$\left(\frac{2^{3n}(2^{2n} - 1)}{2^{3n}(2^{3n} - 1)} \right)^2 = \frac{(2^{2n} - 1)^2}{(2^{3n} - 1)^2}$$

Then we suppose that $I_3(i) \neq I_3(j)$ and in that case, the probability is

$$\frac{2^{3n}2^{2n}(2^n - 1)}{2^{3n}(2^{3n} - 1)} \times \frac{2^{3n}2^{2n}}{2^{3n}(2^{3n} - 1)} = \frac{2^{4n}(2^n - 1)}{(2^{3n} - 1)^2}$$

Finally, we obtain

$$\mathbb{E}(\delta_{i,j}) = \frac{2^{5n} - 2.2^n + 1}{(2^{3n} - 1)^2} = \frac{1}{2^{2n}} + \frac{1}{2^{6n}} - \frac{1}{2^{7n}} + \frac{2}{2^{9n}} - \frac{2}{2^{10n}} + \frac{3}{2^{12n}} - \frac{3}{2^{13n}} + 0(\frac{1}{2^{15n}}).$$

Hence

$$\begin{aligned}\mathbb{E}(\mathcal{N}_{perm}) &= \tfrac{q(q-1)}{2} \tfrac{2^{5n}-2.2^n+1}{(2^{3n}-1)^2} \\ &= \tfrac{q(q-1)}{2}\left(\tfrac{1}{2^n} + \tfrac{1}{2^{6n}} - \tfrac{1}{2^{7n}} + \tfrac{2}{2^{9n}} - \tfrac{2}{2^{10n}} + \tfrac{3}{2^{12n}} - \tfrac{3}{2^{13n}} + 0(\tfrac{1}{2^{15n}})\right)\end{aligned}$$

Suppose now that $\mathcal{O}$ is a G_3^5 network. Then we have

$$I_3(i) \oplus I_3(j) = S_1(i) \oplus S_1(j) \Leftrightarrow f_3([X^1(i), X^2(j)]) = f_3([X^1(j), X^2(j)]).$$

Moreover

$$f_3([X^1(i), X^2(i)]) = f_3([X^1(j), X^2(j)]) \Leftrightarrow \begin{cases} (X^1(i), X^2(i)) = (X^1(j), X^2(j)) \\ \text{or} \\ \begin{cases} (X^1(i), X^2(i)) \neq (X^1(j), X^2(j)) \\ \text{and} \\ f_3([X^1(i), X^2(i)]) = \\ \qquad f_3([X^1(j), X^2(j)]). \end{cases} \end{cases}$$

Since after each round the outputs are pairwise distinct, it is not possible to have after round two the following equality: $(I_3(i), X^1(i), X^2(i)) = (I_3(j), X^1(j), X^2(j))$. The only possibility to have $(X^1(i), X^2(i)) = (X^1(j), X^2(j))$ is when the following conditions are satisfied:

$$\begin{gathered} I_2(i) = I_2(j),\ I_3(i) \neq I_3(j),\ X^1(i) = X^1(j),\ X^2(i) = X^2(j) \\ \text{or} \\ I_2(i) \neq I_2(j),\ I_3(i) \neq I_3(j),\ X^1(i) = X^1(j),\ X^2(i) = X^2(j). \end{gathered}$$

This shows that

$$\begin{aligned}\Pr\left[f_3([X^1(i), X^2(i)]) = f_3([X^1(j), X^2(j)]\right] &= \frac{2^{4n}(2^n - 1)^2 + 2^{4n}(2^n - 1)}{2^{3n}(2^{3n} - 1)} \times \left(\frac{1}{2^n}\right)^2 \\ &= \frac{2^n - 1}{2^{3n} - 1}.\end{aligned}$$

Thus

$$\mathbb{E}(\delta_{i,j}) = \frac{2^n - 1}{2^{3n} - 1} + \left(1 - \frac{2^n - 1}{2^{3n} - 1}\right)\frac{1}{2^n} = \frac{2^{2n} + 2^n - 2}{(2^{3n} - 1)}$$

and

$$\mathbb{E}(\mathcal{N}_{G_3^5}) = \frac{q(q-1)}{2}\frac{2^{2n}+2^n-2}{(2^{3n}-1)} = \frac{q(q-1)}{2}\left(\frac{1}{2^n}+\frac{1}{2^{2n}}-2\frac{1}{2^{3n}}+O(\frac{1}{2^{4n}})\right)$$

Here both expectations are of the same order, thus we need to use the method with the standard deviation as explained in Chap. 5. This means that it is necessary to compute the standard deviation when $\mathcal{O}$ is a random permutation and when $\mathcal{O}$ is a G_3^5 network.

We first compute the standard deviation when $\mathcal{O}$ is a random permutation. Let us recall the "Covariance formula":

$$V(\sum_{i<j}\delta_{i,j}) = \sum_{i<j}V(\delta_{i,j}) + \sum_{i<j,p<l,(i,j)\neq(p,l)}\left[E(\delta_{i,j}\,\delta_{p,l}) - E(\delta_{i,j})\,E(\delta_{p,l})\right].$$

We have

$$\sum_{i<j}V(\delta_{i,j}) = \sum_{i<j}\mathbb{E}(\delta_{i,j}) - (\mathbb{E}(\delta_{i,j}))^2 = \frac{q(q-1)}{2}\left(\frac{1}{2^n}-\frac{1}{2^{2n}}+0(\frac{1}{2^{6n}})\right).$$

We will now compute the second term which is the covariance term and we will see that it is negligible compared to the first one. Since $\mathbb{E}(\delta_{i,j})\,\mathbb{E}(\delta_{p,\ell}) = \frac{1}{2^{2n}}+\frac{2}{2^{7n}}-\frac{2}{2^{8n}}+\frac{4}{2^{10n}}-\frac{4}{2^{11n}}+\frac{1}{2^{12n}}+\frac{4}{2^{13n}}-\frac{5}{2^{14n}}+O(\frac{1}{2^{15n}})$, we just have to compute:

$$\mathbb{E}(\delta_{i,j}\,\delta_{p,\ell}) = \Pr\left[I_3(i)\oplus I_3(j) = S_1(i)\oplus S_1(j) \text{ and } I_3(p)\oplus I_3(\ell) = S_1(p)\oplus S_1(\ell)\right].$$

To exactly compute the covariance term, we need to separate the computation into several cases. We first consider the case where i, j, p, ℓ are pairwise distinct. There are many sub-cases to study. We can have for example the I_3 values all distinct and $I_3(i)\oplus I_3(j)\oplus I_3(p)\oplus I_3(\ell)\neq 0$ or $I_3(i)\oplus I_3(j)\oplus I_3(p)\oplus I_3(\ell) = 0$. We can also have $I_3(i) = I_3(j)$ (hence $S_1(i) = S_1(j)$) and $I_3(p)\oplus I_3(\ell) = S_1(p)\oplus S_1(\ell)\neq 0$. And it is possible to exchange the role of (i, j) and (p, ℓ). Finally, the study of all the possibilities gives

$$\begin{aligned}\mathbb{E}(\delta_{i,j}\,\delta_{p,\ell}) &= \frac{\frac{1}{2^{22n}}-\frac{12}{2^{19n}}+\frac{2}{2^{17n}}+\frac{56}{2^{16n}}-\frac{20}{2^{14n}}-\frac{124}{2^{13n}}+\frac{1}{2^{12n}}+\frac{80}{2^{11n}}+\frac{112}{2^{10n}}-\frac{12}{2^{9n}}-\frac{120}{2^{8n}}+\frac{36}{2^{6n}}}{(2^{3n}(2^{3n}-1)(2^{3n}-2)(2^{3n}-3))^2}\\ &= \frac{1}{2^{2n}}+\frac{2}{2^{7n}}-\frac{2}{2^{8n}}+\frac{4}{2^{10n}}-\frac{4}{2^{11n}}+\frac{1}{2^{12n}}+\frac{12}{2^{13n}}-\frac{13}{2^{14n}}+O(\frac{1}{2^{15n}}).\end{aligned}$$

Thus we obtain

$$\mathbb{E}(\delta_{i,j}\,\delta_{p,\ell}) - \mathbb{E}(\delta_{i,j})\,\mathbb{E}(\delta_{p,\ell}) = \frac{8}{2^{13n}}-\frac{8}{2^{14n}}+O(\frac{1}{2^{15n}}).$$

The dominant term in the covariance term is in $O(\frac{q^4}{2^{13n}})$. Notice that $\frac{q^4}{2^{13n}} \ll \frac{q^2}{2^n} \Leftrightarrow q \ll 2^{6n}$. This is always the case since the maximal number of messages is 2^{3n}. In that case, the covariance term is negligible compared to $\sum_{i<j} \mathbb{E}(\delta_{i,j}) - \mathbb{E}(\delta_{i,j})^2$.

Next we consider the case when in $\{i, j, \ell, p\}$ there are only 3 distinct indices. Since we always have $i \neq j$ and $p \neq \ell$ and the conditions are symmetric in (i, j) and (ℓ, p), we can assume without loss of generality that $\ell = i$ (there are 4 possibilities: $\ell = i$, $\ell = j$, $p = i$ and $p = j$). In that case, we have

$$\mathbb{E}(\delta_{i,j}\,\delta_{p,i}) = \Pr\left[I_3(i) \oplus S_1(i) = I_3(j) \oplus S_1(j) = I_3(p) \oplus S_1(p)\right].$$

Again, there are several cases. For example, it is possible to have $I_3(i) = I_3(j) = I_3(p)$. But we also may have only 2 equalities between $I_3(i)$, $I_3(j)$ and $I_3(p)$, or these values may be pairwise distinct. The study of all the cases shows that

$$\begin{aligned}\mathbb{E}(\delta_{i,j}\,\delta_{p,i}) &= \frac{\frac{1}{2^{16n}} - \frac{6}{2^{13n}} + \frac{3}{2^{11n}} + \frac{10}{2^{10n}} - \frac{12}{2^{8n}} + \frac{4}{2^{6n}}}{(2^{3n}(2^{3n}-1)(2^{3n}-2))^2}\\ &= \frac{1}{2^{2n}} + \frac{3}{2^{7n}} - \frac{3}{2^{8n}} + \frac{6}{2^{10n}} - \frac{6}{2^{11n}} + \frac{4}{2^{12n}} - \frac{3}{2^{13n}} - \frac{1}{2^{14n}} + O(\frac{1}{2^{15n}}).\end{aligned}$$

And we obtain

$$\begin{aligned}\mathbb{E}(\delta_{i,j}\,\delta_{p,i}) - \mathbb{E}(\delta_{i,j})\,\mathbb{E}(\delta_{p,i}) = &\frac{1}{2^{7n}} - \frac{1}{2^{8n}} + \frac{2}{2^{10n}} - \frac{2}{2^{11n}} + \frac{3}{2^{12n}} - \frac{7}{2^{13n}} + \frac{6}{2^{14n}}\\ &+ O(\frac{1}{2^{15n}}).\end{aligned}$$

The dominant term in the covariance term is in $O(\frac{q^3}{2^{7n}})$. Again $\frac{q^3}{2^{7n}} \ll \frac{q^2}{2^n} \Leftrightarrow q \ll 2^{6n}$ and this is always the case since the maximal number of messages is 2^{3n}. In that case, the covariance term is negligible compared to $\sum_{i<j} \mathbb{E}(\delta_{i,j}) - \mathbb{E}(\delta_{i,j})^2$ as well.

The previous study shows that $V(\mathcal{N}_{perm}) \simeq \frac{q^2}{2^n}$ and $\sigma(\mathcal{N}_{perm}) \simeq \frac{q}{2^{n/2}}$.

We now turn to the computation of the standard deviation when $\mathcal{O}$ is a G_3^5 network. First we have

$$\sum_{i<j} V(\delta_{i,j}) = \sum_{i<j} \mathbb{E}(\delta_{i,j}) - (\mathbb{E}(\delta_{i,j}))^2 = \frac{q(q-1)}{2}\left(\frac{1}{2^n} - \frac{4}{2^{3n}} + 0(\frac{1}{2^{4n}})\right).$$

Again we will show that the covariance term is negligible compared to the term computed above.

$$\mathbb{E}(\delta_{i,j})\,\mathbb{E}(\delta_{p,\ell}) = \left(\frac{2^{2n} + 2^n - 2}{(2^{3n} - 1)}\right)^2.$$

In order to simplify the proof, we define

$$p_1 = \Pr\left[f_3([X^1(i), X^2(i)]) = f_3([X^1(j), X^2(j)])\right] = \frac{2^n - 1}{2^{3n} - 1}.$$

Then we have

$$\mathbb{E}(\delta_{i,j}) = p_1 + (1 - p_1)\frac{1}{2^n} \text{ and } (\mathbb{E}(\delta_{i,j}))^2 = p_1^2 + 2(1 - p_1)\frac{1}{2^n} + (1 - p_1)^2\frac{1}{2^{2n}}.$$

The next step is the computation of $\mathbb{E}(\delta_{i,j}\,\delta_{p,\ell})$ when $\mathcal{O}$ is a G_3^5 network.

$$\begin{aligned}\mathbb{E}(\delta_{i,j}\,\delta_{p,\ell}) = \Pr\big[&f_3([X^1(i), X^2(i)]) = f_3([X^1(j), X^2(j)])) \text{ and}\\ &f_3([X^1(p), X^2(p)]) = f_3([X^1(\ell), X^2(\ell)])\big].\end{aligned}$$

First, we suppose that the indices i, j, ℓ, p are pairwise distinct. Then we have to consider several cases.

1. $[X^1(i), X^2(j)] = [X^1(j), X^2(j)]$ and $[X^1(\ell), X^2(\ell)] = [X^1(p), X^2(p)]$. The probability is given by p_1^2.
2. $[X^1(i), X^2(j)] = [X^1(j), X^2(j)]$ and $[X^1(\ell), X^2(\ell)] \neq [X^1(p), X^2(p)]$ or $[X^1(i), X^2(j)] \neq [X^1(j), X^2(j)]$ and $[X^1(\ell), X^2(\ell)] = [X^1(p), X^2(p)]$. In both cases, the probability is $p_1(1 - p_1)\frac{1}{2^n}$.
3. $[X^1(i), X^2(j)] \neq [X^1(j), X^2(j)]$ and $[X^1(\ell), X^2(\ell)] \neq [X^1(p), X^2(p)]$ and $[X^1(i), X^2(j)] = [X^1(\ell), X^2(\ell)]$ and $[X^1(j)), X^2(j)] = [X^1(p), X^2(p)]$ or $[X^1(i), X^2(j)] = [X^1(p), X^2(p)]$ and $[X^1(j)), X^2(j)] = [X^1(\ell), X^2(\ell)]$. Here the probability is $(1 - p_1)p_1^2 \times \frac{1}{2^n}$.
4. $[X^1(i), X^2(j)] \neq [X^1(j), X^2(j)]$ and $[X^1(\ell), X^2(\ell)] \neq [X^1(p), X^2(p)]$ and we are not in case 3. Then the probability is given by

$$\left[(1 - p_1)^2 - 2p_1^2(1 - p_1)\right] \times \frac{1}{2^{2n}}.$$

Finally, we have

$$\begin{aligned}\mathbb{E}(\delta_{i,j}\,\delta_{p,\ell}) = p_1^2 + 2(1 - p_1)p_1 \times \frac{1}{2^n} &+ 2(1 - p_1)p_1^2 \times \frac{1}{2^n}\\ &+ \left[(1 - p_1)^2 - 2p_1^2(1 - p_1)\right] \times \frac{1}{2^{2n}}.\end{aligned}$$

And

$$\mathbb{E}(\delta_{i,j}\,\delta_{p,\ell}) - \mathbb{E}(\delta_{i,j})\,\mathbb{E}(\delta_{p,\ell}) = 2p_1^2(1 - p_1)\left(\frac{1}{2^n} - \frac{1}{2^{2n}}\right).$$

This shows that the dominant term in the covariance term in is $\frac{q^4}{2^{5n}}$. Moreover $\frac{q^4}{2^{5n}} \ll \frac{q^2}{2^n} \Leftrightarrow q \ll 2^{2n}$. In this KPA, since we suppose that q is about $2^{3n/2}$, we obtain that the covariance term is negligible compared to $\sum_{i<j} V(\delta_{i,j})$.

Next we study the case where in $\{i, j, \ell, p\}$ there are only 3 distinct indices. As for a random permutation, we can assume that for example $\ell = i$. Then we have to estimate

$$\Pr\left[f_3([X^1(i), X^2(i)]) = f_3([X^1(j), X^2(j)])) = f_3([X^1(p), X^2(p)])\right].$$

The study is similar to the done when the four indices are pairwise distinct. Then we obtain that the dominant term in the covariance term is in $\frac{q^3}{2^{3n}}$. Since we suppose that we have about $2^{3n/2}$, we obtain that the covariance term is negligible compared to $\sum_{i<j} V(\delta_{i,j})$ Thus we obtain that $V(\mathcal{N}_{G_3^5}) \simeq \frac{q^2}{2^n}$ and $\sigma(\mathcal{N}_{G_3^5}) \simeq \frac{q}{2^{n/2}}$.

This attack is successful when the difference of both expectations is greater than the standard deviations $\sigma(\mathcal{N}_{perm})$ and $\sigma(\mathcal{N}_{G_3^5})$. This is satisfied as soon as $\frac{q^2}{2^{2n}} \geq \frac{q}{2^{n/2}}$, i.e., when $q = O(2^{3n/2})$. We have here a KPA with about $2^{3n/2}$ messages.

8.3.3 *Attacks on 6 Rounds: G_3^6*

8.3.3.1 NCPA with $O(2^{2n})$ Messages and KPA with $O(2^{5n/2})$ Messages

After 6 rounds, the output is $[S_1, S_2, S_3]$ with

$$\begin{aligned} S_1 &= X^4 = X^1 \oplus f_4([X^2, X^3]) = I_1 \oplus f_1([I_2, I_3]) \\ S_2 &= X^5 = X^2 \oplus f_5([X^3, X^4]) \\ S_3 &= X^6 = X^3 \oplus f_6([X^4, X^5]). \end{aligned}$$

We first provide a NCPA with $O(2^{2n})$ messages. The attacker chooses q messages such that $\forall i$, $I^3(i) = 0$. Then the attacker sends them to the oracle $\mathcal{O}$ and receives the output $[S_1(i), S_2(i), S_3(i)]$. Then she will count the number of pairs $(i, j), i < j$ such that $I_2(i) = I_2(j)$ and $I_1(i) \oplus I_1(j) = S_1(i) \oplus S_1(j)$. Suppose that $\mathcal{O}$ is a random permutation.

$$\mathbb{E}(\mathcal{N}_{perm}) \simeq \frac{q^2}{2.2^{2n}} \text{ and } \sigma(\mathcal{N}_{perm}) \simeq \frac{q}{2^n}.$$

The exact computations can be done exactly as for G_3^5. Now if $\mathcal{O}$ is a G_3^5 network, since all the $I_3(i)$ values are equal, $I_2(i) = I_2(j)$ and $X^2(i) = X^2(j)$ and $X^3(i) = X^3(j)$ imply $I_1(i) \oplus I_1(j) = S_1(i) \oplus S_1(j)$. We get

$$\mathbb{E}(\mathcal{N}_{G_3^6}) \simeq \frac{q^2}{2.2^{2n}} + \frac{q^2}{2 \cdot 2^{3n}} \text{ and } \sigma(\mathcal{N}_{G_3^6}) \simeq \frac{q}{2^n}.$$

We can distinguish the two permutations when the difference between the mean values is larger than the standard deviation, i.e., when $\frac{q^2}{2^{3n}} \geq \frac{q}{2^n}$, i.e., for $q \geq 2^{2n}$.

We can obviously transform this NCPA attack into a KPA attack which will succeed as soon as we have $q \geq 2^{\frac{5n}{2}}$.

8.3.3.2 Experimental Results on G_3^6

Some NCPA and KPA attacks against G_3^6 have been implemented for small values of n ($n = 6$ and $n = 8$). The experimental values confirm the theoretical results. The experiments were performed as follows:

- choose randomly an instance of G_3^6
- choose randomly a permutation: for this we use classical balanced Feistel scheme with a large number of rounds (more than 20)
- launch the attack in NCPA with $q = 2^{2n}$, in KPA with $m = 2^{3n}$ ($q = 2^{\frac{5n}{2}}$ also works).
- count the number of plaintext/ciphertext pairs satisfying the relations for the G_3^6 function and for the permutation
- iterate this procedure a large number of times (here 1000 times) to evaluate the mean values and the standard deviations
- compute the mean value and the standard deviation for both the G_3^6 function and the permutation (Table 8.1).

As we can see, the experimental values for $\mathcal{N}_{G_3^6} - \mathcal{N}_{perm}$ are very close to the theoretical expected values ($\frac{q^2}{2 \cdot 2^{4n}}$ in KPA and $\frac{q^2}{2 \cdot 2^{3n}}$ in NCPA). Similarly, the experimental values for ϵ_{perm} are very close to the theoretical expected values ($\frac{q}{\sqrt{2} \cdot 2^{\frac{3n}{2}}}$ in KPA and $\frac{q}{\sqrt{2} \cdot 2^n}$ in NCPA). So these simulations confirm that it is possible to distinguish G_3^6 from a random permutation with the complexity given above.

Table 8.1 Experimental results for KPA and NCPA attacks on G_3^6

Attack	n	$\mathcal{N}_{G_3^6}$	$\mathcal{N}_{perm}$	$\mathcal{N}_{G_3^6} - \mathcal{N}_{perm}$	$\frac{q^2}{2 \cdot 2^{4n}}$	$\sigma_{G_3^6}$	σ_{perm}	$\frac{q}{\sqrt{2} \cdot 2^{\frac{3n}{2}}}$
KPA	6	131006	129011	1995	2048	159	372	362.038
KPA	8	8388308	8355787	32521	32768	2862	2833	2896.309
NCPA	6	2058	2009	49	32	45	44	45.254
NCPA	8	32781	32601	180	128	178	185	182.019

8.3.4 Attacks on 7 Rounds: G_3^7

Here we provide a KPA that will use the maximal number of messages. After 7 rounds, the output is $[S_1, S_2, S_3]$ with

$$\begin{aligned} S_1 &= X^5 = X^2 \oplus f_5([X^3, X^4]) = I_2 \oplus f_2([I_3, X^1]) \oplus f_5([X^3, X^4]) \\ S_2 &= X^6 = X^3 \oplus f_6([X^4, X^5]) \\ S_3 &= X^7 = X^4 \oplus f_7([X^5, X^6]). \end{aligned}$$

The attacker chooses q messages, sends them to the oracle $\mathcal{O}$, and receives the output $[S_1(i), S_2(i), S_3(i)]$. Then she will count the number of pairs $(i, j), i < j$ such that $I_3(i) = I_3(j)$ and $I_2(i) \oplus I_2(j) = S_1(i) \oplus S_1(j)$. Supposc that $\mathcal{O}$ is a random permutation. Computations similar to those previously performed shows that

$$\mathbb{E}(\mathcal{N}_{perm}) \simeq \frac{q^2}{2.2^{2n}} \text{ and } \sigma(\mathcal{N}_{perm}) \simeq \frac{q}{2^n}.$$

When $\mathcal{O}$ is a G_3^7 cipher, then we have

$$\mathbb{E}(\mathcal{N}_{G_3^7}) \simeq \frac{q^2}{2.2^{2n}} + \frac{q^2}{2 \cdot 2^{4n}} \text{ and } \sigma(\mathcal{N}_{G_3^7}) \simeq \frac{q}{2^n}.$$

We can distinguish the two permutations when the difference between both mean values is larger than the standard deviations, i.e., when $\frac{q^2}{2^{4n}} \geq \frac{q}{2^n}$, i.e., for $q \geq 2^{3n}$. Here we have reached the maximal number of messages. The next step will be to attack generators of permutations and not a single permutation.

8.3.5 Attacks on G_3^r Generators for $r \geq 8$

After 8 rounds, we need to attack permutation generators. The aim is to distinguish a random permutation generator from a G_3^r cipher generator. We always suppose that we are testing μ permutations.

8.3.5.1 Attacks on G_3^r Generators with $r = 1 \mod 3$

In that case, the output is given by $[S_1, S_2, S_3]$ with

$$\begin{aligned} S_1 &= X^{r-2} = I_2 \oplus f_2([I_3, X^1]) \oplus f_5([X^3, X^4]) \oplus \ldots f_{r-2}([X^{r-4}, X^{r-3}]) \\ S_2 &= X^{r-1} \\ S_3 &= X^r. \end{aligned}$$

Let $r = 3\ell + 1$. Then $X^{r-2} = I_2 \oplus f_2([I_3, X^1]) \oplus_{t=2}^{\ell} f_{3t-1}([X^{3t-3}, X^{3t-2}])$. The attacker has q messages and sends them to the oracle $\mathcal{O}$ and receives the output $[S_1(i), S_2(i), S_3(i)]$. Then she will count the number of pairs (i, j), $i < j$ such that $I_3(i) = I_3(j)$ and $I_2(i) \oplus I_2(j) = S_1(i) \oplus S_1(j)$. Suppose that $\mathcal{O}$ is a random permutation. Then

$$\mathbb{E}(\mathcal{N}_{perm}) \simeq \mu \frac{q^2}{2.2^{2n}} \text{ and } \sigma(\mathcal{N}_{perm}) \simeq \sqrt{\mu}\frac{q}{2^n}.$$

When $\mathcal{O}$ is a G_3^r cipher:

$$\mathbb{E}(\mathcal{N}_{perm}) \simeq \mu \frac{q^2}{2.2^{2n}} + \mu \frac{q^2}{2\ell n} \text{ and } \sigma(\mathcal{N}_{perm}) \simeq \sqrt{\mu}\frac{q}{2^n}.$$

Thus it is possible to distinguish the two generators when $\mu \frac{q^2}{2\ell n} \geq \sqrt{\mu}\frac{q}{2^n}$, i.e., when $\mu q^2 \geq 2^{(4\ell-2)n}$. When $q = 2^{3n}$, we obtain $\mu = 2^{(4\ell-8)n}$ and the complexity is $\lambda = \mu.q = 2^{(4\ell-5)n}$. We have $3\ell = r - 1$. This gives: $\lambda = 2^{(r+\ell-6)n} = 2^{(r-\lfloor \frac{r}{3} \rfloor-6)n}$.

8.3.5.2 Attacks on G_3^r Generators with $r = 2 \mod 3$

In that case, the output given is $[S_1, S_2, S_3]$ with

$$\begin{aligned}
S_1 &= X^{r-2} = I_3 \oplus f_3([X^1, X^2]) \oplus f_6([X^4, X^5]) \oplus \ldots f_{r-2}([X^{r-4}, X^{r-3}]) \\
S_2 &= X^{r-1} \\
S_3 &= X^r.
\end{aligned}$$

Let $r = 3\ell + 2$. Then $X^{r-2} = I_3 \oplus_{t=1}^{\ell} f_{3t}([X^{3t-2}, X^{3t-1}])$. The attacker has q messages and sends them to the oracle $\mathcal{O}$ and receives the output $[S_1(i), S_2(i), S_3(i)]$. Then she will count the number of pairs (i, j), $i < j$ such that $I_3(i) \oplus I_3(j) = S_1(i) \oplus S_1(j)$. Suppose that $\mathcal{O}$ is a random permutation. Then

$$\mathbb{E}(\mathcal{N}_{perm}) \simeq \mu \frac{q^2}{2.2^{n}} \text{ and } \sigma(\mathcal{N}_{perm}) \simeq \sqrt{\mu}\frac{q}{2^{n/2}}.$$

When $\mathcal{O}$ is a G_3^r cipher:

$$\mathbb{E}(\mathcal{N}_{perm}) \simeq \mu \frac{q^2}{2.2^{n}} + \mu \frac{q^2}{2\ell n} \text{ and } \sigma(\mathcal{N}_{perm}) \simeq \sqrt{\mu}\frac{q}{2^{n/2}}.$$

Thus it is possible to distinguish the two generators when $\mu \frac{q^2}{2\ell n} \geq \sqrt{\mu}\frac{q}{2^{n/2}}$, i.e., when $\mu q^2 \geq 2^{(4\ell-1)n}$. When $q = 2^{3n}$, we obtain $\mu = 2^{(4\ell-7)n}$ and the complexity is $\lambda = \mu.q = 2^{(4\ell-4)n}$. We have $3\ell = r - 2$. This gives: $\lambda = 2^{(r+\ell-6)n} = 2^{(r-\lfloor \frac{r}{3} \rfloor-6)n}$.

8.3.5.3 Attacks on G_3^r Generators with $r = 0 \mod 3$

In that case, the output is given is $[S_1, S_2, S_3]$ with

$$\begin{aligned} S_1 &= X^{r-2} = I_1 \oplus f_1([I_2, I_3]) \oplus f_4([X^2, X^3]) \oplus \ldots f_{r-2}([X^{r-4}, X^{r-3}]) \\ S_2 &= X^{r-1} \\ S_3 &= X^r. \end{aligned}$$

Let $r = 3\ell$. Then $X^{r-2} = I_1 \oplus f_1([I_2, I_3]) \oplus_{t=2}^{\ell} f_{3t-2}([X^{3t-4}, X^{3t-3}])$. The attacker has q messages and sends them to the oracle $\mathcal{O}$ and receives the output $[S_1(i), S_2(i), S_3(i)]$. Then she will count the number of pairs (i, j), $i < j$ such that $I_2(i) = I_2(j)$, $I_3(i) = I_3(j)$, and $I_1(i) \oplus I_1(j) = S_1(i) \oplus S_1(j)$. Suppose that $\mathcal{O}$ is a random permutation. Then

$$\mathbb{E}(\mathcal{N}_{perm}) \simeq \mu \frac{q^2}{2.2^{3n}} \text{ and } \sigma(\mathcal{N}_{perm}) \simeq \sqrt{\mu} \frac{q}{2^{3n/2}}.$$

When $\mathcal{O}$ is a G_3^r cipher:

$$\mathbb{E}(\mathcal{N}_{perm}) \simeq \mu \frac{q^2}{2.2^{3n}} + \mu \frac{q^2}{2\ell n} \text{ and } \sigma(\mathcal{N}_{perm}) \simeq \sqrt{\mu} \frac{q}{2^{3n/2}}.$$

Thus it is possible to distinguish the two generators when $\mu \frac{q^2}{2\ell n} \geq \sqrt{\mu} \frac{q}{2^{3n/2}}$, i.e., when $\mu q^2 \geq 2^{(4\ell-3)n}$. When $q = 2^{3n}$, we obtain $\mu = 2^{(4\ell-9)n}$ and the complexity is $\lambda = \mu.q = 2^{(4\ell-6)n}$. We have $3\ell = r$. This gives: $\lambda = 2^{(r+\ell-6)n} = 2^{(r-\lfloor \frac{r}{3} \rfloor-6)n}$.

8.3.6 *Summary of the Attacks on G_3^r*

In Table 8.2, we provide the results for G_3^d contracting Feistel ciphers.

8.4 Generic Attacks When $k \geq 4$ and $r > k$

This section contains attacks on G_k^r for $k \geq 4$ and $r > k$. For $r \leq k$, the results are given in Sect. 8.2.

Table 8.2 Results on G_3^r. For more than 7 rounds more than one permutation is needed or more than 2^{3n} computations are needed in the best known attacks to distinguish from a random permutation with an even signature

	KPA	NCPA[a]
G_3^1	1	1
G_3^2	1	1
G_3^3	2^n	2
G_3^4	2^n	$2^{n/2}$
G_3^5	$2^{3n/2}$	2^n
G_3^6	$2^{5n/2}$	2^{2n}
G_3^7	2^{3n}	2^{3n}
G_3^8	2^{4n}	2^{4n}
G_3^9	2^{6n}	2^{6n}
G_3^{10}	2^{7n}	2^{7n}
G_3^{11}	2^{8n}	2^{8n}
G_3^{12}	2^{10n}	2^{10n}
$G_3^r, r \geq 12$	$2^{(r+\lfloor \frac{r}{3} \rfloor - 6)}$	$2^{(r+\lfloor \frac{r}{3} \rfloor - 6)}$

[a] Here we do not show CPA, NCPCA and CPCA since for G_3^r, no better attacks are found compared with CPA-1

8.4.1 *Attacks for* $k + t$ *Rounds, with* $1 \leq t < k - 1$

In the NCPA attack, the attacker chooses q messages such that $\forall i$, $I_{t+2}(i) = \cdots = I_k(i) = 0$ and $[I_1(i), \ldots I_t(i)]$ are pairwise distinct. This choice limits the maximal number of plaintext/ciphertext tuples to $\leq 2^{(t+1)n}$. The attacker sends them to the oracle $\mathcal{O}$, and receives the output $[S_1(i), \ldots, S_k(i)]$. Then she will count the number of pairs $(i, j), i < j$ such that $I_{t+1}(i) \oplus I_{t+1}(j) = S_1(i) \oplus S_1(j)$.

When $\mathcal{O}$ is a random permutation, we have

$$\mathbb{E}(\mathcal{N}_{perm}) \simeq \frac{q(q-1)}{2 \cdot 2^n} \text{ and } \sigma(\mathcal{N}_{perm}) \simeq \frac{q}{2^{n/2}}.$$

This can be easily proved using the Covariance Formula and with computations similar to those performed in the previous attacks.

When $\mathcal{O}$ is a G_k^r cipher, the preceding condition appears at random, but we also have the following property:

$$X^1(i) = X^1(j), \ldots, X^t(i) = X^t(j) \Rightarrow S_1(i) \oplus S_1(j) = I_{t+1}(i) \oplus I_{t+1}(jj)$$

since $S_1 = X^{t+1} = L_{t+1} \oplus f^{t+1}([L_{t+2}, \ldots L_k, X^1, \ldots, X^t])$. This gives

$$\mathbb{E}(\mathcal{N}_{G_k^{k+t}}) \simeq \frac{q(q-1)}{2 \cdot 2^n} + \frac{q(q-1)}{2 \cdot 2^{tn}} \text{ and } \sigma(\mathcal{N}_{G_k^{k+t}}) \simeq \frac{q}{2^{n/2}}$$

So

$$|\mathbb{E}(\mathcal{N}_{G_k^{k+t}}) - \mathbb{E}(\mathcal{N}_{perm})| \simeq \frac{q(q-1)}{2 \cdot 2^{tn}}.$$

Thus we distinguish when $\frac{q^2}{2^{tn}} \geq \frac{q}{2^{\frac{n}{2}}}$, i.e., when $q \geq 2^{(t-\frac{1}{2})n}$, which is compatible with the bound given above.

As usual, we are able to transform this attack into a KPA attack which succeeds if $q \geq 2^{(\frac{k+t-2}{2})n}$.

8.4.2 *Attacks for 2k − 1 Rounds*

In that case there is only a KPA attack. the attacker chooses q messages, sends them to the oracle $\mathcal{O}$, and receives the output $[S_1(i), \ldots, S_k(i)]$. Then she will count the number of pairs (i, j), $i < j$ such that $I_k(i) \oplus I_k(j) = S_1(i) \oplus S_1(j)$. When $\mathcal{O}$ is a random permutation, we have

$$\mathbb{E}(\mathcal{N}_{perm}) \simeq \frac{q(q-1)}{2 \cdot 2^n} \text{ and } \sigma(\mathcal{N}_{perm}) \simeq \frac{m}{\sqrt{2^n}}$$

When $\mathcal{O}$ is a G_k^{2k-1} cipher,

$$\mathbb{E}(\mathcal{N}_{G_k^{2k-1}}) \simeq \frac{q(q-1)}{2 \cdot 2^n} + \frac{q(q-1)}{2 \cdot 2^{(k-1)n}} \text{ and } \sigma(\mathcal{N}_{G_k^{2k-1}}) \simeq \frac{q}{\sqrt{2^n}}$$

Since $S_1 = X^k = I_k \oplus f^{2k-1}([X^1, \ldots, X^{k-1}])$, we have that $I_k(i) \oplus I_k(j) = S_1(i) \oplus S_1(j)$ is also implied by the following equations: $X^1(i) = X^1(j)$, $X^2(i) = X^2(j)$, $\cdots$, $X^{k-1}(i) = X^{k-1}(j)$. Thus we can distinguish when $\frac{q^2}{2 \cdot 2^{(k-1)n}} \geq \frac{q}{\sqrt{2^n}}$. This gives $q \geq 2^{(k-\frac{3}{2})n}$.

We can remark that for more than $2k$ rounds we will have to proceed with different attacks, since $X^1(i) = X^1(j), \ldots, X^k(i) = X^k(j)$ implies $i = j$ because we have a permutation.

8.4.3 *Attacks on Generators*

8.4.3.1 Attacks on G_k^r Generators When $r = 2k$

Let μ be the number of permutations that we will use. After $2k$ rounds, the output is given by $[S_1, S_2, \ldots, S_k] = [X^{k+1}, X^{k+2}, \ldots, X^{2k}]$ where $X^{k+1} = X^1 \oplus f_{k+1}([X^2, \ldots, X^k])$. Remember that $X^1 = I_1 \oplus f_1([I_2, \ldots, I_k])$. Let us describe the

KPA attack which concentrates on $S_1 = X^{k+1}$. The attacker counts be the number of pairs $(i, j), i < j$, such that

$$I_2(i) = I_2(j), \ldots, I_k(i) = I_k(j), \quad X^{k+1}(i) \oplus X^{k+1}(j) = I_1(i) \oplus I_1(j). \quad (8.1)$$

There we have necessary $I_1(i) \neq I_1(j)$ and $X^1(i) \neq X^1(j)$. When we are testing μ random permutations,

$$\mathbb{E}(\mathcal{N}_{perm}) \simeq \mu \cdot \frac{q^2}{2 \cdot 2^{kn}} \text{ and } \sigma(\mathcal{N}_{perm}) \simeq \sqrt{\mu} \cdot \frac{q}{2^{\frac{kn}{2}}}$$

With a G_k^{2k} cipher, since $I_2(i) = I_2(j), \ldots, I_k(i) = I_k(j)$, $X^2(i) = X^2(j), \ldots$, $X^k(i) = X^k(j)$ imply (8.1) we have

$$\mathbb{E}(\mathcal{N}_{G_k^{2k}}) = \mu \cdot \frac{q^2}{2 \cdot 2^{kn}} + \mu \cdot \frac{q^2}{2 \cdot 2^{(2k-2)n}} \text{ and } \sigma(\mathcal{N}_{G_k^{2k}}) \simeq \sqrt{\mu} \cdot \frac{q}{2^{\frac{kn}{2}}}$$

Thus we can distinguish the two generators when: $\mu \cdot \frac{q^2}{2^{(2k-2)n}} \geq \sqrt{\mu} \cdot \frac{q}{2^{\frac{kn}{2}}}$, or when $\mu \cdot q \geq 2^{(3k-4)n}$. When $q = 2^{kn}$, we find $\mu = 2^{(k-4)n}$ and $\mu \cdot q = 2^{(2k-4)n}$.

8.4.3.2 Attacks on G_k^r Generators for r Rounds with $r \geq 2k$

It is possible to generalize the attack given above for any $r \geq 2k$. We give here only the main ideas. We concentrate the attack on X^{r-k+1}. In the constraints, there are r conditions and $r - k$ internal variables X^i. We choose conditions number k, $2k$, ... , until we get $\xi = \lfloor \frac{r}{k} \rfloor$ conditions. This gives ξ (internal or external) $\cdot(k - 1)$-multiple equations. When they are satisfied, we have

1. One equation between the input and output variables.
2. φ equations between the output variables where

$$\varphi = (k - 1) - \left(r - \left\lfloor \frac{r}{k} \right\rfloor k\right) = (k - 1) - (r \bmod k)$$

We have μ permutations and the attack proceeds as follows: the attacker counts the number of pairs $(i, j), i < j$, such that these $\varphi + 1$ equations are satisfied. When we are testing a permutation generator, we have

$$\mathbb{E}(\mathcal{N}_{perm}) \simeq \mu \cdot \frac{q(q-1)}{2 \cdot 2^{(\varphi+1)n}} \text{ and } \sigma(\mathcal{N}_{perm}) \simeq \sqrt{\mu} \cdot \frac{m}{2^{(\frac{\varphi+1}{2})n}}$$

With a G_k^r cipher, the $\xi(k-1)$-multiples equations imply the $\varphi+1$ equations described above. This shows that

$$\mathbb{E}(\mathcal{N}_{G_k^r}) \simeq mu \cdot \frac{q(q-1)}{2\cdot 2^{(\varphi+1)n}} + \mu \cdot \frac{q(q-1)}{2\cdot 2^{(k-1)n}} \text{ and } \sigma(\mathcal{N}_{G_k^r}) \simeq \sqrt{\mu}\cdot \frac{m}{2^{(\frac{\varphi+1}{2})n}}$$

We get the condition:

$$\mu \cdot \frac{q^2}{2^{(k-1)n}} \geq \sqrt{\mu}\cdot \frac{q}{2^{(\frac{\varphi+1}{2})n}},$$

$$\mu \cdot q^2 \geq 2^{(2(k-1)\xi-\varphi-1)n}.$$

For the maximal value $q=2^{kn}$, we find $\mu = 2^{(2(k-1)\xi-\varphi-2k-1)n}$ and the complexity is $\lambda = \mu\cdot q = 2^{(2(k-1)\xi-\varphi k-1)n}$. Thus we can write

$$\lambda = 2^{(2(k-1)\lfloor\frac{r}{k}\rfloor+(r \bmod k)-2k)n} = 2^{\left(r+(k-2)\lfloor\frac{r}{k}\rfloor-2k\right)n}.$$

8.4.4 Summary of the Results for r > 4

Table 8.3 gives the results on G_k^r for $r \geq 4$.

Storing a random function of $(k-1)n$ bits to n bits requires a large memory and this may be a practical disadvantage of G_k^r compared with balanced Feistel schemes or Feistel schemes with expanding functions. However if a function generator is used to generate pseudo-random functions, this may not be a problem.

Table 8.3 Results on G_k^r for any $k \geq 4$. For more than $2k$ rounds more that one permutation is needed or more than $2^{(2k-4)n}$ computations are needed in the best known attacks to distinguish from a random permutation with an even signature

	KPA	NCPA[a]
G_k^r, $1 \leq r \leq k-1$	1	1
G_k^k	$2^{\frac{n(k-1)}{2}}$	2
G_k^{k+1}	$2^{\frac{n(k-1)}{2}}$	$2^{\frac{n}{2}}$
G_k^{k+2}	$2^{\frac{k}{2}n}$	$2^{\frac{3}{2}n}$
G_k^{k+3}	$2^{(\frac{k+1}{2})n}$	$2^{\frac{5}{2}n}$
G_k^{k+i}, $1 \leq i < k$	$2^{(\frac{k+i-2}{2})n}$	$2^{(\frac{2i-1}{2})n}$
G_k^{2k}	$2^{(2k-4)n}$	$2^{(2k-4)n}$
G_k^r, $r \geq 2k$	$2^{(r+(k-2)\lfloor\frac{r}{k}\rfloor-2k)n}$	$2^{(r+(k-2)\lfloor\frac{r}{k}\rfloor-2k)n}$

[a] Here we do not show CPA, NCPCA and CPCA since for G_k^r, no better attacks are found compared with CPA-1

References

Hoang, V. T., and P. Rogaway. 2010. On generalized feistel networks. In *Advances in cryptology—CRYPTO 2010*, vol. 6223, ed. T. Rabin. Lecture notes in computer science, 613–630. Berlin: Springer.

Lucks, S. 1996. Faster luby-rackoff ciphers. In *Fast software encrytion—FSE '96*, vol. 1039, ed. D. Gollmann. Lecture notes in computer science, 189–203. Berlin: Springer.

Naor, M., and O. Reingold. 1999. On the construction of pseudo-random permutations: Luby-Rackoff revisited. *Journal of Cryptology* 12: 29–66 Extended abstract. In Proceedings 29th annual ACM symposium on theory of computing, 1997, 189–199.

Patarin, J. 2010. Security of balanced and unbalanced feistel schemes with linear non equalities. In *Cryptology ePrint Archive: Report 2010/293.*

Patarin, J., V. Nachef, and C. Berbain. 2006. Generic attacks on unbalanced feistel schemes with contracting functions. In *Advances in cryptology—ASIACRYPT 2006*, vol. 4284, ed. X. Lai and K. Chen. Lecture notes in computer science, 396–411. Berlin: Springer.

Yun, A., J. H. Park, and J. Lee. 2011. Lai-massey scheme and quasi-feistel networks des. *Codes Cryptography* 58: 45–72.

References

Chapter 9
Generic Attacks on Expanding Feistel Ciphers

Abstract "Generic" Unbalanced Feistel Ciphers with Expanding Functions are Unbalanced Feistel Ciphers with truly random internal round functions from n bits to $(k-1)n$ bits with $k \geq 3$. From a practical point of view, an interesting property of these schemes is that since $n < (k-1)n$ and n can be small (8 bits for example), it is often possible to store these truly random functions in order to design efficient schemes. This was done in the construction of the hash function CRUNCH (Goubin et al. 2008) for example. Attacks on Unbalanced Feistel Ciphers with expanding functions have been first studied by Jutla (1998). Then these attacks were improved and generalized in Patarin et al. (2007). However, in Patarin et al. (2007), some attacks were working only with particular functions (weak keys). This was due to bottleneck in equalities as explained in Sect. 9.3.2. This issue has been addressed in Volte et al. (2010) where the authors created a computer program that systematically analyzes all the possible attacks, reject attacks with bottlenecks and detect the most efficient ones. This led to many new improved attacks by a systematic study of all 2-point and rectangle attacks when $k \leq 7$. The generalization of these improved attacks was done for all k. This chapter is devoted to present the best attacks (KPA and NCPA) on Unbalanced Feistel Ciphers with Expanding Functions. According to the number of rounds, these attacks will be either 2-point attacks or different type of rectangle attacks. As pointed in Jutla (1998) and Patarin et al. (2007), Volte et al. (2010), there are surprisingly much more possibilities for these attacks than for generic balanced Feistel ciphers, generic unbalanced Feistel ciphers with contracting functions, or generalized Feistel ciphers. In fact, this large number of attack possibilities makes the analysis difficult. Many simulations on the attacks are also given, which confirm the theoretical analysis. Security results using the coupling method are given in Hoang and Rogaway (2010).

J. Patarin et al., *Guide to Feistel Ciphers*, Information Security and Cryptography,
https://doi.org/10.1007/978-3-031-99850-8_9

9.1 Notation Definition Properties

We first describe Unbalanced Feistel cipher with Expanding Functions F_k^r. Let us give the definition for the 1-round unbalanced transformation with one expanding function.

Definition 9.1 Let $k \in \mathbb{N}$ and f_1 a function from $\{0, 1\}^n$ to $\{0, 1\}^{(k-1)n}$. The function f_1 is defined as $f_1 = (f_{1,1}, f_{1,2}, \ldots, f_{1,k-1})$, where each function $f_{1,i}$, $1 \le i \le k-1$ is defined from $\{0, 1\}^n$ to $\{0, 1\}^n$. The *1-round unbalanced Feistel network with expanding functions*, denoted $F_k(f_1)$, is a function from $\{0, 1\}^{kn}$ to $\{0, 1\}^{kn}$ defined by (see also Fig. 9.1):

$$\forall (I_1, \ldots, I_k) \in (\{0, 1\}^n)^k,$$
$$F_k(f_1)([I_1, \ldots, I_k]) = [S_1, \ldots, S_k] \iff \begin{cases} \forall j,\ 1 \le j \le k-1,\ S_j = I_{j+1} \oplus f_{1,j}(I_1) \\ S_k = I_1. \end{cases}$$

Proposition 9.1 *For any function f_1 from $\{0, 1\}^n$ to $\{0, 1\}^{(k-1)n}$, $F_k(f)$ is a permutation of $\{0, 1\}^{kn}$.*

Proof We have $I_1 = S_k$ and $\forall i,\ 2 \le j \le k,\ I_j = S_{j-1} \oplus f_{1,j}(S_k)$.

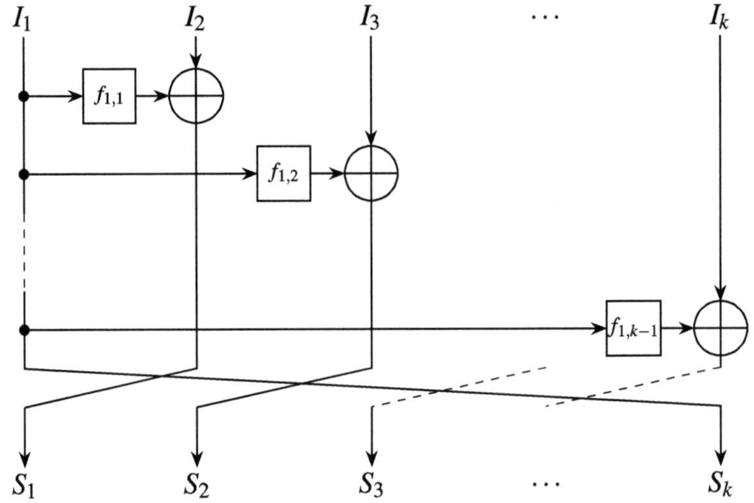

Fig. 9.1 1-round unbalanced Feistel network with expanding function: $F_k(f_1)$

Definition 9.2 Let $k \in \mathbb{N}$, $r \geq 1$ and let $f_1, \ldots, f_r$ be r function from $\{0, 1\}^n$ to $\{0, 1\}^{(k-1)n}$. The *r-round unbalanced Feistel network with expanding functions* associated with $f_1, \ldots, f_r$, denoted $F_k^r(f_1, \ldots, f_r)$ is the function from $\{0, 1\}^{kn}$ to $\{0, 1\}^{kn}$ defined by:

$$F_k^r(f_1, \ldots, f_r) = F_k(f_r) \circ \cdots \circ F_k(f_2) \circ F_k(f_1).$$

Theorem 9.1 *For any functions $f_1, \ldots, f_r$ from $\{0, 1\}^n$ to $\{0, 1\}^{(k-1)n}$, the r-round unbalanced Feistel network with expanding function $F_k^r(f_1, \ldots, f_r)$ is a permutation of $\{0, 1\}^{kn}$*

Proof $F_k^r(f_1, \ldots, f_r)$ is a permutation of $\{0, 1\}^{kn}$ since it is the composition of r permutations of $\{0, 1\}^{kn}$. □

Definition 9.3 Let $r \geq 1$. The *r-round unbalanced Feistel transformation with expanding functions*, denoted F_k^r, maps a r-tuple of functions $(f_1, \ldots, f_r) \in ({\mathscr{F}_{(k-1)n}}^{\mathscr{F}_n})^r$ to the permutation $F_k^r(f_1, \ldots, f_r)$ of $\{0, 1\}^{kn}$ as defined by Definition 9.2.

From unbalanced Feistel networks with expanding functions, we can finally define unbalanced Feistel ciphers with expanding functions, by letting round functions depend on secret keys.

Definition 9.4 Let $r \geq 1$ and let $F = (f_K)$ be a family of functions in ${\mathscr{F}_{(k-1)n}}^{\mathscr{F}_n}$ indexed by a set $\mathscr{K}$. The *r-round unbalanced Feistel cipher with expanding functions* associated with F is the block cipher with key space $\mathscr{K}^r$ and message space $\{0, 1\}^{kn}$ which maps a key $(K_1, \ldots, K_r) \in \mathscr{K}^r$ and a plaintext $[I_1, \ldots, I_k] \in \{0, 1\}^{kn}$ to the ciphertext

$$F_k^r(f_{K_1}, \ldots, f_{K_r})([I_1, \ldots, I_k]).$$

In other words, the permutation of $\{0, 1\}^{kn}$ associated with key $(K_1, \ldots, K_r)$ is the unbalanced Feistel network with expanding functions $F_k^r(f_{K_1}, \ldots, f_{K_r})$.

At each round j, the round function f_j from n bits to $(k-1)n$ bits is defined as $f_j = (f_{j,1}, f_{j,2}, \ldots, f_{j,k-1})$, where each function $f_{j,i}$ is defined from $\{0, 1\}^n$ to $\{0, 1\}^n$. Again, we have to introduce internal variables that are the first n bits of the round entry and denoted by X^{j-1}. We compute $f_j(X^{j-1})$ and obtain $(k-1)n$ bits. Then those bits are xored to the $(k-1)n$ last bits of the round entry and the result is rotated by n bits. We have

$$\begin{aligned}
X^0 &= I_1 \\
X^1 &= I_2 \oplus f_{1,1}(I_1) \\
X^2 &= I_3 \oplus f_{1,2}(I_1) \oplus f_{2,1}(X^1) \\
X^3 &= I_4 \oplus f_{1,3}(I_1) \oplus f_{2,2}(X^1) \oplus f_{3,1}(X^2) \\
&\ldots
\end{aligned}$$

More generally, we can express the X^j recursively:

$$\begin{aligned} \forall \xi < k,\ X^{\xi} &= I_{\xi+1} \oplus_{i=1}^{\xi} f_{i,\xi-i+1}(X^{i-1}) \\ \forall \xi \geq 0,\ X^{k+\xi} &= X^{\xi} \oplus_{i=2}^{k} f_{\xi+i,k-i+1}(X^{\xi+i-1}). \end{aligned}$$

After r rounds ($r \geq k+1$), the output $[S_1, S_2, \ldots, S_k]$ can be expressed by using the internal variables X^j:

$$\begin{aligned} S_k &= X^{d-1} \\ S_{k-1} &= X^{d-2} \oplus f_{d,k-1}(X^{d-1}) \\ S_{k-2} &= X^{d-3} \oplus f_{d-1,k-1}(X^{d-2}) \oplus f_{d,k-2}(X^{d-1}) \\ &\ldots \\ S_{\xi} &= X^{d-1-k+\xi} \oplus_{i=d-k+\xi}^{d-1} f_{i+1,\xi+d-i-1}(X^{i}) \\ &\ldots \\ S_1 &= X^{d-k} \oplus_{i=d-k+1}^{d-1} f_{i+1,d-i}(X^{i}). \end{aligned}$$

Definition 9.5 When it is clear from the context, we will denote the permutation on $\{0,1\}^{kn}$ produced by an expanding Feistel transformation with r rounds, a F_k^r permutation without mentioning the round functions.

We conclude this section by giving the signature of unbalanced Feistel ciphers with expanding functions as a corollary of Theorem 5.1:

Proposition 9.2 *F_k^r has an even signature as soon as $n \geq 2$.*

Proof In the case of our functions F_k^r, the round functions are from $\{0,1\}^n$ to $\{0,1\}^{(k-1)n}$. If $n \geq 2$, F_k^r has always an even signature (however if we change one bit at each round, the Feistel cipher obtained has not always an even signature). □

9.2 Attacks on the First $k + 2$ Rounds

The attacks described in this section are 2-point attacks (except for the first round). These are the most efficient here. When the number of rounds is greater than or equal to $k + 3$, the best attacks known so far are rectangle attacks. They are studied in the next sections.

9.2.1 Attacks on F_k^1

After one round, we have $S_k = I_1$. This leads to the following KPA on F_k^1.

Consider the following KPA-distinguisher D:

1. D makes a query to the oracle, and receives a random plaintext $[I_1, \ldots, I_k]$ together with $[S_1, \ldots, I_k] = \mathcal{O}([I_1, \ldots, I_k])$;
2. is $S_K = I_1$, D outputs 1, otherwise it outputs 0.

The definition of F_k^1 shows that when $\mathcal{O} = F_k^1(f_1, \ldots, f_r)$, D *always* outputs 1. On the other hand, when $\mathcal{O}$ is a random permutation of $\{0, 1\}^{kn}$, then $[S_1, \ldots, S_k]$ is uniformly random in $\{0, 1\}^{kn}$, and the probability that $S_k = I_1$ (and hence that D outputs 1) is exactly 2^{-n}. Therefore, by definition of the advantage (cf. Definition 1.3) we have

$$\mathcal{A}_{F_k^1}(\mathsf{D}) = 1 - \frac{1}{2^n}.$$

Since D makes exactly one query, it follows that

$$\mathcal{A}_{F_k^1}^{\mathrm{KPA}}(1) \geq 1 - \frac{1}{2^n}.$$

Hence, there is a very efficient known-plaintext attack against F_k^1, making only one query and distinguishing a F_k^r permutation from a random permutation with probability negligibly close to one.

9.2.2 2-Point NCPA and KPA on F_k^r, $2 \leq r \leq k$

We first introduce the differential notation for the attacks.

9.2.2.1 Differential Notation for 2-Point Attacks

We use plaintext/ciphertext pairs. The notation $[\mathbf{0}, \mathbf{0}, \ldots, \Delta_{k-1}^0, \Delta_k^0]$ means that the pair of messages (i, j) satisfies $I_1(i) = I_1(j)$, $I_2(i) = I_2(j)$, and $I_s(i) \oplus I_s(j) = \Delta_s^0$, $3 \leq s \leq k$. The differential of the outputs i and j after round r is denoted by $[\Delta_1^r, \Delta_2^r, \ldots, \Delta_{k-1}^r, \Delta_k^r]$. At each round, internal variables are defined by the structure of the scheme. In the attacks, we determine equalities that have to be satisfied by the inputs and the outputs. With a scheme, some equalities on the internal variables on some rounds will allow the differential path to propagate. On an intermediate round, when equalities on the internal variables are needed in order to get a differential characteristic, we use the notation $\mathbf{0}$, to mean that the corresponding internal variables are equal in messages i and j. When we write 0, this means that the differential path propagates without any constraint on the internal functions.

9.2.2.2 Presentation of the Attacks

Let r be such that $2 \le r \le k$.

Consider the following NCPA-distinguisher D:

1. D chooses $I_r(1), I_{r+1}(1), \ldots I_k(1)$ in $\{0,1\}^{(k-r+1)n}$, $I_r(2), I_{r+1}(2), \ldots I_k(2)$ in $\{0,1\}^{(k-r+1)n}$, such that $[I_r(1), I_{r+1}(1), \ldots I_k(1)] \ne [I_r(2), I_{r+1}(2), \ldots I_k(2)]$ and queries $[S_1(1), \ldots, S_k(1)] = \mathcal{O}([0, 0, \ldots, I_r(1), \ldots, I_k(1)])$ and
$$[S_1(2), \ldots, S_k(2)] = \mathcal{O}([0, 0, \ldots, I_r(2), \ldots, I_k(2)]).$$
2. D tests whether $S_k(1) \oplus S_k(2) = I_r(1) \oplus I_r(2)$; if this holds, D outputs 1, otherwise D outputs 0.

With the differential notation, this attack is represented as follows:

By definition of F_k^r, $2 \le r \le k$, and by using the differential path in Table 9.1, it is easy to see that when $\mathcal{O} = F_k^r(f_1, \ldots, f_k)$, D always outputs 1.

When $\mathcal{O}$ is a random permutation of $\{0,1\}^{kn}$, then $[S_1(2), \ldots, S_k(2)]$ is uniformly random in $\{0,1\}^{kn} \setminus \{[S_1(1), \ldots, S_k(1)]\}$. Since there are exactly $2^{(k-1)n}$ possible values of $\{0,1\}^{kn}$ for $[S_1(2), \ldots, S_k(2)]$ in $\{0,1\}^{kn} \setminus \{[S_1(1), \ldots, S_k(1)]\}$ such that $S_k(1) \oplus S_k(2) = I_r(1) \oplus I_r(2)$, D outputs 1 with probability
$$\frac{2^{(k-1)n}}{2^{kn}-1}.$$

Hence, we have, by definition of the advantage,
$$\mathcal{A}_{F_k^r}(\mathsf{D}) = 1 - \frac{2^{(k-1)n}}{2^{kn}-1}.$$

Since D makes exactly two queries, this implies
$$\mathcal{A}_{F_k^r}^{\mathrm{NCPA}}(\mathsf{D}) \ge 1 - \frac{2^{(k-1)n}}{2^{kn}-1}.$$

Table 9.1 Attack on F_k^r with $2 \le r \le k$

Round	0	0	...	0	0	Δ_r^0	...	Δ_k^0
1	0	0	...	0	Δ_r^0	Δ_r^1	...	Δ_k^0
2	0	0	...	Δ_r^0	Δ_{r+1}^0	Δ_{r+1}^0	...	0
⋮	⋮	⋮	⋮	⋮	⋮	⋮	⋮	⋮
$r-1$	Δ_r^0	Δ_{r+1}^0	...	...	...	...	...	0
r	...	...	...	...	...	...	...	Δ_r^0

Hence, there is a very efficient non-adaptive chosen-plaintext attack against F_k^r, making only two queries and distinguishing F_k^r from a random permutation with probability negligibly close to one.

As usual, this attack can be transformed into a KPA with $O(2^{\frac{(k-1)n}{2}})$ random queries.

9.2.3 2-Point NCPA and KPA on F_k^{k+1}

After $k + 1$ rounds, we have

$$S_{k-1} = X^{k-1} \oplus f_{k+1,k-1}(X^k) \quad \text{with} \quad X^{k-1} = I_k \oplus f_{1,k-1}(I_1) \oplus_{i=2}^{k-1} f_{i,k-i}(X^{i-1}).$$

We will use the following NCPA distinguisher.

1. D chooses about $\sqrt{2^n}$ inputs with $I_1, I_2, \ldots, I_{k-1}$ constant, and asks for the corresponding outputs $[S_1(i), S_2(i), \ldots, S_k(i)]$, $1 \le i \le \sqrt{2^n}$. Notice that the conditions on the inputs imply that $I_1, X^1, \ldots, X^{k-2}$ are constant and that
$$\forall i, j, \; X^{k-1}(i) \oplus X^{k-1}(j) = I_k(i) \oplus I_k(j).$$
2. D looks for indexes $i, j, i \neq j$ such that $S_k(i) = S_k(j)$. (Here we notice that $S_k = X^k$ since $r = k + 1$). Since $q \simeq \sqrt{2^n}$, by the birthday paradox, the probability p that D finds such indexes is greater than $\frac{1}{2}$.
3. Then D checks if $S_{k-1}(i) \oplus S_{k-1}(j) = I_k(i) \oplus I_k(j)$; if this holds, D outputs 1, otherwise D outputs 0.

With the differential notation, this attack is represented in Table 9.2.

By definition of F_k^{k+1}, and by using the differential path in Table 9.2, it is easy to see that when $\mathcal{O} = F_k^r(f_1, \ldots, f_k)$, D always outputs 1 with probability at least p. When $\mathcal{O}$ is a random permutation of $\{0, 1\}^{kn}$, with $\sqrt{2^n}$ queries, the probability to have two indexes $i, j, i \neq j$ such that $S_k(i) = S_k(j)$ is again $p > \frac{1}{2}$. Then

Table 9.2 Attack on F_k^{k+1}

Round	0	0	...	0	0	...	0	Δ_k^0
1	0	0	...	0	0	0	...	Δ_k^0
2	0	0	...	0	0	0	...	0
$\vdots$	$\vdots$	$\vdots$	$\vdots$	$\vdots$	$\vdots$	$\vdots$	$\vdots$	$\vdots$
$k-1$	Δ_k^0	Δ_2	0	$\vdots$	$\vdots$	$\vdots$	$\vdots$	$\vdots$
k	$\mathbf{0}$	...	...	...	...	...	...	Δ_k^0
$k+1$	Δ_1^{k+1}	...	...	...	...	...	Δ_k^0	0

$[S_1(j), \ldots, S_{k-1}(j)]$ is uniformly random in $\{0,1\}^{kn} \setminus \{[S_1(i), \ldots, S_{k-1}(i)]\}$. Since there are exactly $2^{(k-2)n}$ possible values of $\{0,1\}^{kn}$ for $[S_1(j), \ldots, S_{k-1}(j)]$ in $\{0,1\}^{kn} \setminus \{[S_1(i), \ldots, S_{k-1}(i)]\}$ such that $S_{k-1}(i) \oplus S_{k-1}(k) = I_k(i) \oplus I_k(j)$, D outputs 1 with probability greater than $p\left(\frac{2^{(k-2)n}}{2^{(k-1)n}-1}\right)$.

Hence, we have, by definition of the advantage,

$$\mathscr{A}_{F_k^{k+1}}(\mathsf{D}) \geq p\Big(1 - \frac{2^{(k-2)n}}{2^{(k-1)n}-1}\Big).$$

and

$$\mathscr{A}^{\mathrm{NCPA}}_{F_k^r}(\mathsf{D}) \geq p\Big(1 - \frac{2^{(k-2)n}}{2^{(k-1)n}-1}\Big).$$

Hence, there is a very efficient non-adaptive chosen-plaintext attack against F_k^{k+1}, making only about $\sqrt{2^n}$ queries and distinguishing F_k^{k+1} from a random permutation with probability negligibly close to p.

As usual, this attack can be transformed into a KPA with $O(2^{\frac{kn}{2}})$ random queries.

9.2.4 2-Point NCPA and KPA on F_k^{k+2}

After $k+2$ rounds, we have

$$S_{k-2} = X^{k-1} \oplus f_{k+1,k-1}(X^k) \oplus f_{k+2,k-2}(X^{k+1})$$

and

$$X^{k-1} = I_k \oplus f_{1,k-1)}(I_1) \oplus_{i=2}^{k-1} f_{i,k-i}(X^{i-1}).$$

We will use the following NCPA distinguisher.

1. D chooses about 2^n inputs with $I_1, I_2, \ldots, I_{k-1}$ constant, and asks for the corresponding outputs $[S_1(i), S_2(i), \ldots, S_k(i)]$, $1 \leq i \leq \sqrt{2^n}$. Notice that the conditions on the inputs imply that $I_1, X^1, \ldots, X^{k-2}$ are constant and that
$$\forall i, j,\ X^{k-1}(i) \oplus X^{k-1}(j) = I_k(i) \oplus I_k(j).$$
2. D looks for indexes $i, j, i \neq j$ such that $S_k(i) = S_k(j)$ and $S_{k-1}(i) = S_{k-1}(j)$. Here, we have the following relations:
$$\begin{aligned} S_k &= X^{k+1} \\ S_{k-1} &= X^k \oplus f_{k+2,k-1}(X^{k+1}) \\ S_{k-2} &= X^{k-1} \oplus f_{k+1,k-1}(X^k) \oplus f_{k+2,k-2}(X^{k+1}). \end{aligned}$$

The X^{k-1} are pairwise distinct but we can get a collision (i, j) for the X^{k+1} variables and the X^k variables. Since $q^2 \geq 2^{2n}$, i.e., $q \geq 2^n$, by the birthday paradox, the probability p that D finds such indexes is greater than $\frac{1}{2}$.
3. Then D checks if $S_{k-1}(i) \oplus S_{k-1}(j) = I_k(i) \oplus I_k(j)$; ; if this holds, D outputs 1, otherwise D outputs 0.

Then using arguments similar to those in the attack on F_k^{k+1}, it is easy to show that

$$\mathscr{A}_{F_k^{k+2}}(\mathsf{D}) \geq p\Big(1 - \frac{2^{(k-3)n}}{2^{(k-2)n} - 1}\Big)$$

and

$$\mathscr{A}_{F_k^r}^{\mathrm{NCPA}}(\mathsf{D}) \geq p\Big(1 - \frac{2^{(k-3)n}}{2^{(k-2)n} - 1}\Big).$$

Hence, there is a very efficient non-adaptive chosen-plaintext attack against F_k^{k+1}, making only about 2^n queries and distinguishing F_k^{k+2} from a random permutation with probability negligibly close to p.

As usual, this attack can be transformed into a KPA with $O(2^{\frac{(k+1)n}{2}})$ random queries and this attack is valid since $2^{\frac{k+1}{2}n} \leq 2^{kn}$ (here $k \geq 3$).

9.3 Rectangle Attacks for $r \geq k+3$

9.3.1 *Notation First Examples*

After $k+3$ rounds, it is still possible to mount 2-point attacks, but rectangle attacks are much better. Moreover, 2-point attacks can be provided up to $2k-1$ rounds, while rectangle attacks are efficient to $3k-1$ rounds. These rectangle attacks use φ-tuples of points where φ is even and there are many more differential path than for 2-point attacks. In φ-point attacks, there is a set of equalities that have to be satisfied by φ-tuples of points (inputs/outputs). The number of this φ-tuples will be different for a random permutation and for permutation produced by a scheme. This is due to the fact, that with a scheme, equalities on the internal variables produced during intermediate rounds will help the equalities to propagate. Thus, the equalities on the outputs will appear more frequently. We now explain what kind of equalities may be satisfied either by the inputs, or by the internal variables, or by the outputs in the case of φ-point attacks. We give them on the inputs. The same definition applies for internal variables and outputs.

Definition 9.6 1. "Quasi-Horizontal equalities" on I_i. This family of equalities includes the following possibilities

a. "Horizontal equalities" on I_i:
$I_i(1) = I_i(3) = \ldots = I_i(\varphi - 1)$ and $I_i(2) = I_i(4) = \ldots = I_i(\varphi)$

b. "Diagonal equalities" on I_i This means that for all j instead of having $I_i(2j-1) = I_i(2j+1)$ and $I_i(2j) = I_i(2j+2)$ (as in horizontal equalities), we have $I_i(2j-1) = I_i(2j+2)$ and $I_i(2j) = I_i(2j+1)$.
c. "Hybrid equalities": for each j, we can have either $I_i(2j-1) = I_i(2j+1)$ and $I_i(2j) = I_i(2j+2)$, or $I_i(2j-1) = I_i(2j+2)$ and $I_i(2j) = I_i(2j+1)$.

2. "Vertical equalities" on I_i:
$I_i(1) = I_i(2), I_i(3) = I_i(4), \ldots, I_i(\varphi-1) = I_i(\varphi)$.
3. "Differential equalities" on I_i:
$I_i(1) \oplus I_i(2) = I_i(3) \oplus I_i(4) = \ldots = I_i(\varphi-1) \oplus I_i(\varphi)$.

Notice that once, we have differential equalities, then for vertical equalities, if we have $I_i(1) = I_i(2)$ the other equalities are automatically satisfied and for quasi-horizontal equalities, we need to have $\frac{\varphi}{2} - 1$ equalities to get the other ones. Figure 9.2 explains the choice of the terms of "vertical" and "quasi-horizontal" equalities. Here $\varphi = 6$. In this the dotted lines represent equalities that are satisfied due to differential equalities and vertical/quasi-horizontal equalities.

We now introduce the notation used for the differential paths. When we write $[0, .\Delta_2^0, \Delta_3^0, \ldots, \Delta_k^0]$, this means that, on the input variables, a vertical equality on the first coordinate and horizontal equalities on the second coordinate. The same notation applies to the internal variable and the output variables. The differential path is constructed such that we always keep the differential equalities. This mean that we need the internal variables to satisfy some equalities. For a vertical equality we will write **0** and for quasi-horizontal equality, we will set •. We will write 0 and . when these equalites propagate without any further assumption.

Definition 9.7 For a given attack, the number of equalitites on the inputs (resp. on the outputs) is denoted by n_I (resp. n_S). The number of equalities on the internal variables needed for the propagation of the differential path is denoted by n_X.

Definition 9.8 The length of a rectangle attack with φ-tuples of points is $\ell = \frac{\varphi}{2}$.

We now give the definition of a valid φ-point attack.

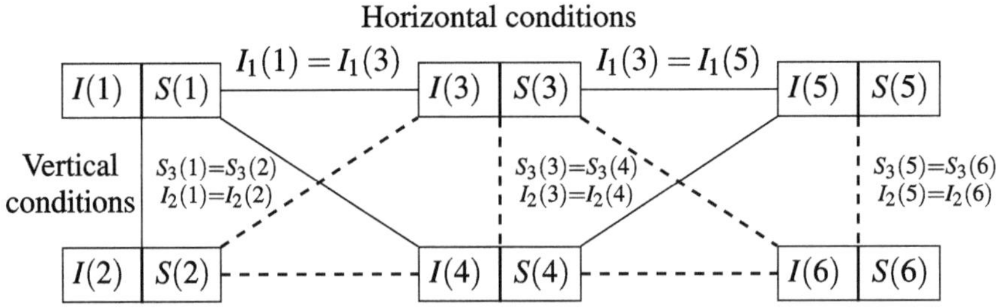

Fig. 9.2 Example of differential equalities for $\varphi = 6$

Definition 9.9 1. The complexity of the attack which is $\frac{n_I + n_X}{2\ell + 2}$ must be smaller than the total number of possible messages, i.e., $\frac{n_I + n_X}{2\ell + 2} \leq k$.
2. There must be less internal conditions than output conditions: $n_X \leq n_S$. This will imply that for a scheme, the number of messages satisfying the condition will be at least twice the number of messages satisfying the condition when testing a random permutation.
3. If $n_X = n_S$ then n_S must be different from the number of final consecutive vertical conditions in the output conditions. If not, it is easy to prove that the output conditions are completely equivalent to the internal conditions. So, the output conditions will not happen more often than for a random permutation.
4. The number of equalities inside the path must be smaller than the number of variables included in them.
5. There is no bottleneck in the equalities, i.e., any subset of equalities must have a greater number of variables. If it is not the case, the attack will only work with very particular functions (weak keys).

9.3.1.1 KPA on F_4^{10}

We now give an example with an attack on F_4^{10}, for which we have vertical conditions on the three first coordinates of the input and quasi-horizontal conditions on the output. In Table 9.3, two kinds of differential paths are provided.

In this attack, for Path 1, we have

$$n_I = 4\ell + 3,\ n_X = 6\ell + 1,\ n_S = 7\ell.$$

Table 9.3 F_4^{10} attack: example of two differential paths

Path 1					Path 2				
Round	0	0	0	Δ_4^0	Round	0	0	0	Δ_4^0
1	0	0	Δ_4^0	0	1	0	0	Δ_4^0	0
2	0	Δ_4^0	0	0	2	0	Δ_4^0	0	0
3	$\bullet\Delta_4^0$	0	0	0	3	$\bullet\Delta_4^0$	0	0	0
4	$\bullet\Delta_1^4$	Δ_2^4	Δ_3^4	$.\Delta_4^0$	4	$\bullet\Delta_1^4$	Δ_2^4	Δ_3^4	Δ_4^0
5	$\bullet\Delta_1^5$	Δ_2^5	$.\Delta_3^5$	$.\Delta_1^4$	5	**0**	Δ_2^5	$.\Delta_3^5$	$.\Delta_1^4$
6	**0**	$.\Delta_2^6$	$.\Delta_5^6$	$.\Delta_1^5$	6	**0**	Δ_3^5	Δ_1^4	0
7	$\bullet\Delta_2^6$	Δ_3^6	Δ_1^5	0	7	$\bullet\Delta_3^5$	Δ_1^4	0	0
8	$\bullet\Delta_1^8$	Δ_2^8	Δ_3^8	$.\Delta_2^6$	8	$\bullet\Delta_1^8$	Δ_2^8	Δ_3^8	$.\Delta_3^5$
9	$\bullet\Delta_1^9$	Δ_2^9	$.\Delta_3^9$	$.\Delta_1^8$	9	$\bullet\Delta_1^9$	Δ_2^9	$.\Delta_3^9$	$.\Delta_1^8$
10	Δ_1^{10}	$.\Delta_2^{10}$	$.\Delta_3^{10}$	$.\Delta_1^9$	10	Δ_1^{10}	$.\Delta_2^{10}$	$.\Delta_3^{10}$	$.\Delta_1^9$

We have to check the validity of the attack. One condition is $6\ell + 1 \leq 7\ell$. This gives $\ell \geq 1$. Since it is better to have the minimum number of points, we choose $\ell = 1$ and we get $\varphi = 4$, with $n_I = 7$, $n_X = 7$ and $n_S = 7$. Thus we have a 4-point attack, whose complexity is $2^{\frac{7n}{2}}$ and we can check that the other conditions to obtain a valid attack are satisfied.

With Path 2, we have

$$n_I = 4\ell + 3,\ n_X = 5\ell + 2,\ n_S = 7\ell$$

Here, we also can choose $\ell = 1$ and the complexity of the attack is $2^{\frac{7n}{2}}$ as well.

9.3.1.2 NCPA on F_3^6

This example gives a NCPA on F_3^6. It is represented in Table 9.4. Here we have $n_I = 3\ell + 2, n_X = 2\ell + 2$ and $n_S = 3\ell + 2$. In order to have $n_X \leq n_S$, the condition on ℓ is $2\ell + 2 \leq 3\ell + 2$. This gives $\ell \geq 1$. If we choose $\ell = 1$, we obtain a 4-point attack. We know transform this KPA in order to get a NCPA as follows. We generate all the possible messages $[I_1, I_2, I_3]$ such that $I_1 = 0$ and the first $n/2$ bits of I_2 are 0. So, we will generate exactly $q = 2^{3n/2}$ messages. We want to know how many 4-tuples of points will verify the input conditions. For the first message there are q possibilities. For the second there are only 2^n possibilities because I_1 and I_2 must be the same as for the first message. For the third point there are again q possibilities, and then there is no more choice for the last point. Therefore there are $q^2 \times 2^n = 2^{4n}$ 4-tuples of points that satisfy all the input conditions. For a F_3^6 scheme, each of these tuples will satisfy at random the 4 internal conditions with a probability equal to $1/2^{4n}$. So, the expected number of 4-tuples that satisfy also the output conditions will be approximately 1. Since there are 5 output conditions, the expected number of 4-tuples that satisfy the input conditions and the output conditions will be much lower for a random permutation. So, this CPA-1 attack will succeed with a high probability. This gives a NCPA with $O(2^{3n/2})$ complexity and $O(2^{3n/2})$ messages. Moreover it is possible to check that all the condition for a valid attack are satisfied.

Table 9.4 Attack on F_3^6

Round	**0**	**0**	Δ_3^0
1	0	Δ_3^0	0
2	$\bullet\Delta_3^0$	0	0
3	$\bullet\Delta_1^3$	Δ_2^3	$.\Delta_3$
4	**0**	**0**	$.\Delta_1^3$
5	0	$.\Delta_1^3$	0
6	$.\Delta_1^3$	0	0

9.3.2 Generation of All Possible Attacks for $k \leq 7$

Here we describe how to generate all the possible attacks. First we choose a value for k, then we increase the value of r, beginning with $r = 1$, until we find no possible attacks. All the attacks (or sometimes only the best attacks when the number is too much important) are put in a specific file corresponding to the values of k and r.

To find an attack, we need to construct all the differential paths. There are two constraints for this construction:

- In the same round, it is not possible to have k vertical conditions, because it leads to a collision between the points.
- In the same round, it is not possible to have k quasi-horizontal conditions, because it also lead to a collision between the points.

When the path is constructed, we look if the attack is valid according to Definition 9.9. Notice that checking that there is no bottleneck is very difficult to carry out without the help of a computer. This last condition was always satisfied in Volte et al. (2010), but not always in Patarin et al. (2007).

Finally, all the possible attacks are sorted in function of their complexity (KPA or NCPA). In Volte et al. (2010), the authors generated all the possible attacks for $k \leq 7$ with the help of a computer and then generalized the attacks for any k, assuming that for $k > 7$, the best attacks are of the same kinds. Before giving these attacks, we first define the different kinds of attacks that can be considered.

9.3.3 Different Kinds of Rectangle Attacks: $R1$, $R2$, $R3$, and $R4$

In rectangle attacks, on the inputs and on the outputs, it is possible to ask for either vertical equalities or quasi-horizontal equalities. It could also be possible to have at the same time vertical and quasi-horizontal equalities on the inputs or outputs. These attacks are not better and we do not consider them here. Thus this leads to define 4 different kinds of attacks.

9.3.3.1 $R1$ Attacks

Here there are vertical conditions on the input and output variables. These attacks are more general than the attacks named R1 in Patarin et al. (2007) since there are more vertical conditions on the inputs and outputs. These attacks were first described by Jutla (1998) and improved in Volte et al. (2010). Thus we have the following conditions:

$$n_I = k\ell + s, \quad n_X = t\ell + w, \quad n_S = k\ell + v.$$

We use t for the number of quasi-horizontal equalities the internal variables and w for vertical conditions on the internal variables. On the inputs (resp. the outputs, the number of vertical conditions is denoted by s (resp. v). Then the number of rounds is given by $s + t + w$. The condition $n_X \leq n_S$ implies that $(k - t)\ell \geq w - v$. In order to avoid weak keys, the number of equations with the internal variables must be smaller than or equal to the number of internal variables. This condition was not always satisfied in Patarin et al. (2007), but it was always verified in Volte et al. (2010). For R1 attacks, it is easy to check that the number of equations is given by $t(k - 1)$ and the number of variables is $k(t + 1) - s - w$. Thus we get the condition: $s + w \leq t + k$. The complexity of such an attack is $2^{\frac{n_I+n_X}{\varphi}n}$. This implies $\frac{n_I+n_X}{\varphi} \leq k$, i.e., $\frac{(k+t)\ell+s+w}{2\ell+2} \leq k$.

9.3.3.2 *R*2 Attacks

Here we have quasi-horizontal conditions on the inputs and vertical conditions on the outputs. This gives

$$n_I = (k + u)\ell, \quad n_X = t\ell + w, \quad n_S = k\ell + v.$$

The number of quasi-horizontal conditions on the inputs variables is denoted by u, the number of vertical conditions on the outputs is denoted by v. On the internal variables t represents the number of quasi-vertical conditions and w the number of vertical conditions. The number of rounds is given by $u + t + w$. The condition $n_X \leq n_S$ is equivalent to $(k - t)\ell \geq w - v$. For $R2$ attacks, it is easy to check that the number of equations is given by $(t + 1)(k - 1)$ and the number of variables is $k(t + 2) - w$. Thus we get the condition: $w \leq t + k + 1$. The complexity of such an attack is $2^{\frac{n_I+n_X}{\varphi}n}$. This implies $\frac{n_I+n_X}{\varphi} \leq k$, i.e., $\frac{(k+t+u)\ell+w}{2\ell+2} \leq k$.

9.3.3.3 *R*3 Attacks

Here we have vertical conditions on the input variables and quasi-horizontal conditions on the output variables. We have

$$n_I = k\ell + s, \quad n_X = t\ell + w, \quad n_S = (k + x)\ell.$$

The number of rounds is given by $s + t + w$. The condition $n_X \leq n_S$ is equivalent to $(k + x - t)\ell \geq w$. For R3 attacks, it is easy to check that the number of equations is given by $(t - x)(k - 1)$ (here we have $t > x$)) and the number of variables is $k(t - x + 1) - s - w$. Thus we get the condition: $w + s + x \leq t + k$. The complexity of such an attack is $2^{\frac{n_I+n_X}{\varphi}n}$. This implies $\frac{n_I+n_X}{\varphi} \leq k$, i.e., $\frac{(k+t)\ell+s+w}{2\ell+2} \leq k$.

9.3.3.4 R4 Attacks

Here we have horizontal conditions on the input and output variables. We have

$$n_L = (k+u)\ell, \quad n_X = t\ell + w, \quad n_S = (k+x)\ell.$$

The number of rounds is given by $u+t+w$. The condition $n_X \leq n_S$ is equivalent to $(k+x-t)\ell \geq w$. For R4 attacks, it is easy to check that the number of equations is given by $(t-x+1)(k-1)$ (here we have $t \geq x$)) and the number of variables is $k(t-x+2)-w$. Thus we get the condition: $w+x \leq t+k+1$. The complexity of such an attack is $2^{\frac{n_I+n_X}{\varphi}n}$. This implies $\frac{n_I+n_X}{\varphi} \leq k$, i.e., $\frac{(k+t+u)\ell+w}{2\ell+2} \leq k$.

9.3.4 *Best KPA Attacks: R1, R2*

Here we provide the best KPA known so far. We will mostly describe one example of $R2$ attacks since for any round there are many possible $R2$ attacks that give the best complexity. It can be noticed that in KPA, there is a symmetry between $R2$ and $R3$ attacks. Thus there always exist $R2$ and $R3$ attacks with the same complexity. Sometimes, it is also possible to have $R1$ attacks with the same complexity. Most of the time, $R4$ attacks are worse. We give attacks from $k+3$ rounds to $3k-1$ rounds since from 1 to $k+2$ rounds, 2-point attacks are most of the time better. In all our attacks, it is easily checked that the conditions to have a valid attack are verified. Moreover, we always look for attacks where the number of points is minimum. The best $R2$ KPA attacks are summarized in Table 9.5.

Table 9.5 Best known KPA on F_k^r, for any $k \geq 3$

r values	n_I	n_X	n_S	ℓ	Complexity
$k+2q \in [k+3; 2k-2]$	$(2k-1)\ell$	$q\ell+q+1$	$k\ell+q+1$	1	$2^{\frac{k+d}{4}n}$
$k+2q+1 \in [k+3; 2k-2]$	$(2k-1)\ell$	$q\ell+q+2$	$k\ell+q+2$	1	$2^{\frac{k+d}{4}n}$
$k+2q \in [2k-1; 3k-2]$	$(2k-1)\ell$	$(q-\lfloor\frac{k-1}{2}\rfloor)\ell+q+\lfloor\frac{k+1}{2}\rfloor$	$k\ell+k-1$	1	$2^{\frac{k+d}{4}n}$
$k+2q+1 \in [2k-1; 3k-2]$	$(2k-1)\ell$	$(q-\lfloor\frac{k-3}{2}\rfloor)\ell+q+\lfloor\frac{k+1}{2}\rfloor$	$k\ell+k-1$	1	$2^{\frac{k+d}{4}n}$
$3k-1$	$k\ell+\ell$	$k\ell+2k-1$	$k\ell+k-1$	k	$2^{(k-\frac{1}{2k+2})n}$

Remark 9.1 1. We have the following $R1$ attacks:

a. When $k+3 \leq r \leq 2k-2$ and $r = k+2q$, we set

$$n_I = (k\ell + k - 1), \quad n_X = q\ell + q + 1, \quad n_S = k\ell + q + 1.$$

It is possible to choose $\ell = 1$ and the complexity is also $2^{\frac{k+r}{4}n}$

b. When $2k-1 \leq r \leq 3k-2$ and $r = k+2q$, we set

$$n_I = k\ell + 2, \quad n_X = q\ell + k + q - 2, \quad n_S = k\ell + k - 1.$$

The complexity is still $2^{\frac{k+r}{4}n}$, but ℓ is greater than 1.

2. In Jutla (1998), Jutla gave a $R1$ attack on $3k-3$ rounds but the complexity that we obtain with a $R2$ attack here is better. It is possible to perform a $R1$ attack on $3k-2$ rounds just by adding a vertical condition on the input variables to the attack on $3k-3$ rounds and the we obtain the same complexity as the one we get with a $R2$ attack. Due to the conditions between the number of equations and internal variables, it is not possible to use the same idea for $3k-1$ rounds. In this last case, we have $R2$ (and of course $R3$) attacks, but no $R1$ attack.

9.3.5 *From KPA Into NCPA*

We now explain how to go from KPA to NCPA. An example was given on F_3^6. As we will see the best NCPA do not always come from the best KPA. We call u the number of quasi-horizontal conditions, s the number of vertical condition on the inputs. Since we suppose that we have either vertical equalities or quasi-horizontal equalities on the inputs but not both for the same attack, we distinguish two cases:

Case $s = 0$: $\overbrace{.\Delta_1 \, .\Delta_2 \, \ldots \, .\Delta_u}^{u\text{"."}} \; \Delta_{u+1} \ldots \, \Delta_{k-u}$

Case $u = 0$: $\underbrace{0\,0 \ldots 0}_{s\text{"0"}} \; \Delta_1 \, \ldots \, \Delta_{k-s}.$

9.3.5.1 For $R1$ Attacks: Vertical Equalities on the Inputs

Suppose that for $1 \leq s \leq q$, we have the following input:

$$\underbrace{0\,0 \ldots 0}_{s} \; \Delta^0_{s+1} \, \Delta^0_{s+2} \cdots \Delta^0_{k-s}.$$

Let $I_{\alpha n}$ be the group $\left((\{0,1\}^n)^\alpha, \oplus\right)$. We will define a subgroup $\overrightarrow{\Delta}$ of $\overrightarrow{E} = (\{0\}^n)^s \times I_{(k-s)n}$, which is itself a subgroup of $I_{kn} = \left((\{0,1\}^n)^k, \oplus\right)$. Let β be the number of packs of n bits used in $\overrightarrow{\Delta}$, then we have $|\overrightarrow{\Delta}| = 2^{\beta n}$. β may not be an integer but we have $\beta \leq k-s$.

Let us choose $d = 2^{bn}$ inputs $I(1), I(2), ..., I(d)$ such that

$$i \neq j \Rightarrow I(j) \oplus I(i) \notin \overrightarrow{\Delta}.$$

Then, the chosen plaintext are of the from $I(i) + \Delta$ with $1 \leq i \leq r$ and $\Delta \in \overrightarrow{\Delta}$. Thus there are exactly $2^{(b+\beta)n}$ chosen plaintexts. We want to have $b+\beta$ as small as possible.

We have to find how many φ-tuples of messages will satisfy the input conditions.

For $M(1)$, it is possible to choose any of the $2^{(b+\beta)n}$ chosen plaintexts, thus there are $2^{(b+\beta)n}$ possibilities.

For $M(2)$, we must choose $M(1)+\Delta$ where $\Delta \in \overrightarrow{\Delta}$, there are $2^{\beta n}$ possibilities.

For $M(3)$, any choice is possible except $M(1)$. If we do not take into account this exception, this gives $2^{(b+\beta)n}$ possibilities.

For $M(4)$, the only choice is $M(3)+\Delta$. This only 1 possibility.

$\vdots$

For $M(\varphi-1)$ there are $2^{(b+\beta)n}$ possibilities.

For $M(\varphi)$, there is only one possibility.

Finally, the number of φ-tuples is given by

$$\left(2^{(b+\beta)n}\right)^{\varphi/2} \times 2^{\beta n} = 2^{(b+\beta)n(\ell+1)+\beta n}.$$

Since we want $b+\beta$to be as small as possible, we need to choose β maximal.

Moreover, each φ-tuple will satisfy the internal conditions with probability $\frac{1}{2^{n \cdot n_X}}$. Thus if we want the conditions to be verified with a non negligible probability, we must have

$$(b+\beta)(\ell+1)+\beta \geq n_X.$$

β is maximal when b is equal to zero, which gives $\beta = \frac{n_X}{\ell+2}$. If this value is smaller than $k-s$, we choose this value and in that case the complexity is $2^{\frac{n_X}{\ell+2}n}$. In the other case, we take $\beta = k-s$ since this is the maximal value and the complexity is $2^{\frac{n_X-k+s}{\ell+1}n}$.

For example, in the attack on F_k^{3k-1} where $k-s=1$, $\ell=k$ and $n_X = k^2+k$, we have $\frac{n_X}{\ell+2} = \frac{k^2+k}{k+2} \geq 1$, thus the NCPA complexity is

$$2^{\frac{k^2+k-1}{k+1}n} = 2^{\left(k-\frac{1}{k+1}\right)n}.$$

9.3.5.2 For $R2$ Attacks: Quasi-Horizontal Equalities on the Inputs

Suppose that there are u quasi-horizontal conditions on the input, i.e., we have

$$.\Delta_1^0 \ .\Delta_2^0 \ \dots \ .\Delta_u^0 \ \Delta_{u+1}^0 \ \dots \ \Delta_k^0.$$

Among the nu first bits, we will choose cn bits and the remaining bits will be supposed to be zero. Among the $(k-u)n$ next bits, we choose βn.

Then we generate all the possible plaintexts with these $(c+\beta)n$ bits. This gives a number of messages equal to $2^{(c+\beta)n}$.

We now count the number of φ-tuples messages satisfying the input conditions.
For $M(1)$, there are $2^{(c+\beta)n}$ possibilities.
For $M(2)$, there are about $2^{(c+\beta)n}$ possibilities.
For $M(3)$, we must keep the cn bits of $M(1)$, thus $2^{\beta n}$ possibilities remain.
For $M(4)$, there is no choice, we take $M(1) \oplus M(2) \oplus M(3)$. This gives 1 possibility.

$\vdots$

For $M(\varphi-1)$, there are about $2^{\beta n}$ possibilities.
For $M(\varphi)$, there is only 1 possibility.
Finally there are $2^{(2(c+\beta)+\ell\beta)n}$ φ-tuples, and for each of them the probability that the internal conditions are satisfied is $\frac{1}{2^{n_X n}}$. The attack will success with a non negligible probability if:

$$n_X = 2(c+\beta) + \ell\beta.$$

Since we want $c+\beta$ to be minimal, β has to be maximal. Since $c \geq 0$, we must have $\frac{n_X}{2+\ell} \geq \beta$. There are two cases:

First case. $\frac{n_X}{2+\ell} \leq k-u$. Then we take $\beta = \frac{n_X}{2+\ell}$ and we have $c = 0$. Thus the complexity is $2^{\frac{n_X}{\ell+2}n}$.

Second case. $\frac{n_X}{2+\ell} \geq k-u$. We take $\beta = k-u$ and we have

$$c+\beta = \frac{n_X - l(k-u)}{2}.$$

Moreover, since $n_I = (k+u)\ell$ we get

$$c+\beta = \frac{n_X - \ell(2k - \frac{n_I}{\ell})}{2} = \frac{n_X + n_I - 2k\ell}{2}.$$

Remembering that $\log_{2^n}(KPA) = (n_X + n_I)/(2l+2)$, we obtain

$$\log_{2^n}(CPA) = \log_{2^n}(KPA) - l(k - \log_{2^n}(KPA)).$$

Table 9.6 KPA to CPA

Conditions		$\log_{2^n}(CPA)$
$u = 0$	$\frac{n_X}{\ell+2} \leq k - s$	$\frac{n_X}{\ell+2}$
s vertical conditions	$\frac{n_X}{\ell+2} > k - s$	$\frac{n_X - k + s}{\ell+1}$
$s = 0$	$\frac{n_X}{\ell+2} \leq k - u$	$\frac{n_X}{\ell+2}$
u horizontal conditions	$\frac{n_X}{\ell+2} > k - u$	$\frac{n_X - \ell(k-u)}{2}$

Example: For attacks F_k^{3k-1} with one quasi-horizontal condition on the first coordinate, we have $u = 1, \ell = k, n_X = (k-1)\ell + 2k - 1$. The complexity of this NCPA is $2^{(k-\frac{1}{2})n}$.

9.3.5.3 Summary of the Results

Best NCPA do not always come from best KPA. However, it is possible to express the NCPA complexity with the KPA complexity, with the following formula: $\log_{2^n}(KPA) = \frac{r + (u+k)\ell + n_X}{2\ell+2}$.

Generally, for all the NCPA, the best choice is to keep the first bits constant and generate all the possible messages with the same first bits (Table 9.6).

9.3.6 *Best NCPA: R_1, R_2. Simulations*

9.3.6.1 NCPA Attacks

In this section, we describe the best NCPA known so far. Again for $k \leq 7$ all the best NCPA have been generated (Volte et al. 2010) and then they have been generalized for any k. The best NCPA are generally $R2$ attacks. Sometimes $R1$ attacks exist with the same complexity. As said previously, it is interesting to note that the best NCPA do not come from the best KPA. We will use the study of NCPA made in Sect. 9.3.5. We will describe NCPA for $k + 3 \leq r \leq 3k - 1$ since for $r \leq k + 2$, the best attacks are 2-point attacks given in Sect. 9.2. Again we will give an example of such an attack for each round. We notice that for the same conditions on the input and output variables, we can find several attacks: the quasi-horizontal and vertical conditions on the internal variables can be displayed differently inside the differential path, but we must respect the conditions between the number of equations and variables at each step of the attack. An example is given below. The best $R2$ NCPA known so far are summarized in Table 9.7.

Table 9.7 Best known NCPA on F_k^r, for any $k \geq 3$

r values	n_L	n_X	n_S	ℓ	Complexity
$k+3$	$k\ell + (k-1)\ell$	$\ell + 3$	$k\ell + 1$	1	$2^{\frac{3n}{2}}$
$k+4$	$k\ell + (k-2)\ell$	$2\ell + 4$	$k\ell + 2$	1	2^{2n}
$k+5$	$k\ell + (k-2)\ell$	$2\ell + 5$	$k\ell + 2$	1	$2^{5n/2}$
$k + 2q \in [k+6; 3k-4]$	$k\ell + (k-q)\ell$	$(q-1)\ell + 2q + 1$	$k\ell + 1$	$q-1$	$2^{\frac{q^2+2}{q+1}n}$
$k + 2q + 1 \in [k+7; 3k-5]$	$k\ell + (k-q)\ell$	$(q-1)\ell + 2q + 2$	$k\ell + 1$	$q-1$	$2^{\frac{q^2+3}{q+1}n}$
$3k-3$	$k\ell + \ell$	$(k-2)\ell + 2k - 2$	$k\ell + 1$	$k-1$	$2^{\frac{(k-1)k}{k+1}n}$
$3k-2$	$k\ell + (k-1)\ell$	$(k - \lfloor\frac{k}{2}\rfloor)\ell + 2k - \lfloor\frac{k-1}{2}\rfloor$	$k\ell + k - 1$	1	$2^{(k-1)n}$
$3k-1$	$k\ell + \ell$	$(k-1)\ell + 2k - 1$	$k\ell + k - 1$	k	$2^{(k-\frac{1}{2})n}$

Remark 9.2 For $r = k+3, k+4, k+5$ and $3k-2$, there exist $R1$ attacks with the same complexity and the same number of points.

9.3.6.2 Experimental Results

We present simulations of these NCPA attacks. For each simulation a random Feistel scheme with 20 rounds is generated, and a F_k^{3k-1} scheme. For both schemes, $2^{(k-1/2)n}$ ciphertext/plaintext pairs are computed, such that they vary only on the last $(k-1/2)n$ bits. After this, all couples of points that satisfy both input and output conditions are extracted and these couples of points are sorted in order to count how many φ-tuples of points match the input and output condition. If there are q couples of points that satisfy all these conditions with $q \geq \varphi/2$, they are counted as if there are $\frac{q!}{(q-\varphi/2)!}$ φ-tuples, because this is the number of φ-tuples that can be taken out these points, by changing the position of the couple of points. Once this is finished, it is possible to compare the number found for each permutation. That often enables to make a distinction between them. See Table 9.8.

9.4 Summary of the Attacks

In Tables 9.9 and 9.10, we give the complexity of the attacks we have found. For $k \leq 7$, since we have generated all the attacks, these are the best possible attacks. Then we have generalized the results for $k > 7$ and we believe that the attacks

Table 9.8 Experimental results for F_k^{3k-1}

n	k	kn	% of success (%)	% of false alarm (%)	# iteration
2	3	6	29,09	0,35	100000
2	4	8	61,6	0,06	10000
2	5	10	98,37	0	10000
2	6	12	99,99	0	10000
2	7	14	100	010000	
2	8	16	100	0	1000
2	9	18	100	0	500
2	10	20	100	0	100
4	3	12	21,15	1,12%	10000
4	4	16	42,5	0	1000
4	5	20	93	0	100
4	6	24	100	0	100
6	3	18	8	1,2	500
8	3	24	2	0	100

Table 9.9 Best known TWO and Rectangle attacks on F_k^d, for any $k \geq 3$

Rounds	KPA	NCPA
F_k^1	1	1
F_k^2	$2^{\frac{n}{2}}$, 2-point	2
F_k^3	2^n, 2-point	2
$F_k^d, 2 \leq d \leq k$	$2^{\frac{d-1}{2}n}$, 2-point	2
F_k^{k+1}	$2^{\frac{k}{2}n}$, 2-point	$2^{\frac{n}{2}}$, 2-point
F_k^{k+2}	$2^{\frac{k+1}{2}n}$, 2-point	2^n, 2-point
F_k^{k+3}	$2^{\frac{2k+3}{4}n}$, R2, R3	$2^{3n/2}$, R2
F_k^{k+4}	$2^{\frac{k+2}{2}n}$, R1, R2, R3	2^{2n}, R2
F_k^{k+5}	$2^{\frac{2k+5}{4}n}$, R2, R3	$2^{5n/2}$, R2
$\vdots$	$\vdots$	$\vdots$
F_k^r, $r = k + 2q$, $3 \leq q \leq k-2$	$2^{\frac{d+k}{4}n}$, R1, R2, R3	$2^{\frac{q^2+2}{q+1}n}$, R2
$F_k^d, r = k + 2q + 1$, $3 \leq q \leq k-3$	$2^{\frac{d+k}{4}n}$, R2, R3	$2^{\frac{q^2+3}{q+1}n}$, R2
$\vdots$	$\vdots$	$\vdots$
F_k^{3k-3}	$2^{(k-\frac{3}{4})n}$, R2, R3	$2^{\frac{(k-1)k}{k+1}n}$, R2
F_k^{3k-2}	$2^{(k-\frac{1}{2})n}$, R1, R2, R3	$2^{(k-1)n}$, R2
F_k^{3k-1}	$2^{(k-\frac{1}{2k+2})n}$, R2, R3,	$2^{(k-\frac{1}{2})n}$, R2

Table 9.10 Best known 2-point and Rectangle attacks on F_3^r

Rounds	KPA	NCPA
F_3^1	1	1
F_3^2	$2^{\frac{n}{2}}$, 2-point	2
F_3^3	2^n, 2-point	2
F_3^4	$2^{\frac{3}{2}n}$, 2-point	$2^{\frac{n}{2}}$, 2-point
F_3^5	2^{2n}, 2-point	2^n, 2-point
F_3^6	$2^{\frac{9}{4}n}$, R2, R3	$2^{\frac{3}{2}n}$, R2
F_3^7	$2^{\frac{5}{2}n}$, R1, R2, R3	2^{2n}, R2
F_3^8	$2^{\frac{23}{8}n}$, R2, R3	$2^{\frac{5}{2}n}$, R2

presented here are also the best possible attacks. For $d \leq k+2$, we have 2-point attacks. For $d \geq k+3$, we have rectangle attacks. As mentioned before, in KPA, there are always R2 and R3 attacks that give the best complexity sometimes there is also a R1 attacks (for $3k-2$ rounds for example). In NCPA, the best complexity is given by R2 attacks, and sometimes R1 attacks.

Problems

9.1 We consider an "extremely Expanding Feistel Cipher", i.e., such that each round from N bits to N bits is built like this: each bit, except the first one, is Xored with a secret function of the first bit. Then, we perform a rotation of one bit on all the bits. Why is this a very bad cipher, whatever the number of rounds like this is done?

References

Goubin, L., Ivasco, M., Jalby, W., Ly, O., Nachef, V., Patarin, J., Treger, J., Volte, E. 2008. CRUNCH. Submission to NIST.

Hoang, V. T., and P. Rogaway. 2010. On generalized feistel networks. In *Advances in cryptology—CRYPTO 2010*, vol. 6223, ed. T. Rabin. Lecture notes in computer science, 613–630. Berlin: Springer.

Hoel, P. G., S. C. Port, and C. J. Stone. 1971. *Introduction to probability theory*. Houghton Mifflin Company.

Jutla, C. S. 1998. Generalized birthday attacks on unbalanced feistel networks. In *Advances in cryptology—CRYPTO 1998*, vol. 1462, ed. H. Krawczyk. Lecture notes in computer science, 186–199. Berlin: Springer.

Patarin, J., V. Nachef, and C. Berbain. 2007. Generic attacks on unbalanced feistel schemes with expanding functions. In *Advances in cryptology—ASIACRYPT 2007*, vol. 4833, ed. K. Kurosawa. Lecture notes in computer science, 325–341. Berlin: Springer.

Volte, E., V. Nachef, and J. Patarin. 2010. Improved generic attacks on unbalanced feistel schemes with expanding functions. In *Advances in cryptology—ASIACRYPT 2010*, vol. 6477, ed. M. Abe. Lecture notes in computer science, 94–111. Berlin: Springer.

Chapter 10
Generic Attacks on Generalized Feistel Ciphers

Abstract Type-1, type-2 and type-3 and alternating Feistel schemes are described by Zhen et al. (1990) (see also Hoang and Rogaway 2010). These generalized Feistel schemes are used in well-known block cipher networks that use generalized Feistel schemes: CAST-256 (type-1), RC-6 (type-2), and BEAR/LION (alternating). Also, type-1 and type-2 Feistel schemes are respectively used in the construction of the hash functions Lesamnta and SHAvite $- 3_{512}$. There exist many kind of attacks on these schemes: impossible differential attacks (Bouillaguet et al. 2012; Luo et al. 2014), boomerang attacks (Choy and Yap 2009), and differential attacks (Nachef et al. 2013). However, the attacks we are going to describe in this chapter generally allow to attack more rounds (or at least the same number of rounds) than for example impossible differential attacks. Moreover in the presented attacks, there is no restriction on the round function, unlike for some impossible differential attacks where it is supposed to be bijective. Security results are given in Hoang and Rogaway (2010).

10.1 Type-1 Feistel Ciphers

10.1.1 Notation Definition

As usual the input is denoted by $[I_1, I_2, \ldots, I_k]$ and the output by $[S_1, S_2, \ldots, S_k]$ where each I_s, S_s is an element of $\{0, 1\}^n$. When there are q messages, $I_s(i)$ represents part s of the input of message number i. The same notation is used for the outputs as well.

After one round, the output is given by $[I_2 \oplus f_1(I_1), I_3, I_4, \ldots, I_k, I_1]$ where f_1 is a function from n bits to n bits (Fig. 10.1).

At round r, the internal variable X^r appears on the first coordinate of the intermediate value. It is easy to check that

$$\begin{cases} X^1 = I_2 \oplus f_1(I_1) \\ \forall r,\ 1 \leq r \leq k-1,\ X^r = I_{r+1} \oplus f_r(X^{r-1}) \\ X^k = I_1 \oplus f_k(X^{k-1}) \\ \forall r > k,\ X^r = X^{r-k} \oplus f_r(X^{r-1}). \end{cases}$$

J. Patarin et al., *Guide to Feistel Ciphers*, Information Security and Cryptography,
https://doi.org/10.1007/978-3-031-99850-8_10

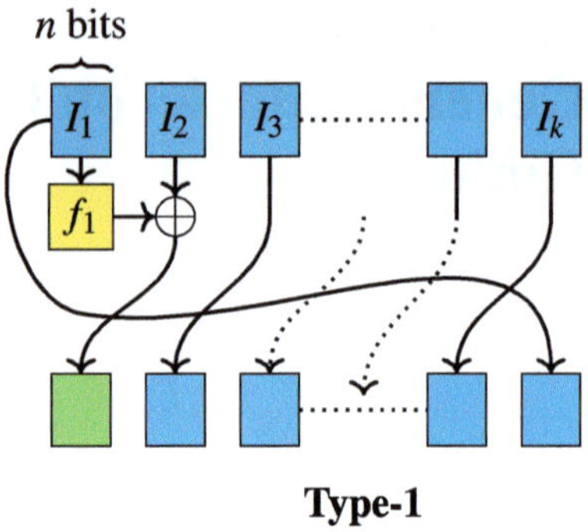

Fig. 10.1 First round for type-1 Feistel Ciphers

The output after round r is given by

$$\begin{cases} 1 \le r \le k-2 : [X^r, I_{r+2}, \dots, I_k, I_1, X^1, \dots, X^{r-1}] \\ r = k-1 : [X^{k-1}, I_1, X^1, \dots, X^{k-2}] \\ r \ge k : [X^r, X^{r-k+1}, X^{r-k+2}, \dots, X^{r-1}]. \end{cases}$$

In particular, when $1 \le r \le k-1$, the input of the round function f_r depends on $I_1, \dots, I_r$. Thus X^r is the xor of I^{r+1} and a function of $I_1, \dots, I_k$, and X^k depends on all the input values.

10.1.2 Simple Attacks on the First Rounds

10.1.2.1 Attacks on the First $k-1$ Rounds

It is easy to see that after rounds r, $1 \le r \le k-1$, we have $S_{k-r+1} = I_1$. This leads to the following KPA. Let r such that $1 \le r \le k-1$. Consider the following KPA-distinguisher D.

1. D makes a query to the oracle and receives a random plaintext $[I_1, \dots, I_k]$ together with $[S_1, \dots, S_k] = \mathcal{O}([I_1, \dots, I_k])$;
2. If $S_{k-r+1} = I_1$, D outputs 1, otherwise, it outputs 0.

When $\mathcal{O}$ is a type-1 Feistel cipher, D always outputs 1. When $\mathcal{O}$ is a random permutation of $\{0, 1\}^{kn}$, then $[S_1, \dots, S_k]$ is uniformly random in $\{0, 1\}^{kn}$, and the probability that $S_{k-r+1} = I_1$ is 2^{-n}. Therefore, by the definition of the advantage, we have

$$\mathscr{A}(\mathsf{D}) = 1 - \frac{1}{2^n}.$$

Since D makes exactly one query, it follows that

$$\mathscr{A}^{\mathrm{KPA}}(\mathsf{D}) \geq 1 - \frac{1}{2^n}.$$

This attack is able to distinguish a permutation produced by a Type 1 Feistel cipher when the round functions are from n bits to n bits and randomly chosen, from a truly random permutation of $\mathscr{P}_{kn}$ with a high probability using one random plaintext/ciphertext pair. This attack is a KPA with $q = 1$ plaintext/ciphertext pair.

10.1.2.2 NCPA on r Rounds with $k \leq r \leq 2k - 2$

Let r such that $k \leq r \leq 2k - 2$. Consider the following NCPA-distinguisher D.

1. D chooses $I_{r-k+2}, I_{r-k+3}, \dots I_k$ in $\{0, 1\}^{(2k-r-1)n}$, and asks for the value of $\mathscr{O}([0, 0, \dots, I_{r-k+2}, I_{r-k+3}, \dots I_k]) = [S_1, \dots, S_k]$.
2. D chooses $I'_{r-k+2}, I'_{r-k+3}, \dots I'_k$ in $\{0, 1\}^{(2k-r-1)n}$, such that

$$[I_{r-k+2}, I_{r-k+3}, \dots I_k] \neq [I'_{r-k+2}, I'_{r-k+3}, \dots I'_k]$$

and asks for the value of

$$\mathscr{O}([0, 0, \dots, I'_{r-k+2}, I'_{r-k+3}, \dots I'_k]) = [S'_1, \dots, S'_k].$$

3. D tests whether $S_2 \oplus S'_2 = I_{r-k+2} \oplus I'_{r-k+2}$.
4. If this holds, D outputs 1, otherwise D outputs 0.

When $\mathscr{O}$ is a type-1 Feistel cipher, D always outputs 1, since after r rounds $k \leq r \leq 2k - 2$, the output is given by $[X^r, X^{r-k+1}, X^{r-k+2}, \dots, X^{r-1}]$. When $\mathscr{O}$ is a random permutation of $\{0, 1\}^{kn}$, then $[S_1, \dots, S_k]$ is uniformly random in $\{0, 1\}^{kn}$, and the probability that $S_2 \oplus S'_2 = I_{r-k+2} \oplus I'_{r-k+2}$ is 2^{-n}. Therefore, by the definition of the advantage, we have

$$\mathscr{A}(\mathsf{D}) = 1 - \frac{1}{2^n}.$$

Since D makes exactly two queries, it follows that

$$\mathscr{A}^{\mathrm{KPA}}(\mathsf{D}) \geq 1 - \frac{1}{2^n}.$$

This attack is able to distinguish a permutation produced by a Type 1 Feistel cipher when the round functions are from n bits to n bits and randomly chosen, from a truly gives a NCPA with $q = 2$ plaintext/ciphertext pairs.

As usual, these attacks can be transformed into KPA, but there exist attacks with better complexities, as we will see below.

10.1.3 NCPA and KPA Using the Expectation

10.1.3.1 KPA on r Rounds with $k \leq r \leq 3k - 2$

The attacks make use of the following differential path described in Table 10.1.

Complexity in $O(2^{n/2})$ for $k \leq r \leq 2k - 1$.

For the KPA on k rounds, the attacker chooses q messages, sends them to the oracle $\mathcal{O}$, and receives the outputs $[S_1(i), S_2(i), \ldots, S_k(i)]$, $1 \leq i \leq q$. Then he will count the number of pairs (i, j), $i < j$ such that $S_k(i) = S_k(j)$ and $I_1(i) \oplus I_1(j) = S_1(i) \oplus S_1(j)$. Suppose that $\mathcal{O}$ is a random permutation. We obtain that

$$\mathbb{E}(\mathcal{N}_{perm}) \simeq \frac{q^2}{2.2^{2n}}.$$

When $\mathcal{O}$ is type-1 Feistel cipher, Table 10.1 shows that when $X^{k-1}(i) = X^{k-1}(j)$, then the inputs and outputs will automatically satisfy the right conditions, which may also appear at random. This shows that

$$\mathbb{E}(\mathcal{N}_{Type1_k^k}) \simeq \frac{q^2}{2.2^n} + \frac{q^2}{2 \cdot 2^{2n}}.$$

When $q \simeq 2^{n/2}$, $\mathbb{E}(\mathcal{N}_{Type1_k^k})$ is much greater than $\mathbb{E}(\mathcal{N}_{perm})$ and the attack succeeds. The KPA on $r = k + t$ rounds with $1 \leq t \leq k - 2$ is quite similar, the conditions

Table 10.1 Differential path for KPA on r rounds with $k \leq r \leq 3k - 2$

round	Δ_1^0	Δ_2^0	Δ_3^0	$\ldots$	$\ldots$	Δ_{k-1}^0	Δ_k^0
1							Δ_1^0
2						Δ_1^0	
$\vdots$	$\vdots$	$\vdots$	$\vdots$	$\vdots$	$\vdots$	$\vdots$	$\vdots$
$k-1$	**0**	Δ_1^0					
k	Δ_1^0						0
$k+1$						0	Δ_1^0
$\vdots$	$\vdots$	$\vdots$	$\vdots$	$\vdots$	$\vdots$	$\vdots$	$\vdots$
$2k-2$		0	Δ_1^0				
$2k-1$	**0**	Δ_1^0					
$2k$	Δ_1^0						0
$2k+1$						0	Δ_1^0
$\vdots$	$\vdots$	$\vdots$	$\vdots$	$\vdots$	$\vdots$	$\vdots$	$\vdots$
$3k-2$		0	Δ_1^0				

are given by $S_{k-t}(i) = S_{k-t}(j)$ and $I_1(i) \oplus I_1(j) = S_{k-t-1}(i) \oplus S_{k-t-1}1(j)$. Again, this gives a successful KPA with $2^{n/2}$ queries.

On $2k-1$ rounds, the attacker chooses q messages, sends them to the oracle $\mathcal{O}$, and receives the outputs $[S_1(i), S_2(i), \ldots, S_k(i)]$, $1 \le i \le q$. Then he will count the number of pairs (i, j), $i < j$ such that $I_1(i) \oplus I_1(j) = S_2(i) \oplus S_2(j)$. Suppose that $\mathcal{O}$ is a random permutation. We obtain that

$$\mathbb{E}(\mathcal{N}_{perm}) \simeq \frac{q^2}{2.2^n}.$$

When $\mathcal{O}$ is type-1 Feistel cipher, Table 10.1 shows that when $X^{k-1}(i) = X^{k-1}(j)$, the conditions are satisfied. But the conditions may also appear at random. This shows that $\mathbb{E}(\mathcal{N}_{Type1^k_k}) \simeq 2\mathbb{E}(\mathcal{N}_{perm})$ and the attack succeeds as soon as q is about $2^{n/2}$.

Complexity in $O(2^n)$ for $2k \le r \le 3k-2$.
The attacks are very similar to those provided for $k \le r \le 2k-2$ with the same conditions. We obtain that

$$\mathbb{E}(\mathcal{N}_{perm}) \simeq \frac{q^2}{2.2^{2n}}$$

With a type-1 Feistel scheme, if we have $X^{k-1}(i) = X^{k-1}(j)$ and $X^{2k-1}(i) = X^{2k-1}(j)$ as shown in Table 10.1, then the conditions will be satisfied. However, the conditions may also appear at random. Again, $\mathbb{E}(\mathcal{N}_{Type1^k_k}) \simeq 2\mathbb{E}(\mathcal{N}_{perm})$ and the attack succeeds as soon as q is about 2^n.

10.1.3.2 NCPA on r Rounds with $2k-1 \le r \le 4k-3$

The attacks use the differential path shown in Table 10.2.

The NCPA are very similar to the KPA performed previously. The attacker chooses q values $I_k(i)$, $1 \le i \le q$ pairwise distinct and asks for the value of $\mathcal{O}([0, 0, \ldots, 0, \ldots I_k(i)]) = [S_1, \ldots, S_k]$. This will lead to a NCPA with $2^{n/2}$ messages when $2k-1 \le r \le 3k-2$ and with 2^n messages for $3k-1 \le r \le 4k-3$.

10.1.4 NCPA and KPA Using the Standard Deviation

10.1.4.1 KPA for $3k-1 \le r \le k^2+2k-1$

In Table 10.3, we give the general pattern of the differential path used for KPA. All the attacks are very similar, we explain the attack for $r = tk-2$. The conditions after $tk-2$ rounds ($t \ge 4$) are given by

Table 10.2 Differential path for NCPA on r rounds with $2k-1 \leq r \leq 4k-3$

Round	0	0	0	$\ldots$	$\ldots$	0	0	Δ_k^0
1	0	0	0	0	0	0	Δ_k^0	0
$\vdots$	$\vdots$	$\vdots$	$\vdots$	$\vdots$	$\vdots$	$\vdots$	$\vdots$	$\vdots$
$k-2$	0	Δ_k^0	0	0	0	0	0	0
$k-1$	Δ_k^0	0	0	0	0	0	0	0
k	Δ_1^k	0	0	0	0	0	0	Δ_k^0
$k+1$	Δ_1^{k+1}					0	Δ_1^0	Δ_1^k
$\vdots$	$\vdots$	$\vdots$	$\vdots$	$\vdots$	$\vdots$	$\vdots$	$\vdots$	$\vdots$
$2k-3$	Δ_1^{2k-3}	0	Δ_k^0	$\ldots$	$\ldots$	Δ_1^{2k-6}	Δ_1^{2k-5}	Δ_1^{2k-4}
$2k-2$	**0**	Δ_k^0	Δ_1^k	$\ldots$	$\ldots$	Δ_1^{2k-5}	Δ_1^{2k-4}	Δ_1^{2k-3}
$2k-1$	Δ_k^0							0
$2k$							0	Δ_k^0
$\vdots$	$\vdots$	$\vdots$	$\vdots$	$\vdots$	$\vdots$	$\vdots$	$\vdots$	$\vdots$
$3k-3$	Δ_1^{3k-3}	0	Δ_k^0	$\ldots$	$\ldots$	Δ_1^{3k-6}	Δ_1^{3k-5}	Δ_1^{3k-4}
$3k-2$	**0**	Δ_k^0	Δ_1^k	$\ldots$	$\ldots$	Δ_1^{3k-5}	Δ_1^{3k-4}	Δ_1^{3k-3}
$3k-1$	Δ_k^0							0
$3k$							0	Δ_k^0
$\vdots$	$\vdots$	$\vdots$	$\vdots$	$\vdots$	$\vdots$	$\vdots$	$\vdots$	$\vdots$
$4k-3$	Δ_1^{4k-3}	0	Δ_k^0	$\ldots$	$\ldots$	Δ_1^{4k-6}	Δ_1^{4k-5}	Δ_1^{4k-4}

Table 10.3 Differential characteristic used in KPA

Round	Δ_1^0	Δ_2^0	Δ_3^0	...	Δ_{k-1}^0	Δ_k^0
1				...		Δ_1^0
2				...	Δ_1^0	
$\vdots$						
$k-1$	**0**	Δ_1^0		...		
k	Δ_1^0			...		0
$k+1$				...		Δ_1^0
$\vdots$						
$tk-2$		0	Δ_1^0	...		
$tk-1$	**0**	Δ_1^0		...		
tk	Δ_1^0			...		0
$\vdots$						
$(t+1)k-2$		0	Δ_1^0	...		

$$\begin{cases} S_2(i) = S_2(j) \\ I_1(i) \oplus I_1(j) = S_3(i) \oplus S_3(j). \end{cases} \tag{10.1}$$

We count the number of indices (i, j) such that these conditions are satisfied. Then $\mathbb{E}(\mathcal{N}_{perm}) \simeq \frac{q^2}{2.2^{2n}}$. For $\mathbb{E}(\mathcal{N}_{perm})$, the conditions appear at random or because some conditions are satisfied by the internal variables and we get $\mathbb{E}(\mathcal{N}_{Type1_k^r}) \simeq \frac{m^2}{2.2^{2n}} + O(\frac{m^2}{2^{(t-1)n}})$. The O function comes from the conditions $\mathbf{0}$ that we impose on the differential path. Both standard deviations satisfy $\sigma(\mathcal{N}_{perm}) \simeq \sqrt{\mathbb{E}(\mathcal{N}_{perm})}$ and $\sigma(\mathcal{N}_{Type1_k^r}) \simeq \sqrt{\mathbb{E}(\mathcal{N}_{Type1_k^r})} \simeq \sqrt{\mathbb{E}(\mathcal{N}_{perm})}$ when $t \geq 4$. This means that we can distinguish between a random permutation and a type-1 Feistel scheme as soon as $\frac{q^2}{2^{(t-1)n}} \geq \frac{q}{2^n}$. This gives the condition $q \geq 2^{(t-2)n}$. Since the maximal number of messages is 2^{kn}, these attacks work for $t - 2 \leq k$ and then with $r = k + 2$, we can attack up to $r = (k+2)k - 2 = k^2 + 2k - 2$ rounds.

10.1.4.2 NCPA on r Rounds with $4k - 2 \leq r \leq k^2 + k - 1$

The differential path is shown in Table 10.4.

Again all the attacks are quite similar. We explain a NCPA on $tk - 1$ rounds (with $t \geq 4$) in Table 10.5, where we choose the messages such that I_1 takes only one value for all messages. Here, we have $m \geq 2^{(t-2)n}$. Since the maximal number of messages

Table 10.4 Differential path used in NCPA on type-1 Feistel schemes

Round	$\mathbf{0}$	Δ_0^2	Δ_0^3	...	Δ_0^{k-1}	Δ_0^k
1	Δ_0^2	Δ_0^3		...		0
2				...	0	Δ_0^2
⋮						
$k-1$		0	Δ_0^2	...		
k	$\mathbf{0}$	Δ_0^2		...		
$k+1$	Δ_0^2				...	0
$k+2$				...	0	Δ_0^2
⋮						
$2k-1$		0	Δ_0^1	...		
$2k$	$\mathbf{0}$	Δ_0^2		...		
⋮						
$tk-1$		0	Δ_0^2	...		
tk	$\mathbf{0}$	Δ_0^2		...		
⋮						
$(t+1)k-1$		0	Δ_0^2	...		

Table 10.5 Type-1 Feistel scheme: NCPA on $tk - 1$ rounds

Differential	$\mathbb{E}(\mathcal{N}_{perm})$	$\mathbb{E}(\mathcal{N}_{Type1^r_k})$	σ	q
$\Delta_2^{tk-2} = 0$ $\Delta_3^{tk-2} = \Delta_2^0$	$\frac{q^2}{2.2^{2n}}$	$\frac{q^2}{2.2^{2n}} +$ $O(\frac{q^2}{2^{(t-1)n}})$	$\frac{q}{\sqrt{2}2^n}$	$2^{(t-2)n}$

is $2^{(k-1)n}$, these attacks work as long as $t - 2 \leq k - 1$. Thus with $rtk + 1$, we can attack up to $r = (k+1)k - 1 = k^2 + k - 1$ rounds.

10.1.5 Summary of the Results

The complexities of the best known attacks so far are provided in Table 10.6.

10.1.6 Signature of Type-1 Feistel Ciphers

The proofs are very similar to those given for classical Feistel schemes (Sect. 2.2) and for unbalanced Feistel schemes (Propositions 8.2 and 9.2).

Table 10.6 Complexities of the attacks on type-1 Feistel schemes

r Rounds	KPA	r rounds	CPA-1	r	CPA-1
$1 \to k-1$	1	1			
$k \to 2k-1$	$2^{n/2}$	$\vdots$	1	$\vdots$	
$2k \to 3k-2$	2^n	$k-1$			
$\vdots$		k		$pk-(p-2)$	
$rk-2$	$2^{(r-2)n}$	$\vdots$	2	$\vdots$	$2^{(p-2)n}$
$rk-1$	$2^{(r-3/2)n}$	$2k-2$		$(p+1)k-p$	
rk		$2k-1$			
$\vdots$	$2^{(r-1)n}$	$\vdots$	$2^{n/2}$	$\vdots$	
$(r+1)k-2$		$3k-2$			
$\vdots$		$3k-1$		k^2+1	
k^2+2k-2	2^{kn}	$\vdots$	2^n	$\vdots$	$2^{(k-1)n}$
		$4k-3$		k^2+k-1	

Theorem 10.1 *Let Ψ_1 be a type-1 Feistel scheme. If $n \geq 2$, Ψ_1 has an even signature.*

Proof It is enough to prove Theorem 10.1 for one round since the composition of even permutations is an even permutation. Let f be a function from $\{0, 1\}^n$ to $\{0, 1\}^n$. Let $[I_1, \ldots, I_k] \in \{0, 1\}^{kn}$. Then for one round, we have: $\Psi_1(f)[I_1, \ldots, I_k] = [I_2 \oplus f(I_1), I_3, \ldots, I_k, I_1]$. We can write $\Psi(f) = \sigma \circ \Psi_1'(f)$, where $\Psi_1'(f)[I_1, \ldots, I_k] = [I_1, I_2 \oplus f(I_1), I_3, \ldots, I_k]$ and σ is a rotation of n bits. It is easy to check that $\Psi_1'(f) \circ \Psi_1'(f) = id$, where id is the identity function. So in $\Psi_1'(F)$ we have only cycles with 1 or 2 elements, and the signature is given by the number of cycles with 2 elements modulo 2. The number of cycles with 2 elements is exactly $\frac{2^n.\alpha}{2}$, where α is the number of values I_1 such that $f(I_1) \neq 0$. Thus, when $n \geq 2$, the signature of $\Psi_1'(f)$ is even. In the proof of Theorem 5.1, it is shown that a rotation of n bits has an even signature. This completes the proof that a type-1 Feistel scheme has an even signature. □

10.2 Type-2 Feistel Ciphers

10.2.1 Notation Definition

Here k is even. After one round, the output is given by $[I_2 \oplus f_{1,1}(I_1), I_3, I_4 \oplus f_{1,2}(I_3), \ldots, I_k \oplus f_{1,\frac{k}{2}}(I_{k-1}), I_1]$ where each $f_{1,s}$, $1 \leq s \leq \frac{k}{2}$ is a function from n bits to n bits (Fig. 10.2).

10.2.2 KPA

Since the arguments are similar to those given for previous scheme, we will only give the conditions that have to be satisfied.

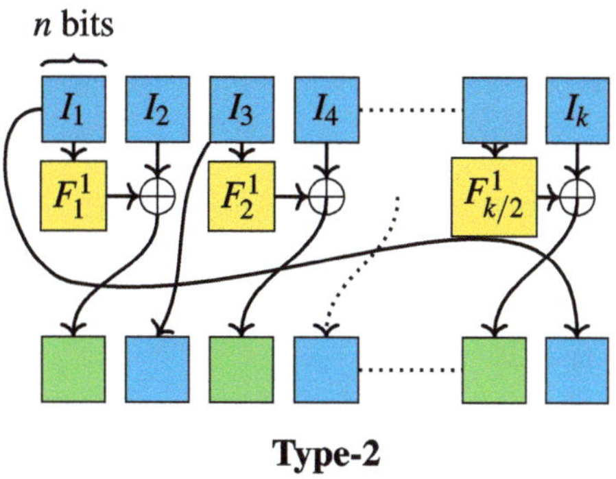

Fig. 10.2 First round for type-2 Feistel Cipher

10.2.2.1 Attacks on the First Round

It is easy to see that after one round we have $S_k = I_1$. This gives a KPA with one message.

10.2.2.2 Overview of Attacks on r Rounds with $2 \le r \le 2k + 2$

Tables 10.7 and 10.8 represent a KPA on $2k + 2$ rounds.

We explain how to get attacks on intermediate rounds. After $2r$ rounds, $r \ge 1$, we have in Table 10.9.

Where $1 \le s \le k$ and $s \equiv 2 - 2r \pmod{k}$.

In this attack, $q = 2^{(r-1)n}$. Thus, for $r = k + 1$, we have reached the maximal number of rounds with $2^{(k-1)n}$ messages.

After $2r + 1$ rounds, $r \ge 1$, the attack is represented in Table 10.10:

Where $1 \le t \le k$ and $t \equiv 1 - 2r \pmod{k}$.

Table 10.7 Differential path used in KPA on type-2 Feistel ciphers

round	$\mathbf{0}$	Δ_2^0	Δ_3^0	Δ_4^0	...	Δ_{k-3}^0	Δ_{k-2}^0	Δ_{2k-1}^0	Δ_k^0
1	Δ_2^0				...				0
2					...			$\mathbf{0}$	Δ_2^0
3					...		0	Δ_2^0	
⋮									
$k-1$		0	Δ_2^0		...				
k	$\mathbf{0}$	Δ_2^0			...				
$k+1$	Δ_2^0				...				0
$k+2$					...			$\mathbf{0}$	Δ_2^0
⋮									
$2k-1$		0	Δ_2^0		...				
$2k$	$\mathbf{0}$	Δ_2^0			...				
$2k+1$	Δ_2^0				...				0
$2k+2$					...				Δ_2^0

Table 10.8 Type-2 Feistel cipher: KPA on $2k + 2$ rounds

Differential	$\mathbb{E}(\mathcal{N}_{perm})$	$\mathbb{E}(\mathcal{N}_{scheme})$	σ	q
$\Delta_1^0 = 0$ $\Delta_k^k = \Delta_2^0$	$\frac{q^2}{2.2^{2n}}$	$\frac{q^2}{2.2^{2n}} + O(\frac{q^2}{2^{(k+1)n}})$	$\frac{q}{\sqrt{22^n}}$	2^{kn}

Table 10.9 Type-2 Feistel cipher: KPA on $2r$ rounds

Differential	$E(\mathcal{N}_{perm})$	$E(\mathcal{N}_{scheme})$	σ	q
$\Delta_1^0 = 0$ $\Delta_s^{2r} = \Delta_2^0$	$\frac{q^2}{2.2^{2n}}$	$\frac{q^2}{2.2^{2n}} + O(\frac{q^2}{2^{rn}})$	$\frac{q}{\sqrt{22^n}}$	$2^{(r-1)n}$

Table 10.10 Type-2 Feistel cipher: KPA on $2r + 1$ rounds

Differential	$\mathbb{E}(\mathcal{N}_{perm})$	$\mathbb{E}(\mathcal{N}_{scheme})$	σ	q
$\Delta_1^0 = 0$ $\Delta_t^{2r+1} = \Delta_2^0$ $\Delta_{t-1}^{2r+1} = 0$	$\frac{q^2}{2.2^{3n}}$	$\frac{q^2}{2.2^{3n}} + O(\frac{q^2}{2^{(r+1)n}})$	$\frac{q}{\sqrt{22^n}}$	$2^{(r-\frac{1}{2})n}$

10.2.3 NCPA

10.2.3.1 Attacks on r Rounds with $2 \le r \le k$

The distinguisher D chooses $I_r, \ldots I_k$ in $\{0,1\}^{(k-r-1)n}$, and asks for the value of $f[0, 0, \ldots, \mathrm{i}_r, \ldots I_k] = [S_1, \ldots, S_k]$. Then it chooses $I'_r, \ldots I'_k$ in $\{0,1\}^{(k-r-1)n}$, such that $[I_r, I_{r+1}, \ldots I_k] \neq [I'_r, I'_{r+1}, \ldots I'_k]$ and asks for the value of $f[0, 0, \ldots, I'_r, \ldots I'_k] = [S'_1, \ldots, S'_k]$. Finally, D tests if $S_k \oplus S'_k = I_r \oplus I'_r$and if it is, ϕ outputs 1, otherwise ϕ outputs 0.

It is easy to see that with a random permutation, the probability to output 1 is $1/2^n$ while it is equal to 1 with a type-2 Feistel cipher. Therefore, this gives a NCPA with $q = 2$.

10.2.3.2 Attacks on r Rounds with $k + 1 \le r \le 2k + 1$

For NCPA, we can impose conditions on a given number of input variables. We give in Tables 10.11 and 10.12 an example of an attack on $2k - 1$ rounds for which we consider messages where I_1, I_2, I_3 are given constant values. Then we will generalize.

For round $2k - 2$, the attack is represented in Table 10.13.

More generally, if we suppose that for the input variables, we have $I_1, \ldots, I_r$ are constants ($r \le k - 1$), we can perform the same kind of attacks. It is easy to check that we can attack up to $2k - r + 2$ rounds and we need exactly $2^{(k-r)n}$ messages. In order to get the best CPA-1 for each round, we will change the conditions on the input variables. For example, for $k + 1$, $k + 2$ and $k + 3$ rounds, we choose $I_1, \ldots I_{k-1}$ to be constant values, then we will have $I_1, \ldots I_{k-2}$ constants, and so on.

Table 10.11 Differential path used in NCPA on type-2 Feistel ciphers

Round	0	0	0	Δ_4^0	Δ_5^0	Δ_6^0	...	Δ_{k-3}^0	Δ_{k-2}^0	Δ_{k-1}^0	Δ_k^0
1	0	0	Δ_4^0				...				0
2	0	Δ_4^0					...				0
3	Δ_4^0						...				0
4							...			**0**	Δ_4^0
5							...		0	Δ_4^0	
$\vdots$											
k			**0**	Δ_4^0			...				
$k+1$		0	Δ_4^0				...				
$k+2$	**0**	Δ_4^0					...				
$k+3$	Δ_4^0						...				0
$\vdots$											
$2k-2$					**0**	Δ_4^0	...				
$2k-1$				0	Δ_2^0		...				

Table 10.12 Type-2 Feistel cipher: NCPA on $2k-1$ rounds

Differential	$\mathbb{E}(\mathcal{N}_{perm})$	$\mathbb{E}(\mathcal{N}_{scheme})$	σ	q
$\Delta_4^{2k-1}=0$ $\Delta_5^{2k-1}=\Delta_4^0$	$\frac{q^2}{2.2^{2n}}$	$\frac{q^2}{2.2^{2n}}+O(\frac{q^2}{2^{(k-2)n}})$	$\frac{q}{\sqrt{2}2^n}$	$2^{(k-3)n}$

Table 10.13 Type-2 Feistel cipher: NCPA on $r=2k-2$ rounds

Differential	$\mathbb{E}(\mathcal{N}_{perm})$	$\mathbb{E}(\mathcal{N}_{scheme})$	σ	q
$\Delta_6^{2k-2}=0$	$\frac{q^2}{2.2^n}$	$\frac{q^2}{2.2^n}+O(\frac{q^2}{2^{(k-3)n}})$	$\frac{q}{\sqrt{2}2^{\frac{n}{2}}}$	$2^{(k-\frac{7}{2})n}$

10.2.4 Summary of the Results

Table 10.14 summarizes the complexities for type-2 Feistel schemes. Flow chart illustrating a data encoding process. The top section is labeled "kn bits" and is divided into two parts: "n" and "(k–1)n". The chart includes blue rectangles representing data blocks, a yellow diamond indicating a decision or process step, and a green rectangle at the bottom. Arrows connect these elements, showing the flow of data through the process. A circle with a plus sign represents a summation or combination point. The diagram visually explains the relationship and transformation of data blocks.

Table 10.14 Complexities of the attacks on type-2 Feistel schemes

r	KPA	NCPA
1	1	1
2	$2^{n/2}$	2
$3 \le r \le k$	$2^{\frac{r-2}{2}n}$	2
$k+1$	$2^{(k-1/2)n}$	$2^{n/2}$
$k+2$	$2^{\frac{k}{2}n}$	$2^{n/2}$
$k+3 \le r \le 2k+1$	$2^{\frac{r-2}{2}n}$	$2^{(r-k-2)n}$
$2k+2$	2^{kn}	2^{kn}

10.2.5 Signature of Type-2 Feistel Ciphers

We show that generalized Feistel schemes have an even signature. The proofs are very similar to those given for classical Feistel schemes (Sect. 2.2) and for unbalanced Feistel schemes (Propositions 8.2 and 9.2).

Theorem 10.2 *Let Ψ_2 be a type-2 Feistel scheme. If $n \ge 2$, Ψ_2 has an even signature.*

Proof Again it is enough to prove Theorem 10.2 for one round since the composition of even permutations is an even permutation. Here we know that k is even. Let $f_1, \ldots, f_{k/2}$ be $k/2$ functions from $\{0,1\}^n$ to $\{0,1\}^n$. Then for one round, we have:

$$\Psi_2(f_1, \ldots, f_{k/2}) = \sigma \circ \sigma^2 \circ \Psi_1'(f_{k/2}) \circ \sigma^2 \circ \Psi'(f_{k/2-1}) \circ \ldots \circ \sigma^2 \circ \Psi'(f_1)$$

$$\Psi_2(f_1, \ldots, f_{k/2}) = \sigma^3 \circ \Psi_1'(f_{k/2}) \circ \sigma^2 \circ \Psi'(f_{k/2-1}) \circ \ldots \circ \sigma^2 \circ \Psi'(f_1).$$

Since Ψ_2 is the composition of even permutation, it is also an even permutation. □

10.3 Type-3 Feistel Ciphers

10.3.1 Notation Definition

After one round, the output is given by $[I_2 \oplus f_{1,1}(I_1), I_3 \oplus f_{1,2}(I_2), I_4 \oplus f_{1,3}(I_3), \ldots, I_k \oplus f_{1,k-1}(I_{k-1}), I_1]$ where each $f_{1,s}$, $1 \le s \le k-1$ is a function from n bits to n bits (Fig. 10.3).

We will present our attacks when k is even. For k odd, the computations are similar. The results are summarized in Table 10.19.

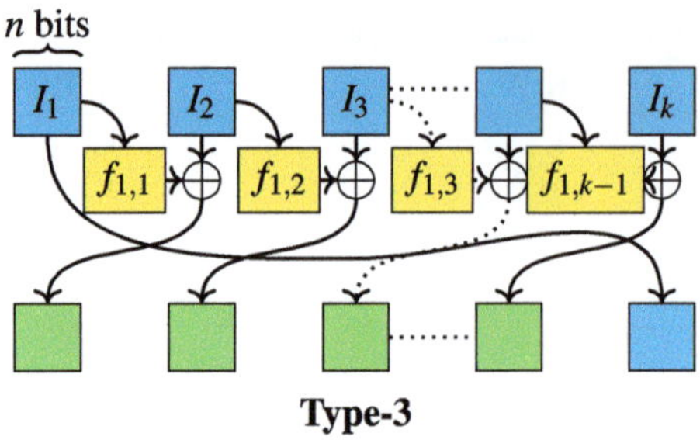

Fig. 10.3 First round for type-3 Feistel ciphers

10.3.2 KPA

Since the arguments are similar to those used for type-1 and type-2 Feistel ciphers, we simplify the presentation of the attacks.

For one round, we need one message, we just have to check if $I_1 = S_k$. With a random permutation, this happens with probability $\frac{1}{2^n}$ and with a scheme with probability one. The attack on r rounds with $2 \le r \le k$ proceeds as follows. The attacker waits until he gets 2 messages such that $I_1(1) = I_1(2), \ldots, I_{r-1}(1) = I_{r-1}(2)$. Then he tests if $I_{r-1}(1) \oplus I_{r-1}(2) = S_k(1) \oplus S_k(2)$. With a random permutation, this happens with probability $\frac{1}{2^n}$ and with a scheme with probability one. Moreover, from the birthday paradox, with $2^{\frac{(r-1)n}{2}}$ messages, the probability to get 2 messages with the given conditions is rather high. Table 10.15 shows a KPA on $k+4$ rounds, when $4 \le \frac{k}{2} + 1$.

For this KPA on $k+4$ rounds, we have in Table 10.16.

Table 10.15 Differential paths used for KPA on type-3 Feistel ciphers

Round	0	...	0	0	0	0	Δ_0^k
1	0	...	0	0	0	Δ_k^0	0
2	0	..	0	0	Δ_k^0		0
3	0	..	0	Δ_k^0			0
$\vdots$							
$k-1$	Δ_k^0	...					0
k		...	**0**	**0**	**0**	**0**	Δ_k^0
$k+1$		...	0	0	0	Δ_k^0	
$k+2$		...	0	0	Δ_k^0		
$k+3$		...	0	Δ_k^0			
$k+4$		...	Δ_k^0				

Table 10.16 Type-3 Feistel cipher: KPA on $r = k + 4$ rounds

Differential	$\mathbb{E}(\mathcal{N}_{perm})$	$\mathbb{E}(\mathcal{N}_{scheme})$	σ	q
$\Delta_1^0 = 0$ $\Delta_2^0 = 0$ $\vdots$ $\Delta_{k-1}^0 = 0$ $\Delta_k^0 = \Delta_{k-5}^{k+4}$	$\frac{q^2}{2.2^{kn}}$	$\frac{q^2}{2.2^{kn}} + O(\frac{q^2}{2^{(k+3)n}})$	$\frac{q}{\sqrt{2}2^{\frac{kn}{2}}}$	$2^{(\frac{k}{2}+3)n}$

Since $q = 2^{(\frac{k}{2}+3)n}$, it is possible to perform the same kind of attack for $k + r$ rounds, with $r \leq \frac{k}{2} + 1$. We can attack up to $k + \frac{k}{2} + 1$ rounds. For $k + \frac{k}{2} + 1$, the maximal number of messages, i.e., 2^{kn}, is needed.

10.3.3 NCPA

For NCPA, it is easy to see that after one round, one message is sufficient. We just have to check if $S_k = I_1$. For 2 rounds, the attacker chooses 2 messages such that $I_1(1) = I_1(2)$ and checks if $S_k(1) \oplus S_k(2) = I_2(1) \oplus I_2(2)$. With a random permutation this happens with probability $\frac{1}{2^n}$, but with a scheme, the probability is one. Thus, it is possible to distinguish between the two permutations with only 2 messages. More generally, for r rounds with $r \leq k$, the attacker chooses 2 messages such that $I_s(1) = I_s(2)$ for $1 \leq s \leq k - 1$ and he checks if $S_k(1) \oplus S_k(2) = I_d(1) \oplus I_d(2)$. With a random permutation this happens with probability $\frac{1}{2^n}$, but with a scheme, the probability is one. Thus, with 2 messages, it is easy to attack up to k rounds.

For $k + 1$ rounds, the attacker chooses q messages such that $I_1, I_2, \ldots, I_{k-1}$ have a constant value. The NCPA is described in Tables 10.17 and 10.18.

Table 10.17 Differential paths used for NCPA on type-3 Feistel ciphers

Round	**0**	**0**	...	**0**	**0**	**0**	**0**	Δ_k^0
1	0	0	...	0	0	0	Δ_k^0	0
2	0	0	..	0	0	Δ_k^0		0
3	0	0	..	0	Δ_k^0			0
$\vdots$								
$k-1$	Δ_k^0		...					0
k			...				**0**	Δ_k^0
$k+1$			...				Δ_k^0	

Table 10.18 Type-3 Feistel cipher: NCPA on $k+1$ rounds

Differential	$\mathbb{E}(\mathcal{N}_{perm})$	$\mathbb{E}(\mathcal{N}_{scheme})$	σ	q
$\Delta_{k-1}^{k+1} = \Delta_k^0$	$\frac{q^2}{2.2^n}$	$\frac{q^2}{2.2^n} + O(\frac{q^2}{2^n})$	$\frac{q}{\sqrt{2}2^{\frac{n}{2}}}$	$2^{\frac{n}{2}}$

10.3.4 Summary of the Results

Table 10.19 gives KPA and NCPA complexities.

10.3.5 Signature of Type-3 Feistel Ciphers

We show that generalized Feistel schemes have an even signature. The proofs are very similar to those given for classical Feistel schemes (Sect. 2.2) and for unbalanced Feistel schemes (Propositions 8.2 and 9.2).

Theorem 10.3 *Let Ψ_3 be a type-3 Feistel scheme. If $n \geq 2$, Ψ_3 has an even signature.*

Proof As previously, it is enough to prove Theorem 10.3 for one round since the composition of even permutations is an even permutation. Let $f_1, \ldots, f_{k-1}$ be $k-1$ functions from $\{0,1\}^n$ to $\{0,1\}^n$. Then for one round, we have:

$$\Psi_3(f_1, \ldots, f_{k-1}) = \sigma^2 \circ \sigma^{-1} \circ \Psi_1'(f_1) \circ \sigma^{-1} \circ \Psi_1'(f_2) \circ \ldots \circ \sigma^{-1} \circ \Psi_1'(f_{k-1}) \circ \sigma^{k-2}$$

$$\Psi_3(f_1, \ldots, f_{k-1}) = \sigma \circ \Psi_1'(f_1) \circ \sigma^{-1} \circ \Psi_1'(f_2) \circ \ldots \circ \sigma^{-1} \circ \Psi_1'(f_{k-1}) \circ \sigma^{k-2}.$$

This decomposition shows that Ψ_3 is an even permutation. □

Table 10.19 Complexities of the attacks on type-3 Feistel ciphers

r	KPA	NCPA
1	1	1
2	$2^{n/2}$	2
3	2^n	2
$\vdots$		
k	$2^{(k-1)n/2}$	2
$k+1$	$2^{\frac{k}{2}n}$	$2^{n/2}$
$k+2 \leq r \leq k + \lfloor\frac{k}{2}\rfloor + 1$	$2^{(r-\lfloor\frac{k}{2}\rfloor-1)n}$	$2^{(r-\lfloor\frac{k}{2}\rfloor-1)n}$

References

Bouillaguet, C., O. Dunkelman, P. A. Fouque, and G. Leurent. 2012. New insights on impossible differential cryptanalysis. In *Selected areas in cryptography—SAC '12*, vol. 7118, ed. A. Miri and S. Vaudenay. Lecture notes in computer science, 243–259. Berlin: Springer.

Choy, J., and H. Yap. 2009. Impossible boomerang attacks for block cipher structures. In *Advances in information and computer security—IWSEC '09*, vol. 5824, ed. T. Takagi and M. Mambo. Lecture notes in computer science, 22–37. Berlin: Springer.

Hoang, V. T., and P. Rogaway. 2010. On generalized feistel networks. In *Advances in cryptology—CRYPTO 2010*, vol. 6223, ed. T. Rabin. Lecture notes in computer science, 613–630. Berlin: Springer.

Luo, Y., Z. Wu, X. Lai, and G. Gong. 2014. A unified method for finding impossible differentials of block cipher structures. *Information Science* 263: 211–220.

Nachef, V., E. Volte, and J. Patarin. 2013. Differential attacks on generalized feistel schemes. In *CANS 2013*, vol. 8257, ed. M. Abdalla, C. Nita-Rotaru, and R. Dahab. Lecture notes in computer science, 1–19. Berlin: Springer.

Zhen, Y., T. Matsumo, and H. Imai. 1990. On the construction of block ciphers provably secure and not relying on any unproved hypotheses. In *Advances in cryptology—CRYPTO '89*, vol. 435, ed. G. Brassard. Lecture notes in computer science, 461–480. Berlin: Springer.

Chapter 11
Quantum Generic Attacks on Feistel Schemes

Abstract In this chapter we present generic quantum attacks on Feistel schemes. This is a field where a large gap exists between security results and generic attacks, and it is interesting to note that, even though we deal with secret key cryptography, quantum attacks on Feistel schemes are better than classical ones. Nevertheless, these attacks require adversarial queries that are in quantum superposition, and we may wonder about the realism of such a scenario.

11.1 Introduction

The power of quantum computing lies in its ability to process information differently than classical computing. In classical computing, a program processes the value of an n-bit register 1 bit at a time, whereas, in quantum computing, a program operates on an n-qubit register, handling 2^n values simultaneously.

The main differences between bits and qubits are that qubits can exist in a superposition of states (superpositions of 0 and 1 for example), and that qubits can be entangled, allowing the state of one qubit to depend on another, regardless of physical distance.

These superposition and entanglement properties enable quantum computers to perform certain computations much more efficiently than classical computers.

We will not present in this book the basic definitions and results about qubits and quantum computations, but instead we will show how to use some of these results (for example, Simon's quantum algorithm to find a period) on Feistel schemes.

When some measurements are done, at the end or during a quantum computation, part or all the quantum superposition is lost, and from the probability distribution of the results we will be able to solve some problem faster than with non-quantum computers. However, for some other problems we will not obtain any advantage.

When we study a quantum algorithm, we assume that the loss of superposition, or the loss of entanglement, is done exactly when we perform some measurements, and with the expected theoretical probability distribution. However, it is extremely difficult to avoid spontaneous loss of quantum properties (we speak of decoherence or quantum errors) and it is not clear at present (April 2025) if powerful and useful

J. Patarin et al., *Guide to Feistel Ciphers*, Information Security and Cryptography,
https://doi.org/10.1007/978-3-031-99850-8_11

quantum computers will really appear soon or not due to these problems. (Some quantum error corrections are possible, but it is not clear if powerful quantum error corrections on thousands of useful for quantum computations qubits will appear soon, and if a lot of quantum computations can be done without decoherence.)

In this chapter, we present attacks that are either QCPA (quantum chosen plaintext attacks) or QCCA (quantum chosen plaintext and ciphertext attacks). In these attacks, the hardware that computes the Feistel Cipher will be in quantum superposition. These attacks are therefore very different from Shor's famous quantum attacks on RSA, for example, where the factorization of the public modulo is obtained by a quantum algorithm (and not by using a superposition of inputs to an algorithm that uses the secret key).

We can imagine, for example, that the secret keys, or secret functions, used by the Feistel scheme are inside a tamper-resistant hardware, such that it is not possible for an adversary to read these secret values, but it is however possible for this adversary to put the software that use this hardware in a quantum superposition and to run the software. Another example, maybe more realistic, is that the Feistel scheme is written in white box, i.e., there is a public code of the Feistel computation, but in this public code the secret values are obfuscated. In this case, the quantum attack will make quantum superpositions of such public codes. A final example is the case where quantum computing and quantum memories have become prevalent, and data in superposition needs to be stored encrypted at rest, as it is currently done for classical data. In this scenario, encryption algorithms would have to manipulate superposition of messages, and may allow an attacker to mount QCPAs and QCCAs.

Remark 11.1 In this chapter, we will not present quantum known plaintext attacks. In the classical setting, KPA means that an adversary is given plaintext/ciphertext pairs for random plaintexts. In the quantum setting, a message becomes an element of the unit sphere of a complex vector space with finite dimension. Thus, it is not entirely clear how this classical KPA concept could be translated, and which probability distributions on the message space would be reasonable to consider.

The attacks we present will be compared with classical, i.e., non-quantum attacks, in Table 11.1 when random internal functions are used, and in Table 11.2 when random internal permutations are used, from 1 to 6 rounds. For more rounds the best known quantum complexities have the same complexities as non-quantum attacks.

We can notice that Tables 11.1 and 11.2 have identical complexities, except for KPA non-quantum attacks with 3 rounds.

These quantum attacks on 3 and 4 rounds will be based on Simon's algorithm to find a period. On 5 rounds they will be based on quantum algorithms to find a collision, and on 6 rounds they will be based on Childs and Eisenberg's algorithm.

Table 11.1 Number of queries and number of computations to distinguish Feistel schemes with random internal functions from random permutations (best known attacks)

	KPA	CPA	CCA	QCPA	QCCA
ψ^1	1	1	1	1	1
ψ^2	$2^{n/2}$	2	2	2	2
ψ^3	$2^{n/2}$	$2^{n/2}$	3	n (*)	3
ψ^4	2^n	$2^{n/2}$	$2^{n/2}$	$2^{n/2}$	n (*)
ψ^5	$2^{3n/2}$	2^n	2^n	$2^{2n/3}$	$2^{2n/3}$
ψ^6	2^{2n}	2^{2n}	2^{2n}	$2^{8n/5}$	$2^{8n/5}$

(*): We need n queries and about n^3 computations

Table 11.2 Number of queries and number of computations to distinguish Feistel schemes with random internal permutations from random permutations (best known attacks)

	KPA	CPA	CCA	QCPA	QCCA
ψ^1	1	1	1	1	1
ψ^2	$2^{n/2}$	2	2	2	2
ψ^3	2^n(+)	$2^{n/2}$	3	n (*)	3
ψ^4	2^n	$2^{n/2}$	$2^{n/2}$	$2^{n/2}$	n (*)
ψ^5	$2^{3n/2}$	2^n	2^n	$2^{2n/3}$	$2^{2n/3}$
ψ^6	2^{2n}	2^{2n}	2^{2n}	$2^{8n/5}$	$2^{8n/5}$

(*): We need n queries and about n^3 computations. (+): Worse complexity than for Feistel with internal functions

11.2 Simon's Algorithm to Find a Period

Let f be a function from $\{0, 1\}^n$ to $\{0, 1\}^n$ with a period s (for a commutative group law, for example $\oplus$). So we have: for all $x \in \{0, 1\}^n$:

$$f(x \oplus s) = f(x).$$

The problem is to find s.

With classical (i.e., non-quantum) algorithms, the complexity to find s will be in $\mathcal{O}(2^{n/2})$ queries to the function f, and a similar number of computations (via the birthday paradox), and it is possible to prove that it is not possible to have a better complexity. However, on a quantum computer, via the algorithm of Daniel Simon, only $\mathcal{O}(n)$ queries to the function f will be enough, and then we will have to solve a linear system of $n - 1$ equations with n unknowns. This can be done after at most $\mathcal{O}(n^3)$ computations by Gaussian reductions, for example. In this book, we will not give details on how Simon algorithm work, but we will use the result.

Remark 11.2 Initially the group Law in Simon's Problem (finding a period s) was the XOR. A generalization was obtained by Peter Shor when the group law is the addition modulo of an integer n, with a similar complexity to find a period s. Then, a similar (polynomial) complexity was obtained for any finite commutative group law (to find a period s). However, when the group law is not commutative, no polynomial quantum algorithm is known in general to find a period s.

11.3 QCPA Attack on Ψ^3

Let us denote, as usual, Ψ^3 a 3-round balanced Feistel cipher with internal functions (or permutations) f_1, f_2, f_3. We have seen that:

$$\Psi^3[L, R] = [S, T] \iff \begin{cases} S = R \oplus f_2\left(L \oplus f_1(R)\right), \\ T = L \oplus f_1(R) \oplus f_3(S). \end{cases}$$

Let α and β be two distinct elements of $\{0, 1\}^n$. Let b be an element of $\{0, 1\}$, and let f be the following function from $\{0, 1\}^{n+1}$ to $\{0, 1\}^n$:

$$\begin{aligned} f_{\alpha,\beta} : \{0, 1\} \times \{0, 1\}^n &\longrightarrow \{0, 1\}^n \\ (0, L) &\longmapsto S(L, \alpha) \oplus \beta = \alpha \oplus f_2\left(L \oplus f_1(\alpha)\right) \oplus \beta, \\ (1, L) &\longmapsto S(L, \beta) \oplus \alpha = \beta \oplus f_2\left(L \oplus f_1(\beta)\right) \oplus \alpha. \end{aligned}$$

We will see that $1||L \oplus f_1(\alpha) \oplus f_1(\beta)$ is a period of f. Indeed, one has

$$\begin{aligned} f(0, L \oplus f_1(\alpha) \oplus f_1(\beta)) &= \alpha \oplus f_2(L \oplus f_1(\alpha) \oplus f_1(\beta) \oplus f_1(\alpha)) \oplus \beta \\ &= \alpha \oplus f_2(L \oplus f_1(\beta)) \oplus \beta \\ &= f(1, L). \end{aligned}$$

Similarly,

$$\begin{aligned} f(1, L \oplus f_1(\alpha) \oplus f_1(\beta)) &= \beta \oplus f_2(L \oplus f_1(\alpha) \oplus f_1(\beta) \oplus f_1(\beta)) \oplus \alpha \\ &= \beta \oplus f_2(L \oplus f_1(\alpha)) \oplus \alpha \\ &= f(0, L). \end{aligned}$$

Simon's algorithm will be able to find this period on Ψ^3 after $\mathcal{O}(n)$ queries and $\mathcal{O}(n^3)$ computations, while on a random permutation, no such period generally exists. This gives a QCPA distinguisher between Ψ^3 and a random permutation on $2n$ bits with $\mathcal{O}(n)$ queries and $\mathcal{O}(n^3)$ computations.

Reference: This attack comes from Kuwakado and Morii (2010) for internal permutations, and from Kaplan et al. (2016) for internal functions.

11.4 QCCA Attack on Ψ^4

In this section, we describe an attack by Ito et al. (2019). Let us denote, as usual, Ψ^4 a 4-round balanced Feistel cipher with internal functions (or permutations) f_1, f_2, f_3, f_4. We have seen that:

$$\Psi^4[L, R] = [S, T] \iff \begin{cases} S = L \oplus f_1(R) \oplus f_3[R \oplus f_2(L \oplus f_1(R))], \\ T = R \oplus f_2(L \oplus f_1(R)) \oplus f_4(S), \end{cases}$$

and

$$(\Psi^4)^{-1}[S, T] = [L, R] \iff \begin{cases} L = S \oplus f_3(T \oplus f_4(S)) \oplus f_1(R), \\ T = R = T \oplus f_4(S) \oplus f_2[S \oplus f_3(T \oplus f_4(S))]. \end{cases}$$

Let G be either a Ψ^4 or a random permutation. We want to find a distinguisher. Let β be 0 or 1, and let α_0 and α_1 be two elements of $\{0, 1\}^n$.

We define the function f from $\{0, 1\} \times \{0, 1\}^n$ to $\{0, 1\}^n$ like this. f takes (β, x) as input. First, we compute

$$G[x, \alpha_\beta] = [S, T].$$

Then we compute

$$G^{-1}[S, T \oplus \alpha_0 \oplus \alpha_1] = [L, R].$$

Finally,

$$f[x, \alpha_\beta] = R \oplus \alpha_\beta.$$

Theorem 11.1 (Ito et al. 2019) *When G is a Ψ^4 permutation, this function f has this property:*

$$\begin{aligned} f(\beta, x) = &\alpha_0 \oplus \alpha_1 \oplus f_2(x \oplus f_1(\alpha_\beta)) \\ &\oplus f_2[x \oplus f_1(\alpha_\beta) \oplus f_3(\alpha_\beta \oplus f_2(x \oplus f_1(\alpha_\beta))] \\ &\oplus f_3[\alpha_\beta \oplus \alpha_0 \oplus \alpha_1 \oplus f_2(x \oplus f_1(\alpha_\beta))]. \end{aligned}$$

Theorem 11.2 (Ito et al. 2019) *When G is a Ψ^4 permutation, this function f has this property:*

$$f(\beta, x) = f(\beta \oplus 1, x \oplus f_1(\alpha_0) \oplus f_1(\alpha_1)).$$

So f has the period $s = (1, f_1(\alpha_0) \oplus f_1(\alpha_1))$.

Since Theorem 11.2 is generally not true when G is a random permutation, and from Simon's algorithm, we see that we have here a QCCA distinguisher for Ψ^4 with a quantum complexity of $\mathcal{O}(n)$ queries and $\mathcal{O}(n^3)$ computations.

11.5 QCPA Attack on Ψ^5

In this section, we first describe a non-quantum attack that distinguishes a 5-round Feistel scheme from a $2n$ bits random permutation with a complexity of 2^n presented in Patarin (2001). Then, we describe a quantum chosen plaintext attack that distinguishes a 5-round Feistel scheme from a $2n$ bits random permutation with a complexity of $2^{2n/3}$ from Chartouny et al. (2023). This attack will rely on the following theorem:

Theorem 11.3 (Zhandry 2015) *Let f be a random function with domain size M and codomain size N. Assume $M = \Omega(N^{1/2})$. Then the quantum query complexity of finding a collision with constant probability is $\Theta(N^{1/3})$.*

11.5.1 Classical Attack on Ψ^5

We can choose messages $[L_i, R_i]$ and $[L_j, R_j]$ such that $R_i = R_j$, $\forall i, j$. Then, we can check whether S_i is equal to S_j and $L_i \oplus L_j$ is equal to $T_i \oplus T_j$, i.e., we count the number of (i, j) such that:

$$\begin{cases} S_i = S_j, \\ L_i \oplus L_j = T_i \oplus T_j. \end{cases}$$

For a 5-round Feistel scheme, we have two times more such collision than for truly random permutations. Indeed, for a truly random permutation if $R_i = R_j$, the numbers of (i, j), $1 \le i < j \le m$, where $m \simeq 2^n$, such that $S_i = S_j$ and $T_i \oplus T_j = L_i \oplus L_j$ is approximately $\dfrac{m(m-1)}{2} \dfrac{1}{2^{2n}}$.

However, for a 5-round Feistel scheme, we have

$$\begin{cases} T_i = L_i \oplus f_1(R_i) \oplus f_3(R_i \oplus f_2(L_i \oplus f_1(R_i))) \oplus f_5(S_i), \\ T_j = L_j \oplus f_1(R_j) \oplus f_3\left(R_j \oplus f_2(L_j \oplus f_1(R_j))\right) \oplus f_5(S_j). \end{cases}$$

Let us suppose that $R_i = R_j$ and $S_i = S_j$, hence, $T_i \oplus T_j = L_i \oplus L_j$ is equivalent to

$$f_3(R_i \oplus f_2(L_i \oplus f_1(R_i))) = f_3\left(R_j \oplus f_2(L_j \oplus f_1(R_j))\right),$$

which can occur either if $f_2(L_i \oplus f_1(R_i)) = f_2\left(L_j \oplus f_1(R_j)\right)$ with an approximate probability of $\frac{1}{2^n}$ or if these values are distinct but when XORed with R, they have the same images by f_3 with an approximate probability of $\frac{1}{2^n}$.

Hence, $\forall (i, j)$, $i < j$, the probability that $S_i = S_j$ and $T_i \oplus T_j = L_i \oplus L_j$ when $R_i = R_j$ is approximately $\frac{m(m-1)}{2} \frac{2}{2^{2n}}$.

Thus we have two times more such collision for a 5-round Feistel scheme compared to a truly random permutation. (This can also be demonstrated with the H-coefficient technique, see Patarin 1991, p. 148, value of h_5.)

Therefore, we will be able to distinguish the 5-round Feistel scheme from truly random permutations when $\frac{m(m-1)}{2} \frac{2}{2^{2n}} \geq 1$, i.e., when m is about greater or equal to 2^n. We see that here the complexity is in 2^n by searching for collision of the form of $S_i \,||\, L_i \oplus T_i$ (birthday paradox).

Note that there are several attacks on Ψ_5 with the same complexity but we choose this one to be able to detect collisions in quantum.

11.5.2 Quantum Attack on Ψ^5

In quantum we detect collisions, when they exist, faster than on classical computers. In fact, we use the same attack in quantum to detect these collisions.

We apply Theorem 11.3 with $M = 2^n$ and $N = 2^{2n}$ ($M = \Omega(N^{1/2})$). Hence, the quantum complexity to detect such a collision is in $N^{1/3} = 2^{2n/3}$ (unlike 2^n as seen before). Therefore, we can distinguish 5-round Feistel schemes from truly random permutations with a quantum complexity of $2^{2n/3}$.

11.6 QCPA Attack on Ψ^6

In this section, we start by describing classical CPAs against ψ^6 with internal functions and internal permutations. We will then derive a QCPA against ψ^6 with a time and query complexity of about $2^{8n/5}$. These results first appeared in Chartouny et al. (2024).

11.6.1 Preliminaries

11.6.1.1 Notations

Let $\mathscr{F}_{m,n}$ be the set of all functions from $\{0, 1\}^m$ to $\{0, 1\}^n$. When $m = n$, the set of all functions from $\{0, 1\}^n$ to $\{0, 1\}^n$ will be denoted by $\mathscr{F}_n$.

11.6.1.2 A Useful Lemma

In this paragraph, we state a Chernoff-type bound for binary random variables that are moderately dependent, and do not necessarily share a common bound. This result can be seen as a slight variant of Chen et al. (2014, Lemma 5).

Lemma 11.1 *Let* $\mathbf{A} = (A_i)_{1\leq i\leq q}$ *be a sequence of random variables taking value in* $\{0, 1\}$. *Assume that, for any* $1 \leq i \leq q$, *there exists* $0 \leq \varepsilon_i < 1$ *such that, for any binary sequence* $(a_1, \ldots, a_{i-1})$, *one has*

$$\Pr[A_i = 1 \mid (A_1, \ldots, A_{i-1}) = (a_1, \ldots, a_{i-1})] \leq \varepsilon_i.$$

Then, for any $\delta > 0$, *one has*

$$\Pr\left[\sum_{i=1}^{q} A_i \geq (1+\delta)m\right] \leq e^{-\frac{\delta^2}{2+\delta}m},$$

where $m = \sum_{i=1}^{q} \varepsilon_i$.

The proof of this lemma can straightforwardly be derived from the one of Chen et al. (2014, Lemma 5).

11.6.1.3 6-Round Feistel Scheme

We now provide detailed equations of the Feistel network for the initial 6 rounds. Let $M = L||R$, then

$$\text{1 round: } \begin{cases} R, \\ L \oplus f_1(R) = X. \end{cases} \qquad \text{4 rounds: } \begin{cases} Z, \\ Y \oplus f_4(Z) = U. \end{cases}$$

$$\text{2 rounds: } \begin{cases} X, \\ R \oplus f_2(X) = Y. \end{cases} \qquad \text{5 rounds: } \begin{cases} U, \\ Z \oplus f_5(U) = S. \end{cases}$$

$$\text{3 rounds: } \begin{cases} Y, \\ X \oplus f_3(Y) = Z. \end{cases} \qquad \text{6 rounds: } \begin{cases} S, \\ U \oplus f_6(S) = T. \end{cases}$$

11.6.1.4 Alternative Notation

We will also introduce variables X^i such that for each round k we have

$$X^k = X^{k-2} \oplus f_k(X^{k-1}). \tag{11.1}$$

Hence, $X^1 = L$, $X^0 = R$, $X^1 = X$, $X^2 = Y$, $X^3 = Z$, $X^4 = U$, $X^5 = S$.

11.6.1.5 Theorems

Let $[L_i, R_i]$ be known inputs with $1 \leq i \leq m$ and let

$$[S_i, T_i] = \Psi^6(f_1, \ldots, f_6)([L_i, R_i]).$$

We assume that $[L_i, R_i]$ values are pairwise distinct, i.e., we cannot have $L_i = L_j$ and $R_i = R_j$ with $i \neq j$.

Theorem 11.4 *If we have internal functions or internal permutations, and if* $i \neq j$, *we cannot have* $X_i^{k-1} = X_j^{k-1}$ *and* $X_i^k = X_j^k$.

Proof This theorem can be proved by induction on k using Eq. (11.1).

Theorem 11.5 *If we have internal permutations, and if* $i \neq j$ *we cannot have*

$$\begin{cases} X_i^{k-2} = X_j^{k-2}, & (*) \\ X_i^k = X_j^k. & (**) \end{cases}$$

Proof By definition, we have that $(**)$ is equivalent to $X_i^{k-2} \oplus f_k(X_i^{k-1}) = X_j^{k-2} \oplus f_k(X_j^{k-1})$ and so $f_k(X_i^{k-1}) = f_k(X_j^{k-1})$ (thanks to Equation $(*)$). Since f_k is a permutation, f_k is injective, we get $X_i^{k-1} = X_j^{k-1}$. Hence, we have

$$\begin{cases} X_i^{k-2} = X_j^{k-2}, \\ X_i^{k-1} = X_j^{k-1}. \end{cases}$$

From Theorem 11.4, this is not possible.

11.6.2 Classical Attacks on 6-Round Feistel Schemes

In this section, we described a four points classical attack on 6-round Feistel schemes with internal permutations. We identified an attack as detailed below with a complexity of $\mathcal{O}(2^{2n})$. Moreover, we made a detailed analysis of a four points classical attack on 6-round Feistel schemes with internal functions presented in Patarin (2001). Note that this attack does not need the full code book.[1] Besides, its time complexity is also in $\mathcal{O}(2^{2n})$, which is much smaller than exhaustive key search.

---

[1] In particular, we can distinguish a 6-round Feistel scheme from an even random permutation.

11.6.2.1 The Attack

Let $[L_i, R_i] \in \{0,1\}^{2n}$. We generate m messages $[L_i, R_i]$ with $1 \le i \le m$ and $m = \mathcal{O}(2^{2n})$. We then compute $[S_i, T_i] = \Psi^6(f_1, \dots, f_6)([L_i, R_i])$. We look if among these values we can find four pairwise distinct indices denoted by $1, 2, 3, 4$ such that these eight equations are satisfied:

$$(\mathcal{S}) \begin{cases} R_1 = R_3, \\ R_2 = R_4, \\ S_1 = S_2, \\ S_3 = S_4, \\ L_1 \oplus L_4 = L_2 \oplus L_3, \\ L_1 \oplus S_1 = L_3 \oplus S_3, \quad (\Longleftrightarrow L_2 \oplus S_2 = L_4 \oplus S_4) \\ R_1 \oplus T_1 = R_2 \oplus T_2, \\ R_3 \oplus T_3 = R_4 \oplus T_4. \quad (\Longleftrightarrow T_1 \oplus T_4 = T_2 \oplus T_3). \end{cases}$$

We also have $R_1 \neq R_2$, $S_1 \neq S_3$ and $T_1 \neq T_2$. We give a visual representation of these eight equations in Fig. 11.1.

For a 6-round Feistel scheme with internal permutations, we will show that we have about two times more such collisions than for truly random permutations. Similarly, for a 6-round Feistel scheme with internal functions, we show that we have 12 times more such collisions than for a truly random permutation. This will give us an attack of $\mathcal{O}(2^{2n})$ in order to distinguish a random permutation from a 6-round Feistel scheme with internal functions or permutations.

Remark 11.3 This result was already known for internal functions (Patarin 2001). In fact, when we have internal functions, different kinds of attacks are known with a complexity of $\mathcal{O}(2^{2n})$: "four points" attacks as described in this paper, or "two points" attacks. However, this result is new for internal permutations. Note also that for internal permutations four points attacks are in $\mathcal{O}(2^{2n})$, but "two points" attacks are in $\mathcal{O}(2^{3n})$.

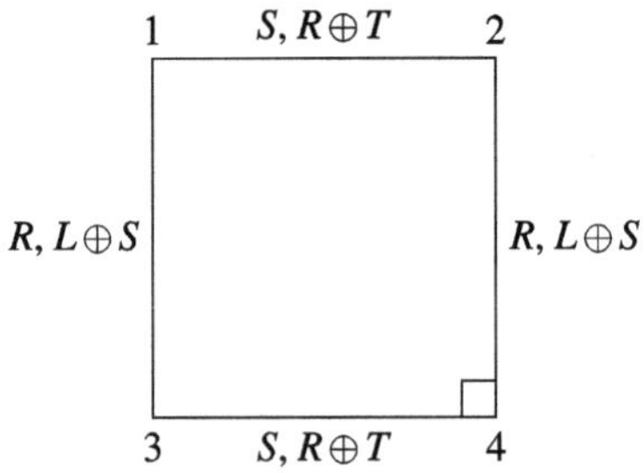

Fig. 11.1 Representation of the eight equations of the system $(\mathcal{S})$. The small square in the right-hand corner below represents $L_1 \oplus L_2 \oplus L_3 \oplus L_4 = 0$

11.6.2.2 Case of 6-Round Feistel with Internal Permutations

Theorem 11.6 *Let us assume that*

$$(1)\begin{cases} R_1 = R_3, \\ R_2 = R_4, \\ S_1 = S_2, \\ L_1 \oplus L_4 = L_2 \oplus L_3. \end{cases}$$

Let us also assume that

$$(2)\begin{cases} X_1 = X_4, \\ Y_1 = Y_2, \\ Z_1 = Z_3, \\ U_1 = U_4. \end{cases}$$

Then, we will necessarily have for a 6-round Feistel schemes with internal functions (and hence also for internal permutations):

$$(3)\begin{cases} S_3 = S_4, \\ L_1 \oplus S_1 = L_3 \oplus S_3, \\ R_1 \oplus T_1 = R_2 \oplus T_2, \\ R_3 \oplus T_3 = R_4 \oplus T_4. \end{cases}$$

Note that the equations of System (2) *are compatible with Theorems 11.4 and 11.5 of Sect. 11.6.*

Proof Suppose we have System (1) and System (2), let us show that System (3) is verified.

From $X_1 = X_4$ we get $X_2 = X_3$. In fact, $X_1 = X_4$ is equivalent to $L_1 \oplus f_1(R_1) = L_4 \oplus f_1(R_4)$. Since $R_1 = R_3$, $R_4 = R_2$ and $L_1 \oplus L_4 = L_2 \oplus L_3$, we get $L_2 \oplus f_1(R_2) = L_3 \oplus f_1(R_3)$, i.e., $X_2 = X_3$.

Similarly, from $Y_1 = Y_2$ we get $Y_3 = Y_4$. Indeed, $Y_1 = Y_2$ means $R_1 \oplus f_2(X_1) = R_2 \oplus f_2(X_2)$. Since $R_1 = R_3$, $R_2 = R_4$, $X_1 = X_4$ and $X_2 = X_3$, we get $R_3 \oplus f_2(X_3) = R_4 \oplus f_2(X_4)$, i.e., $Y_3 = Y_4$.

Also, from $Z_1 = Z_3$ we get $Z_2 = Z_4$. In fact, $Z_1 = Z_3$ means $X_1 \oplus f_3(Y_1) = X_3 \oplus f_3(Y_3)$. Since $X_1 = X_4$, $X_3 = X_2$, $Y_1 = Y_2$, and $Y_3 = Y_4$, we obtain $X_2 \oplus f_3(Y_2) = X_4 \oplus f_3(Y_4)$, i.e., $Z_2 = Z_4$.

Moreover, from $U_1 = U_4$ we obtain $U_2 = U_3$. In fact, $U_1 = U_4$ is equivalent to $Y_1 \oplus f_4(Z_1) = Y_4 \oplus f_4(Z_4)$. Since $Z_1 = Z_3$, $Z_4 = Z_2$, $Y_1 = Y_2$ and $Y_4 = Y_3$, we get $Y_2 \oplus f_4(Z_2) = Y_3 \oplus f_4(Z_3)$, i.e., $U_2 = U_3$.

Similarly, from $S_1 = S_2$ we get $S_3 = S_4$. Indeed, $S_1 = S_2$ means $Z_1 \oplus f_5(U_1) = Z_2 \oplus f_5(U_2)$. Since $Z_1 = Z_3$, $Z_2 = Z_4$, $U_1 = U_4$ and $U_2 = U_3$ we get $Z_3 \oplus f_5(U_3) = Z_4 \oplus f_5(U_4)$ which means $S_3 = S_4$.

Now from $S_1 = S_2$ we get

$$U_1 \oplus U_2 = T_1 \oplus T_2, \tag{11.2}$$

because $T_1 = U_1 \oplus f_6(S_1)$ and $T_2 = U_2 \oplus f_6(S_2)$.

Similarly from $S_3 = S_4$ we get $U_3 \oplus U_4 = T_3 \oplus T_4$, i.e.:

$$T_3 \oplus T_4 = U_1 \oplus U_2, \tag{11.3}$$

because $U_4 = U_1$ and $U_3 = U_2$.

Similarly from $U_1 = U_4$ we get $S_1 \oplus S_4 = Z_1 \oplus Z_4$, i.e.:

$$S_1 \oplus S_3 = Z_1 \oplus Z_2, \tag{11.4}$$

since $S_4 = S_3$ and $Z_4 = Z_2$.

Similarly from $Z_1 = Z_3$ we get $U_1 \oplus U_3 = Y_1 \oplus Y_3$, i.e.:

$$U_1 \oplus U_2 = Y_1 \oplus Y_3, \tag{11.5}$$

since $U_3 = U_2$.

Similarly from $Y_1 = Y_2$ we get:

$$Z_1 \oplus Z_2 = X_1 \oplus X_2. \tag{11.6}$$

Similarly from $X_1 = X_4$ we get $Y_1 \oplus Y_4 = R_1 \oplus R_4$, i.e.:

$$Y_1 \oplus Y_3 = R_1 \oplus R_2, \tag{11.7}$$

because $Y_4 = Y_3$ and $R_4 = R_2$.

Similarly from $R_1 = R_3$ we get $X_1 \oplus X_3 = L_1 \oplus L_3$, i.e.:

$$X_1 \oplus X_2 = L_1 \oplus L_3, \tag{11.8}$$

because $X_3 = X_2$.

From Eqs. (11.2), (11.5) and (11.7) we obtain

$$T_1 \oplus T_2 = R_1 \oplus R_2 \quad (= Y_1 \oplus Y_3).$$

From Eqs. (11.3), (11.5) and (11.7) we obtain $R_1 \oplus R_2 = T_3 \oplus T_4$. Since $R_1 = R_3$ and $R_2 = R_4$ we get

$$R_3 \oplus R_4 = T_3 \oplus T_4.$$

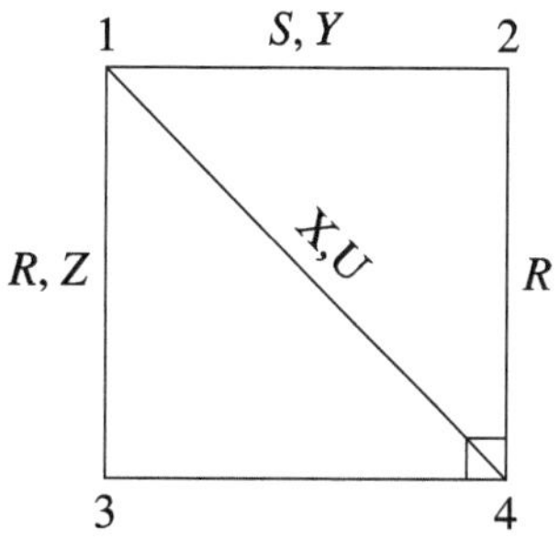

Fig. 11.2 Representation of our eight starting equations. The small square in the right-hand corner below represents $L_1 \oplus L_2 \oplus L_3 \oplus L_4 = 0$

Also from Eqs. (11.4), (11.6) and (11.8) we obtain

$$S_1 \oplus S_3 = L_1 \oplus L_3 \quad (= Z_1 \oplus Z_2 = X_1 \oplus X_2).$$

Therefore, we have proved that the System (3) is verified.

From Theorem 11.6, we see that we will have at least about two times more solutions of the System $(\mathscr{S})$ when we have a 6-round Feistel scheme with internal permutations than when we have a random permutation. In fact, for a 6-round Feistel scheme, the System $(\mathscr{S})$ can appear randomly or from the equations of Fig. 11.2. In both cases we have 8 equations on four indices. This gives an attack in $\mathscr{O}\left(\frac{q^4}{2^{8n}}\right)$, i.e., when $q \approx \mathscr{O}(2^{2n})$.

11.6.2.3 Case of 6-Round Feistel with Internal Functions

This attack also works on 6-round Feistel with internal functions. In fact, instead of about two more times solutions, we will have here about 12 more times solutions. These solutions come from random solutions of the System $(\mathscr{S})$ plus these 11 families listed below. In fact, in Theorem 11.6, we can replace the System (2) with any system of these 11 families and still prove in a similar way that System (3) is verified. This still gives an attack in $\mathscr{O}(2^{2n})$.

$$\text{Family 1:} \begin{cases} X_1 = X_2, \\ Y_1 = Y_3, \\ Z_1 = Z_2, \\ U_1 = U_3. \end{cases} \qquad \text{Family 2:} \begin{cases} X_1 = X_2, \\ Y_1 = Y_3, \\ Z_1 = Z_2, \\ U_1 = U_4. \end{cases}$$

$$\text{Family 3:} \begin{cases} X_1 = X_2, \\ Y_1 = Y_3, \\ Z_1 = Z_4, \\ U_1 = U_3. \end{cases} \qquad \text{Family 4:} \begin{cases} X_1 = X_2, \\ Y_1 = Y_4, \\ Z_1 = Z_2, \\ U_1 = U_3. \end{cases}$$

$$\text{Family 5: } \begin{cases} X_1 = X_2, \\ Y_1 = Y_4, \\ Z_1 = Z_2, \\ U_1 = U_4. \end{cases} \qquad \text{Family 6: } \begin{cases} X_1 = X_2, \\ Y_1 = Y_4, \\ Z_1 = Z_3, \\ U_1 = U_4. \end{cases}$$

$$\text{Family 7: } \begin{cases} X_1 = X_4, \\ Y_1 = Y_2, \\ Z_1 = Z_3, \\ U_1 = U_4. \end{cases} \qquad \text{Family 8: } \begin{cases} X_1 = X_4, \\ Y_1 = Y_2, \\ Z_1 = Z_4, \\ U_1 = U_3. \end{cases}$$

$$\text{Family 9: } \begin{cases} X_1 = X_4, \\ Y_1 = Y_3, \\ Z_1 = Z_2, \\ U_1 = U_3. \end{cases} \qquad \text{Family 10: } \begin{cases} X_1 = X_4, \\ Y_1 = Y_3, \\ Z_1 = Z_2, \\ U_1 = U_4. \end{cases}$$

$$\text{Family 11: } \begin{cases} X_1 = X_4, \\ Y_1 = Y_3, \\ Z_1 = Z_4, \\ U_1 = U_3. \end{cases}$$

Note that these families are compatible with Theorem 11.4 of Sect. 11.6. Note also that Family (7) is the same as System (2).

Remark 11.4 Family (7) is the only family that is compatible with the case of 6-round Feistel schemes with internal permutations because of Theorem (3).

Remark 11.5 We found an even number of solutions since if we exchange indices 1 and 3 and indices 2 and 4, we also obtain a solution of $(\mathscr{S})$. Alternatively, it would have been possible to impose that index 1 is strictly less than index 3.

Remark 11.6 Instead of the eight equations of $(\mathscr{S})$ we may want to compute solutions such that

$$(\mathscr{S}') \begin{cases} R_1 = R_2, \\ R_3 = R_4, \\ S_1 = S_3, \\ S_2 = S_4, \\ T_1 \oplus T_4 = T_2 \oplus T_3, \\ T_1 \oplus R_1 = T_3 \oplus R_3, \\ S_1 \oplus L_1 = S_2 \oplus L_2, \\ L_1 \oplus L_2 \oplus L_3 \oplus L_4 = 0. \end{cases}$$

We use here the fact that the inverse of a Feistel scheme is another Feistel scheme when we permute the left and right parts, i.e., there is a symmetry R/S and L/T. However $(\mathscr{S}')$ is just the same as $(\mathscr{S})$ when we permute the indices 1 and 2.

11.6.3 Complexity of Classical Attacks on 6-Round Feistel Schemes

In this section, we will show that the complexity of the attack for 6-round Feistel schemes with random functions or with random permutations is in $\mathcal{O}(2^{2n})$.

First, we select indices 1 and 2 such that:

$$\begin{cases} S_2 = S_1, \\ R_2 \oplus T_2 = R_1 \oplus T_1. \end{cases}$$

We expect to find $\frac{m^2}{2^{2n}}$ possibilities with $m = 2^{2n}$. Indeed, it's possible to do this in $\mathcal{O}(m)$ computations and $\mathcal{O}(m)$ memory by storing all the m values $(S_i, R_i \oplus T_i)$ in a hash table and looking for collisions.

Moreover, we find index 3 such that:

$$\begin{cases} R_3 = R_1, \\ L_3 \oplus S_3 = L_1 \oplus S_1. \end{cases}$$

We expect to find $\frac{m}{2^{2n}}$ possibilities with $m = 2^{2n}$. In fact, it is possible to do this in $\mathcal{O}(1)$ computations and $\mathcal{O}(m)$ memory by storing all the m values $(R_i, L_i \oplus S_i)$ in a hash table and looking for $(R_1, L_1 \oplus S_1)$.

Finally, we look for index 4 such that:

$$\begin{cases} R_4 = R_2, \\ L_4 = L_1 \oplus L_2 \oplus L_3. \end{cases}$$

It is possible to do this in $\mathcal{O}(1)$, and in this case, we don't need a hash table.

At the end we have at most m choices of pairwise distinct indices $(1, 2, 3, 4)$. Among these, we keep those that give $R_1 \neq R_2$ and $L_1 \neq L_3$. Therefore, the total complexity is in $\mathcal{O}(2^{2n})$ computations and $\mathcal{O}(2^{2n})$ memory.

11.6.4 Quantum Attack on Ψ^6

In this section, we start by recalling in Sect. 11.6.5 Ambainis' quantum collision-finding algorithm given in Ambainis (2004). We then recall the quantum algorithms for subset finding in Sect. 11.6.6 and carefully give the time complexity of the algorithm. We describe the data structures from Ambainis' algorithm in Sect. 11.6.7. Finally, we apply it to the problem of distinguishing 6-round Feistel schemes from a random permutation in Sect. 11.6.8.

11.6.5 Recalling Ambainis' Algorithm

In this section, we explain how to solve the k-distinctness problem for a fixed value of k. Specifically, given N values $x_1, \ldots, x_N$ that belong to a set $\mathscr{X}$ of size M, are there k distinct indices $1 \leq i_1, \ldots, i_k \leq N$ such that $x_{i_1} = \cdots = x_{i_k}$? In Ambainis (2004), proves the following result.

Theorem 11.7 (Ambainis 2004) *Let $r \geq k$, $r = o(N)$. There is a quantum algorithm that solves element k-distinctness in time $\widetilde{\mathcal{O}}\left(\max\left(\frac{N^{k/2}}{r^{(k-1)/2}}, r\right)\right)$, with $\mathcal{O}\left(\max\left(\frac{N^{k/2}}{r^{(k-1)/2}}, r\right)\right)$ queries, using $\widetilde{\mathcal{O}}(r)$ qubits of memory.*

We will give a rough overview of the algorithm and the necessary conditions that are required to prove Theorem 11.7, and in particular with respect to the time complexity of the algorithm.

Remark 11.7 We will only describe the quantum algorithm that solves the k-distinctness problem in the specific case where only a single k-collision exists. Ambainis gives a generic method to derive an algorithm that solves the general problem using the restricted one as a subroutine and without changing the complexity.

From now on, we assume that our goal is to find the unique tuple of indices that form a k-collision. From a high level, the algorithm considers the set $\mathscr{T}$ of tuples of the form $(\mathbf{S}, x_{\mathbf{S}}, y)$ where

- $\mathbf{S} = \{y_1, \ldots, y_r\}$ is a subset of $\{1, \ldots, N\}$ of size r,
- $x_{\mathbf{S}} = (x_{y_1}, \ldots, x_{y_r}) \in \mathscr{X}^r$,
- $y \in \{1, \ldots, N\} \setminus \mathbf{S}$.

It then works as a quantum walk over the set of all these tuples by taking the following steps:

1. Build a uniform superposition of all tuples $(\mathbf{S}, x_{\mathbf{S}}, y)$.
2. Repeat the following procedure $t_1 = \mathcal{O}\left((N/r)^{k/2}\right)$ times:
    a. If there exists k distinct indices $i_1, \ldots, i_k$ in $\mathbf{S}$ such that $x_{i_1} = \cdots = x_{i_k}$, apply the conditional phase flip.
    b. Perform $t_2 = \mathcal{O}\left(\sqrt{r}\right)$ times the following quantum walk step:
        i. Apply a diffusion operator that will create a superposition of all y values such that $y \in \{1, \ldots, N\} \setminus \mathbf{S}$.[2]
        ii. Add y to $\mathbf{S}$ and modify the corresponding $x_{\mathbf{S}}$ vector accordingly.

[2] We stress that this superposition is not uniform, which is critical in the proof of Theorem 11.7. However, the corresponding amplitudes will not be relevant to our discussion, which is why we chose to ignore them for the sake of simplicity.

iii. In order to map the current vector back to $\mathscr{T}$, we need to remove an element from $\mathbf{S}$ and to shrink the vector x accordingly. This is done by creating a superposition, over all possible values of $y' \in \mathbf{S}$, of all tuples of the form $(\mathbf{S} \setminus \{y'\}, x_{\mathbf{S}\setminus\{y'\}}, y')^2$.

3. Measure the final state and check if $\mathbf{S}$ contains a collision.

Theorem 11.7 states that this algorithm indeed finds a collision with constant probability. Moreover, it has a *query* complexity of r to create the initial superposition, and $t_1 \times t_2$ for the second part of the algorithm, which leads to an optimal value of $r = \mathcal{O}\left(N^{k/(k+1)}\right)$. In the following sections, we discuss the time complexity of the algorithm in more detail.

11.6.6 Childs and Eisenberg's Algorithm

We start by noting that Ambainis' algorithm does not necessarily deal with collisions. Indeed, it is a quantum search algorithm that finds a k-tuple of elements in a set that satisfies a specific relation. Moreover, it does so by relying on a quantum walk over a r-tuple of elements in this same set. In particular, the query complexity of Ambainis' algorithm could be generalized to *any* relation that involves exactly k indices, using exactly the same procedure, and with the same *query* complexity (Childs and Eisenberg 2005). A critical part of the analysis from Ambainis (2004) is that the *time* complexity of the collision-finding algorithm actually matches its query complexity. In order to achieve this, Ambainis relies on a clever combination of data structures in order to satisfy the following constraints:

- All manipulation of the data structures can be made in a reversible way.
- The time complexity of the search, insertion, and deletion operations has to be (poly-)logarithmic.
- The time complexity of *checking if the relation is satisfied* has to be smaller than $t_2 = \mathcal{O}(\sqrt{r})$.

The last point stems from the fact that step 2.a of the algorithm (the conditional phase flip) should not be more time-consuming than step 2.b (t_2 iterations of the quantum walk). This remark calls for the following definition.

Definition 11.1 A k-ary relation $\mathscr{R}$ over $\mathscr{X}^k$ is said to be (f, g, r)-testable if there exists a data structure to represent subsets of $\{1, \ldots, |\mathscr{X}|\}$ such that the following properties hold:

- It uses $\widetilde{\mathcal{O}}(s)$ qubits of memory to store a subset of elements of size $s \leq r + 1$.
- Insertion and deletion of an element is (poly)-logarithmic and requires a single query.

- Testing if there exists a tuple $(i_1, \ldots, i_k)$ of k distinct elements in a set of size r such that $(x_{i_1}, \ldots, x_{i_k}) \in \mathcal{R}$ takes a time at most $f(r)$, and requires at most $g(r)$ queries.

Then, one has the following result.

Theorem 11.8 (Childs and Eisenberg 2005) *Let $r \geq k$, $r = o(N)$, and let $\mathcal{R}$ be a (f, g, r)-testable k-ary relation over $\mathcal{X}^k$ for some functions f and g. There is a quantum algorithm that outputs a k-tuple $(i_1, \ldots, i_k)$ of distinct elements such that $(x_{i_1}, \ldots, x_{i_k}) \in \mathcal{R}$ with $\mathcal{O}\left(\max\left(\left(\frac{N}{r}\right)^{k/2}(\sqrt{r} + g(r)), r\right)\right)$ queries using $\widetilde{\mathcal{O}}(r)$ qubits of memory.*

Proof Childs and Eisenberg proved this theorem in Childs and Eisenberg (2005).

Corollary 11.1 *This quantum algorithm takes $\widetilde{\mathcal{O}}\left(\max\left(\left(\frac{N}{r}\right)^{k/2}(\sqrt{r} + f(r)), r\right)\right)$ in time.*

Remark 11.8 In Ambainis (2004), proves that the k-collision relation is actually $\left(\widetilde{\mathcal{O}}(1), 0, r\right)$-testable when $r = o(N)$.

11.6.7 On Data Structures

In Ambainis (2004), Ambainis' shows that the actual running time of the quantum algorithm is in $\widetilde{\mathcal{O}}(N^{k/k+1})$. In this section, we will recall the data structures that Ambainis relies on for the k-distinctness problem to achieve this running time.

11.6.7.1 Required Operations

The required operations are as follows:

1. Adding y to **S** and storing x_y.
2. Removing y from **S** and erasing x_y.
3. Checking if **S** contains the solution to perform the conditional phase flip.
4. Applying the diffusion operator.
5. Storing the same set **S** in the same way independently of how **S** was created.
6. Using classical subroutines that must terminate in fixed time t.

The data structures used combine a hash table and a skip list. In the following, v is set as a variable that tracks the count of distinct $x \in [M]$ for which the set **S** contains $i_1, \ldots, i_k$ such that $x_{i_1} = \cdots = x_{i_k}$.

11.6.7.2 Hash Table

The hash table is used for storing pairs (i, x_i) and retrieving them when searching for x_i corresponding to a given i.

The hash table consists of r buckets, each equipped with memory for $\lceil \log N \rceil$ entries and $\lfloor \log r \rfloor$ counters. To determine the placement of entries, the set $\{1, \ldots, N\}$ is hashed into the r buckets using the function $h(i) = \lfloor i \cdot r/N \rfloor + 1$. The j-th bucket organizes pairs (i, x_i) for $i \in \mathbf{S}$ such that $h(i) = j$, in ascending order of i. In cases where there are more than $\lceil \log N \rceil$ entries with $h(i) = j$, only $\lceil \log N \rceil$ of them are stored in the bucket. Ambainis demonstrates that the probability of this occurrence is small.

As for the counters, denoted as $d_1, \ldots, d_{\lfloor \log r \rfloor}$, the counter d_1 in the j-th bucket counts the number of $i \in \mathbf{S}$ such that $h(i) = j$. Additionally, d_l for $l > 1$ is used only if j is divisible by 2^l. In such instances, it counts the number of $i \in \mathbf{S}$ satisfying $j - 2^l + 1 \leq h(i) \leq j$.

11.6.7.3 Skip List

The skip list is used for organizing the x_i in ascending order. Consequently, when adding a new element i to $\mathbf{S}$, we can verify whether x_i matches any of the x_j values for $j \in \mathbf{S}$. For a more detailed understanding of how this skip list operates, please refer to Pugh (1990); Ambainis (2004).

Each element $i \in \mathbf{S}$ is assigned a random level l_i ranging from 0 to $l_{\max} = \lceil \log N \rceil$. To choose the level independently for each i, we define the levels using $l_{\max}$ functions $h_1, h_2, \ldots, h_{l_{\max}} : [N] \to \{0, 1\}$. If $h_1(i) = \cdots = h_l(i) = 1$ but $h_{l+1}(i) = 0$ then $i \in [N]$ is associated to level l. If $h_1(i) = \cdots = h_{l_{\max}}(i) = 1$, it implies that the element $i \in [N]$ is associated with level $l_{\max}$. The reason for assigning the level in this manner is due to a drawback in the straightforward implementation.

11.6.7.4 Insertion and Deletion

To include i in the set $\mathbf{S}$, we first query the value x_i. Then, we calculate $h(i)$ and add (i, x_i) into the $h(i)$-th bucket. If the bucket already contains some entries, we may move some of them to ensure that, after the insertion of (i, x_i), the entries are still in the order of increasing i. Following this, we increment the counter d_1 for the $h(i)$-th bucket and the counter d_l for the $(\lceil \frac{h(i)}{2^l} \rceil \cdot 2^l)$-th bucket, for each $l \in \{2, \ldots, \lfloor \log r \rfloor\}$. Following that, we modify the skip list as follows: first, run a search to locate the last element preceding i. Then, for every level $l \in \{0, \ldots, l_i\}$ assign j_l to be the level-l pointer of i_l. Set the level-l pointer of i to be equal to j_l and the level-l pointer of i_l to be equal to i. Once the update is finished, we use the skip list to identify the smallest j where $x_j = x_i$. We then use level-0 pointers to determine whether the count of j

with $x_j = x_i$ is less than k, exactly k, or exceeds k. If there are exactly k such j, we increase v by 1.

An element i can be deleted from **S** by running this procedure in reverse.

11.6.7.5 Checking for k-Collisions

To find k-collisions in **S**, we simply verify whether $v > 0$.

11.6.7.6 Diffusion Transform

Based on Grover's result (Grover 1996), Ambainis demonstrates that the algorithm requires approximately $\widetilde{\mathcal{O}}(1)$ time for the diffusion process.

11.6.7.7 Uniqueness

S is always stored in the same way. The values $i \in \mathbf{S}$ are consistently hashed to buckets by h, and within each bucket, the entries are arranged in increasing order of i. The counters, responsible to count the number of entries in the buckets, are uniquely defined by the set **S**. Similarly, the configuration of the skip list is also uniquely established.

11.6.7.8 Guaranteed Running Time

Ambainis demonstrates that the probability of lookup, insertion, or deletion of an element taking more than $\widetilde{\mathcal{O}}(1)$ steps is negligible for any set **S**. In such instances, the algorithm is adjusted for lookup, insertion, or deletion, causing them to abort after $\widetilde{\mathcal{O}}(1)$ steps. It is highlighted in Ambainis (2004) that this modification has no significant impact on the entire quantum search algorithm.

11.6.8 Application to the Case of Feistel Networks

We use the same type of data structure mentioned in Sect. 11.6.7 but with a few small adjustments.

In the following, v is set as a variable that tracks the count of distinct solutions of the System ($\mathcal{S}$) that are included in **S**.

11.6.8.1 Skip List

For a Feistel scheme with 6 rounds, we need two skip lists instead of one. Specifically, the first one is used to store $S||R \oplus T$. The second skip list stores $R||L \oplus S$.

11.6.8.2 Insertion and Deletion

To include i in the set $\mathbf{S}$, we first query the value $O(L, R)$.[3] Then, we calculate $h(i)$ and add $(i, (S_i, T_i))$ into the $h(i)$-th bucket. After that, we update the skip list and then we update the value of v by adding the value of tuple of indices that satisfy the System $(\mathscr{S})$ and include the new index i. This can take the form of $(i, j, k, l) \in \mathbf{S}^4$ that satisfies the System $(\mathscr{S})$, or $(j, i, k, l) \in S^4$ that satisfies the System $(\mathscr{S})$, or $(j, k, i, l) \in \mathbf{S}^4$ that satisfies the System $(\mathscr{S})$, or $(j, k, l, i) \in \mathbf{S}^4$ that satisfies the System $(\mathscr{S})$. For example, to verify if (i, j, k, l) is a solution, it suffices to check whether $S_j||R_j \oplus T_j = S_i||R_i \oplus T_i$ and $R_k||L_k \oplus S_k = R_i||L_i \oplus S_i$ and $R_l||L_l = R_j||L_i \oplus L_j \oplus L_k$. Another instance is checking if (j, i, k, l) is a solution, for which the verification involves ensuring that $S_i||R_i \oplus T_i = S_j||R_j \oplus T_j$ and $R_l||L_l \oplus S_l = R_j||L_j \oplus S_j$ and $S_k||T_k = S_l||T_i \oplus T_j \oplus T_l$.

An element i can be deleted from $\mathbf{S}$ by running this procedure in reverse.

11.6.8.3 Checking for k-Collisions

To find collisions in $\mathbf{S}$, we simply verify whether $v > 0$.

11.6.8.4 Guaranteed Running Time

We need a guaranteed running time when testing if a new entry can be query 1, 2, 3, of a tuple of solution. Let ℓ_1 (resp. ℓ_2) be the length of the longest $[S, R \oplus T]$ (resp., $[R, L \oplus S]$) line,[4] where we say that we have a $[A, B]$ line of length ℓ if there exists pairwise distinct indices $i_1, \ldots, i_\ell$ such that $(A_{i_1}, B_{i_1}) = \cdots = (A_{i_\ell}, B_{i_\ell})$. In this case, the complexity of checking if a new entry triggers the completion of a new tuple is $\widetilde{\mathscr{O}}(\ell_1 \ell_2)$. However, for our attack to work, we need to be able to guarantee that the insertion or deletion of an element runs in time $\widetilde{\mathscr{O}}(1)$, except with a small probability over the uniformly random sampling of the $2n$-bit permutation or the 6 n-bit random functions. In Sect. 11.6.9, we provide a complete proof of the fact that $\ell_1, \ell_2 = \widetilde{\mathscr{O}}(1)$ with a probability that is close to 1, which is sufficient to conclude the analysis of our quantum attack. While Sect. 11.6.9 is quite long, the ideas behind the proof are quite simple, and we briefly outline them here. In the case of a random permutation,

[3] The oracle O is implemented either with a uniformly random permutation P or a 6-round Feistel scheme Ψ.

[4] This is defined over the full message space.

the experiment is close to the classical balls-into-bins problem (with a moderate dependence between throws), which explains why the length of a line is logarithmic with overwhelming probability. In the case of a Feistel network, collisions can come from two possible avenues:

- purely random collisions that occur from a degree of freedom during the evaluation of the Feistel network;
- structural collisions that are forced due to internal collisions with previous queries.

The first category behaves close to a standard balls-into-bins problem and also yields lines of logarithmic length with overwhelming probability. The second type of collisions, although less numerous, are more difficult to analyze. However, the intuition behind our result is that, in such a structural collision brings at least two independent equations per query involved in the collision. Hence, the average number of such collisions is close to 1, which means that it will be logarithmic with a probability that is also close to 1. This is sufficient to ensure that Child and Eisenberg's algorithm will find solutions to the System (11.1) with a non-negligible probability, and with a query complexity of $O(2^{8n/5})$ and a time complexity of $\widetilde{\mathcal{O}}(2^{8n/5})$.

11.6.9 On the Maximum Line Length

11.6.9.1 Permutation Case

In this section, we consider the random variable ℓ, which corresponds to the length of the longest $[R, L \oplus S]$ line (the other case can be treated similarly by symmetry). Let us start by fixing any arbitrary ordering over $\{0, 1\}^n$ such that

$$\{0, 1\}^n = \{L_1, \ldots, L_{2^n}\}.$$

For any $A, B \in \{0, 1\}^n$, let $\ell_{A,B}$ be the number of indices $i \in \{1, \ldots, 2^n\}$ such that $P(L_i||A) = (B \oplus L_i)||*$. Then, one has $\ell = \max_{A,B}(\ell_{A,B})$ and, for any c,

$$\Pr[\ell \geq c] \leq \sum_{A,B} \Pr[\ell_{A,B} \geq c].$$

Let us fix any $A, B \in \{0, 1\}^n$ and, for all $i = 1, \ldots, 2^n$, let $\mathcal{Y}_i$ be the random variable that is equal to 1 if $P(L_i||A) = (B \oplus L_i)||*$, and to 0 otherwise. It is easy to see that

$$\ell_{A,B} = \sum_{i=1}^{2^n} \mathcal{Y}_i.$$

Let us fix any $i \in \{1, \dots, 2^n\}$, and any $(y_1, \dots, y_{i-1}) \in \{0, 1\}^{i-1}$. Then, one has

$$\Pr[\mathscr{Y}_i = 1 \mid (\mathscr{Y}_1, \dots, \mathscr{Y}_{i-1}) = (y_1, \dots, y_{i-1})] \leq \varepsilon_i,$$

where

$$\varepsilon_i = \frac{2^n}{2^{2n} - i + 1} = \frac{1}{2^n} + \frac{i-1}{2^{3n} - (i-1)2^n}.$$

Hence, one has

$$1 \leq 1 + \frac{(2^n - 1)}{2^{2n+1}} \leq \sum_{i=1}^{2^n} \varepsilon_i \leq 1 + \frac{1}{2^{n+1} - 2} \leq \frac{3}{2}.$$

Let us now apply Lemma 11.1 to $(Y_i)_{1 \leq i \leq 2^n}$. One has

$$\Pr\left[\ell_{A,B} \geq \frac{3}{2}(1+\delta)\right] \leq \Pr\left[\ell_{A,B} \geq (1+\delta)\sum_{i=1}^{2^n} \varepsilon_i\right] \leq e^{-\frac{\delta^2}{2+\delta}\sum_{i=1}^{2^n}\varepsilon_i} \leq e^{-\frac{\delta^2}{2+\delta}}.$$

Using the fact that $n \geq 1$ and $e \geq 2$, and taking $\delta = \frac{9n}{2}$, we get

$$\Pr\left[\ell_{A,B} \geq 9n\right] \leq \Pr\left[\ell_{A,B} \geq \frac{3}{2}(1 + \frac{9n}{2})\right] \leq e^{-3n} \leq 2^{-3n}.$$

Hence, after summing over all possible values for A, B, we finally have

$$\Pr[\ell \geq 9n] \leq \frac{1}{2^n}.$$

11.6.9.2 Feistel Case

Let us fix $n \geq 14$. We will split the study into two parts, depending on the number of fresh inputs of the queries that appear in the line. Let ℓ' be the length of the longest free line in $[R, L \oplus S]$ (i.e., a line such that, for each involved query, the intermediate input of *at least one function* among f_3, f_4, and f_5 is fresh). Let N be the number of queries that are part of a $[R, L \oplus S]$ line and whose intermediate inputs of f_3, f_4, and f_5 all collide with another query from the same line. It is clear that $\ell \leq \ell' + N$. We will prove that $\Pr[\ell' \geq 4n] \leq \frac{1}{2^{3n}}$, and $\Pr[N \geq 4n] \leq \frac{1+\frac{6}{2^n}}{4n}$. Overall, this will prove that

$$\Pr[\ell \geq 8n] \leq \Pr[\ell' \geq 4n] + \Pr[N \geq 4n] \leq \frac{1}{2^{3n}} + \frac{1+\frac{6}{2^n}}{4n}.$$

First scenario
Let us fix any $A, B \in \{0,1\}^n$, $i \in \{1,\ldots,2^n\}$ and any $(y_1,\ldots,y_{i-1}) \in \{0,1\}^{i-1}$. For any set $I = \{i_1,\ldots,i_k\}$ of $4n$ pairwise distinct indices in $\{1,\ldots,2^n\}$, we let $\ell_{A,B}(I)$ denote the event

$$\Psi^6(L_{i_j}||A) = (B \oplus L_{i_j})|| * \qquad \text{for } i = 1,\ldots,4n.$$

If we rewrite this equality in terms of the intermediate variables X_{i_j} and Y_{i_j}, this equality becomes

$$f_3\left(Y_{i_j}\right) \oplus f_5\left(Y_{i_j} \oplus f_4\left(X_{i_j} \oplus f_3\left(Y_{i_j}\right)\right)\right) = B \oplus f_1(A). \tag{11.9}$$

Let us fix $j \in \{1,\ldots,4n\}$, and let us assume that Eq. (11.9) holds for all $j' < j$. The event "Eq. (11.9) holds for the j-th query" is included in the following set of events:

1. the input to function f_5 is fresh (i.e., it does not collide with any previous inputs to f_5), and Eq. (11.9) holds;
2. there exists $j' < j$ such that

$$\begin{aligned} Y_{i_j} \oplus f_4\left(X_{i_j} \oplus f_3\left(Y_{i_j}\right)\right) &= Y_{i_{j'}} \oplus f_4\left(X_{i_{j'}} \oplus f_3\left(Y_{i_{j'}}\right)\right), \\ f_3(Y_{i_j}) &= f_3(Y_{i_{j'}}) \qquad \text{(from Eq. (11.9))}, \end{aligned}$$

   but $X_{i_j} \oplus f_3\left(Y_{i_j}\right)$ is fresh;
3. there exists $j', j'' < j$ such that

$$\begin{aligned} Y_{i_j} \oplus f_4\left(X_{i_j} \oplus f_3\left(Y_{i_j}\right)\right) &= Y_{i_{j'}} \oplus f_4\left(X_{i_{j'}} \oplus f_3\left(Y_{i_{j'}}\right)\right), \\ X_{i_j} \oplus f_3\left(Y_{i_j}\right) &= X_{i_{j''}} \oplus f_3\left(Y_{i_{j''}}\right), \\ f_3(Y_{i_j}) &= f_3(Y_{i_{j'}}) \qquad \text{(from Eq. (11.9))}, \end{aligned}$$

   but Y_{i_j} is fresh.

Overall, in all cases, at least one value will be chosen uniformly at random in $\{0,1\}^n$. Indeed, one has $X_{i_j} = L_{i_j} \oplus f_1(A)$ and $Y_{i_j} = A \oplus f_2(X_{i_j})$. This means that the X_i values are always pairwise distinct,[5] and the Y_i values are chosen uniformly and independently at random. We now consider all the probabilities of all three cases occurring.

Case 1: Since the input to f_5 is fresh, this case occurs with a probability lower than

$$\frac{1}{2^n}.$$

[5] Recall that $X_i = L_i \oplus f_1(R_i) = L_i \oplus f_1(A)$. This means, in a $[R, L \oplus S]$ line, all L and X values have to be pairwise distinct.

Case 2: There are at most $j-1$ possible choices for j'. Since the input to f_4 is fresh, the first equation holds with a probability 2^{-n}. For the second equation, two cases can occur: either Y_{i_j} is fresh (which means that the second equation occurs with a probability 2^{-n}), or Y_{i_j} is not fresh, which occurs with a probability at most $(j-1)/2^n$. Overall, the probability of this case occurring is lower than

$$\frac{(j-1)+(j-1)^2}{2^{2n}} \leq \frac{2(4n)^2}{2^{2n}} \leq \frac{1}{2 \cdot 2^n},$$

since $2(4n)^2 \leq 2^n/2$ when $n \geq 14$.

Case 3: One has $X_{i_j} \oplus f_3\left(Y_{i_j}\right) = X_{i_{j''}} \oplus f_3\left(Y_{i_{j''}}\right)$. This means that the first equation can be rewritten as

$$Y_{i_j} \oplus f_4\left(X_{i_{j''}} \oplus f_3\left(Y_{i_{j''}}\right)\right) = Y_{i_{j'}} \oplus f_4\left(X_{i_{j'}} \oplus f_3\left(Y_{i_{j'}}\right)\right),$$

which happens with a probability 2^{-n} over the choice of Y_{i_j}. Since Y_{i_j} is fresh, the third equation will also occur with a probability at most 2^{-n}. Overall, this case happens with probability smaller than

$$\frac{(j-1)^2}{2^{2n}} \leq \frac{1}{2 \cdot 2^n}.$$

Hence, the probability that Eq. (11.9) holds for the j-th query is smaller than $2/2^n$. Thus, taking a product over j yields

$$\Pr[\ell_{A,B}(I) \geq 4n] \leq \frac{2^{4n}}{2^{4n^2}}.$$

Summing over all possible choices for I, A, B yields

$$\Pr[\ell' \geq 4n] \leq \frac{2^{(4n+2)n+4n}}{(4n)! \cdot 2^{4n^2}} = \frac{2^{6n}}{(4n)!}.$$

One has $(4n)! \geq (2n)^{2n} \cdot (2n)! \geq 2^{2n} n^{3n}$, which gives

$$\Pr[\ell' \geq 4n] \leq \frac{2^{4n}}{n^{3n}} \leq \left(\frac{3}{n}\right)^{3n} \leq \frac{1}{2^{3n}}$$

since $n \geq 14$.

Second scenario

Here, we want to upper bound the probability that there exist four indices i_1, i_2, i_3, i_4 such that:

- all indices belong to the same $[R, L \oplus S]$ line;
- $Y_{i_4} = Y_{i_3}$;

- $X_{i_4} \oplus f_3(Y_{i_4}) = X_{i_2} \oplus f_3(Y_{i_2})$ (i.e., $Z_{i_4} = Z_{i_2}$);
- $Y_{i_4} \oplus f_4(X_{i_4} \oplus f_3(Y_{i_4})) = Y_{i_1} \oplus f_4(X_{i_1} \oplus f_3(Y_{i_1}))$ (i.e., $U_{i_4} = U_{i_1}$).
- all indices are the first indices (for an arbitrary ordering of the queries) that satisfies these conditions; this means that all queries i_1, i_2, i_3 have at least one input to f_3, f_4, or f_5 that is fresh.

In particular, these conditions imply that

- $i_4 \neq i_3, i_2, i_1$;
- $f_3(Y_{i_4}) = f_3(Y_{i_3}) = f_3(Y_{i_1})$ since $U_{i_4} = U_{i_1}$ and $L_{i_4} \oplus S_{i_4} = L_{i_1} \oplus S_{i_1}$, which corresponds to Eq. (11.9);
- $i_3 \neq i_2$ and $i_2 \neq i_1$ since otherwise the queries would be identical due to the bijectivity of Feistel networks and the fact that the collision of two successive intermediate values implies a full-state collision (but it may happen that $i_3 = i_1$);
- $Y_{i_4} \neq Y_{i_2}$, since otherwise $X_{i_4} = X_{i_2}$, which is impossible as $i_4 \neq i_2$.

We now divide our analysis into several sub-cases.

1. All four indices are pairwise distinct. The equality $Y_{i_4} = Y_{i_3}$ happens with probability 2^{-n} due to the fact that all queries belong to the same line, which implies that all X values are pairwise distinct, and thus all Y values are uniformly random and independent. Now, the equality $X_{i_4} \oplus f_3(Y_{i_4}) = X_{i_2} \oplus f_3(Y_{i_2})$ becomes $X_{i_4} \oplus f_3(Y_{i_3}) = X_{i_2} \oplus f_3(Y_{i_2})$ and three possible cases can occur:

   a. Y_{i_3} is fresh, which means that the equality occurs with probability 2^{-n}. Since $Y_{i_4} = Y_{i_3} \neq Y_{i_1}$, the equality $f_3(Y_{i_4}) = f_3(Y_{i_1})$ occurs with a probability 2^{-n}. Let us now study the equality $Y_{i_4} \oplus f_4(X_{i_4} \oplus f_3(Y_{i_4})) = Y_{i_1} \oplus f_4(X_{i_1} \oplus f_3(Y_{i_1}))$, which is equivalent to $Y_{i_3} \oplus f_4(X_{i_2} \oplus f_3(Y_{i_2})) = Y_{i_1} \oplus f_4(X_{i_1} \oplus f_3(Y_{i_1}))$. Due to the fact that $Y_{i_3} \neq Y_{i_2}$, this is impossible in the case where both inputs to f_4 are equal. Hence, the inputs have to be different (i.e., $Z_{i_1} \neq Z_{i_2}$) and the collision happens with a probability smaller than 2^{-n}. Finally, we need to consider the condition $L_{i_1} \oplus S_{i_1} = L_{i_2} \oplus S_{i_2} = L_{i_3} \oplus S_{i_3}$. We start by studying the equality

   $$f_3(Y_{i_1}) \oplus f_5(U_{i_1}) = f_3(Y_{i_2}) \oplus f_5(U_{i_2}).$$

   If $U_{i_1} \neq U_{i_2}$, this equality happens with a probability 2^{-n}. On the contrary, it means that $Z_{i_2} \neq Z_{i_1}$, i.e., and $U_{i_1} = U_{i_2}$ becomes $Y_{i_1} \oplus Y_{i_2} = f_4(Z_{i_1}) \oplus f_4(Z_{i_2})$, which is also equal to $Y_{i_3} \oplus Y_{i_1}$. This would imply that $Y_{i_2} = Y_{i_3}$, which is impossible. For the equality

   $$f_3(Y_{i_1}) \oplus f_5(U_{i_1}) = f_3(Y_{i_3}) \oplus f_5(U_{i_3}),$$

   we need to consider three sub-cases.

      i. If U_{i_3} is fresh, it occurs with probability 2^{-n}.
      ii. $U_{i_3} = U_{i_1}$: this is equivalent to $Y_{i_3} \oplus f_4(Z_{i_3}) = Y_{i_1} \oplus f_4(Z_{i_1})$, and in that case $L_{i_1} \oplus S_{i_1} = L_{i_3} \oplus S_{i_3}$ is automatically satisfied as $f_3(Y_{i_3}) =$

$f_3(Y_{i_4}) = f_3(Y_{i_1})$. Note that $Z_{i_3} = X_{i_3} \oplus f_3(Y_{i_3}) \neq X_{i_1} \oplus f_3(Y_{i_1}) = Z_{i_1}$ since $f_3(Y_{i_4}) = f_3(Y_{i_3}) = f_3(Y_{i_1})$ and $X_{i_1} \neq X_{i_3}$ (because $i_1 \neq i_3$). Moreover, one has $X_{i_4} \oplus X_{i_2} = f_3(Y_{i_3}) \oplus f_3(Y_{i_2}) = Z_{i_3} \oplus X_{i_3} \oplus Z_{i_2} \oplus X_{i_2}$, which implies that $Z_{i_3} \neq Z_{i_2}$ since $i_3 \neq i_4$. Thus, this case also happens with probability 2^{-n}.

iii. $U_{i_3} = U_{i_2}$: as in the previous case, it happens with a probability smaller than 2^{-n}.

b. $Y_{i_3} = Y_{i_2}$: this implies that $Y_{i_4} = Y_{i_2}$, which is impossible.

c. $Y_{i_3} = Y_{i_1}$: this happens with probability 2^{-n}, and also implies that $f_3(Y_{i_4}) = f_3(Y_{i_1})$, which automatically yields $L_{i_4} \oplus S_{i_4} = L_{i_1} \oplus S_{i_1}$. Moreover, since $Y_{i_1} (= Y_{i_4}) \neq Y_{i_2}$, the equality $X_{i_4} \oplus f_3(Y_{i_3}) = X_{i_2} \oplus f_3(Y_{i_2})$ becomes $X_{i_4} \oplus f_3(Y_{i_1}) = X_{i_2} \oplus f_3(Y_{i_2})$ and happens with probability 2^{-n}. Let us now study the equality $Y_{i_4} \oplus f_4(X_{i_4} \oplus f_3(Y_{i_4})) = Y_{i_1} \oplus f_4(X_{i_1} \oplus f_3(Y_{i_1}))$, which is equivalent to $f_4(X_{i_2} \oplus f_3(Y_{i_2})) = f_4(X_{i_1} \oplus f_3(Y_{i_1}))$ (i.e., $f_4(Z_{i_1}) = f_4(Z_{i_2})$). Due to the previous equality, one necessarily has $X_{i_2} \oplus f_3(Y_{i_2}) \neq X_{i_1} \oplus f_3(Y_{i_1})$, since otherwise we would have $X_{i_1} = X_{i_4}$. Hence, the equality also occurs with probability 2^{-n}. Since $Y_{i_1} \neq Y_{i_2}$, the equality $f_4(X_{i_2} \oplus f_3(Y_{i_2})) = f_4(X_{i_1} \oplus f_3(Y_{i_1}))$ gives $Z_{i_1} \neq Z_{i_2}$. Let us now consider the equality $f_3(Y_{i_1}) \oplus f_5(U_{i_1}) = f_3(Y_{i_2}) \oplus f_5(U_{i_2})$. If $U_{i_1} \neq U_{i_2}$, this happens with probability 2^{-n}. On the contrary, it means that $Y_{i_1} \oplus Y_{i_2} = f_4(Z_{i_1}) \oplus f_4(Z_{i_2}) = 0$, which is impossible. For the equality

$$f_3(Y_{i_1}) \oplus f_5(U_{i_1}) = f_3(Y_{i_3}) \oplus f_5(U_{i_3}),$$

we need to consider three sub-cases:

i. If U_{i_3} is fresh, it occurs with probability 2^{-n}.

ii. $U_{i_3} = U_{i_1}$: this is equivalent to $f_4(Z_{i_3}) = f_4(Z_{i_1})$, and in that case $L_{i_1} \oplus S_{i_1} = L_{i_3} \oplus S_{i_3}$ is automatically satisfied. Note that $Z_{i_3} = X_{i_3} \oplus f_3(Y_{i_3}) \neq X_{i_1} \oplus f_3(Y_{i_1}) = Z_{i_1}$ since $f_3(Y_{i_4}) = f_3(Y_{i_3}) = f_3(Y_{i_1})$ and $X_{i_1} \neq X_{i_3}$ (because $i_1 \neq i_3$). Moreover, one has $X_{i_4} \oplus X_{i_2} = f_3(Y_{i_3}) \oplus f_3(Y_{i_2}) = Z_{i_3} \oplus X_{i_3} \oplus Z_{i_2} \oplus X_{i_2}$, which implies that $Z_{i_3} \neq Z_{i_2}$ since $i_3 \neq i_4$. Thus, this case also happens with probability 2^{-n}.

iii. $U_{i_3} = U_{i_2}$: as in the previous case, it happens with a probability smaller than 2^{-n}.

2. One has $i_3 = i_1$. Then, the equality $Y_{i_4} = Y_{i_3}$ happens with probability 2^{-n}. However, in this case, $f_3(Y_{i_4}) = f_3(Y_{i_1})$ is automatic. As in the previous case, the equality $X_{i_4} \oplus f_3(Y_{i_4}) = X_{i_2} \oplus f_3(Y_{i_2})$ becomes $X_{i_4} \oplus f_3(Y_{i_1}) = X_{i_2} \oplus f_3(Y_{i_2})$, with $Y_{i_1} \neq Y_{i_2}$ due to the fact that $Y_{i_1} = Y_{i_3} = Y_{i_4} \neq Y_{i_2}$. Hence, this equality occurs with a probability 2^{-n}. Let us now study the equality $Y_{i_4} \oplus f_4(X_{i_4} \oplus f_3(Y_{i_4})) = Y_{i_1} \oplus f_4(X_{i_1} \oplus f_3(Y_{i_1}))$, which is equivalent to $f_4(X_{i_2} \oplus f_3(Y_{i_2})) = f_4(X_{i_1} \oplus f_3(Y_{i_1}))$ (i.e., $f_4(Z_{i_1}) = f_4(Z_{i_2})$). Due to the fact that $X_{i_4} \oplus X_{i_2} =$

$f_3(Y_{i_1}) \oplus f_3(Y_{i_2})$, both inputs to f_4 have to be different (i.e., $Z_{i_1} \neq Z_{i_2}$) and the collision happens with a probability smaller than 2^{-n}. Finally, we need to consider the condition $L_{i_1} \oplus S_{i_1} = L_{i_2} \oplus S_{i_2}$. Since $f_4(Z_{i_1}) = f_4(Z_{i_2})$ and $Y_{i_1} \neq Y_{i_2}$, one has $U_{i_1} \neq U_{i_2}$, which means that this equation happens with probability 2^{-n}.

Final bound
Overall, the number of possible choices for the indices of the first case is at most $2^{2n} \times 2^{3n}$: there are 2^n choices for i_1, and then 2^n choices for the three remaining indices since the R value has to be constant. Hence, the probability of the first case adds a query to a free line is at most $6/2^n$, which is also the average number of such a tuple of indices. Similarly, for the second case, there are at most $2^{2n} \times 2^{2n}$ possible choices for the indices, which means that the average number of such triples of queries is at most 1. Markov's inequality yields

$$\Pr[N \geq 4n] \leq \frac{1 + \frac{6}{2^n}}{4n}.$$

Remark 11.9 Although the current bound on ℓ in the Feistel case is sufficient to prove that the attack works, we point out that the actual bound should be much sharper. Using a careful analysis of N, it should be possible to prove that $\Pr[N \geq 4n] \leq \widetilde{\mathcal{O}}\left(\frac{1}{2^n}\right)$. We support this conjecture by providing computer simulations in Table 11.3. As expected, the number of occurrences of the three-point structure from case 2 never grows beyond $2n$. In fact, for $n = 7$ we have an average of 0.9638 and a standard deviation of 1.0334. For $n = 8$ we have an average of 1.0034 and a standard deviation of 1.0226.

Table 11.3 Number of occurrences of the three-point structure from case 2, for $n = 7, 8$, and 10 000 trials

Number of solutions	n = 7	n = 8
0	3982	3743
1	3521	3594
2	1660	1832
3	616	608
4	157	174
5	54	39
6	7	9
7	1	1
8	2	0

11.7 Security Results Against Quantum Attacks

Very few results are known and are present on Feistel schemes against quantum attack. More precisely the only result that we are aware of is: For all non adaptative QCPA the advantage is always lower or equal to $3\sqrt{\frac{q^6}{2^n}} + 6\sqrt{\frac{q^5}{2^n}}$ where q is the number of quantum queries and $2n$ is the size of inputs and outputs.

Therefore for non-adaptive QCPA the quantum query complexity is always greater or equal to $\mathcal{O}(2^{n/6})$.

References

Ambainis, A. 2004. Quantum walk algorithm for element distinctness. In *45th symposium on foundations of computer science (FOCS 2004)*. Proceedings, 22–31. Rome, Italy: IEEE Computer Society.

Chartouny, M., B. Cogliati, and J. Patarin. 2024. Classical and quantum generic attacks on 6-round feistel schemes. In *Selected areas in cryptography—SAC 2024—31st international conference, revised selected papers, Part II*, ed. M. Eichlseder and S. Gambs. volume 15517 of Lecture notes in computer science, 85–111. Berlin: Springer.

Chartouny, M., J. Patarin, and A. Toulemonde. 2023. Quantum cryptanalysis of 5 rounds feistel schemes and benes schemes. In *Codes, cryptology and information security—4th international conference, C2SI 2023, Proceedings*, ed. S. E. Hajji, S. Mesnager, and E. M. Souidi. volume 13874 of Lecture notes in computer science, 196–203. Berlin: Springer.

Chen, S., R. Lampe, J. Lee, Y. Seurin, and J. Steinberger. 2014. Minimizing the two-round even-mansour cipher. In *Advances in cryptology—CRYPTO 2014*, ed. J. A. Garay and R. Gennaro, 39–56. Berlin: Springer.

Childs, A. M., and J. M. Eisenberg. 2005. Quantum algorithms for subset finding. *Quantum Information and Computation* 5 (7): 593–604.

Grover, L. K. 1996. A fast quantum mechanical algorithm for database search. In *Proceedings of the twenty-eighth annual ACM symposium on theory of computing*, 212–219.

Ito, G., A. Hosoyamada, R. Matsumoto, Y. Sasaki, and T. Iwata. 2019. Quantum chosen-ciphertext attacks against feistel ciphers. In *Topics in cryptology—CT-RSA 2019—The cryptographers' track at the RSA conference 2019, proceedings*. Volume 11405 of Lecture notes in computer science, 391–411. Berlin: Springer.

Kaplan, M., G. Leurent, A. Leverrier, and M. Naya-Plasencia. 2016. Breaking symmetric cryptosystems using quantum period finding. In *Advances in cryptology—CRYPTO 2016—36th annual international cryptology conference, Proceedings, Part II*, ed. M. Robshaw and J. Katz. Volume 9815 of Lecture notes in computer science, 207–237. Berlin: Springer.

Kuwakado, H., and M. Morii. 2010. Quantum distinguisher between the 3-round feistel cipher and the random permutation. In *IEEE international symposium on information theory, ISIT 2010, Proceedings*, 2682–2685. IEEE.

Patarin, J. 1991. *Etude des generateurs de permutations pseudo-aleatoires bases sur le schema du d. E. S.* Ph.D. thesis.

Patarin, J. 2001. eneric attacks on feistel schemes. In *Advances in cryptology—ASIACRYPT 2001, 7th international conference on the theory and application of cryptology and information security, Proceedings*, ed. C. Boyd. Volume 2248 of Lecture notes in computer science, 222–238. Berlin: Springer.

Patarin, J. 2001. Generic attacks on feistel schemes. In *Advances in cryptology—ASIACRYPT 2001, 7th international conference on the theory and application of cryptology and information security, Proceedings*. Volume 2248 of Lecture notes in computer science, 222–238. Berlin: Springer.

Pugh, W. 1990. Skip lists: a probabilistic alternative to balanced trees. *Communications of the ACM* 33 (6): 668–676.

Zhandry, M. 2015. A note on the quantum collision and set equality problems. *Quantum Information and Computation* 15 (7–8): 557–567.

Part III
DES and Other Specific Feistel Ciphers

Chapter 12
DES and Variants: 3*DES*, *DES* − *X*

Abstract In this chapter, we will describe briefly the *DES* schemes and its main variants. Then we will present the known crytpanalysis results on these schemes.

12.1 Description

The data encryption standard, or *DES*, is the first secret-key algorithm that was completely published. Indeed, previously, algorithms were kept secret, even though, according to Kerckhoff's principle, they had to be secure after publication (when the keys are kept secret but the algorithm becomes public). *DES* was described in 1977 and was adopted as a standard by the National Institute of Standards and Technology or NIST. It was the most widely used secret-key algorithm between 1977 and 1990, with a huge number of civil applications. Since it uses a secret key with only 56 bits, *DES* is not currently enough secure, but some variants of *DES* (like 3*DES* for example) that use larger keys are still used, particularly in the banking sector with smart cards.

12.1.1 General Description of DES

DES is a Feistel cipher with 16 rounds (see Fig. 12.1) that encrypts blocks of 64 bits with a 56-bit key. The 16-round functions $F_1, \ldots, F_{16}$ are from 32 bits to 32 bits. They are described below. These functions are not pseudo-random but they can be computed very quickly. Thus in some way, the construction principle of *DES* is not based on a generic Feistel cipher (random Feistel cipher) with 6 rounds but 16 rounds and functions that can be computed rapidly: the relatively large number of rounds is supposed to compensate the relative weakness of the functions. The functions $F_1, \ldots, F_{16}$ have been chosen with care as we will see.

J. Patarin et al., *Guide to Feistel Ciphers*, Information Security and Cryptography,
https://doi.org/10.1007/978-3-031-99850-8_12

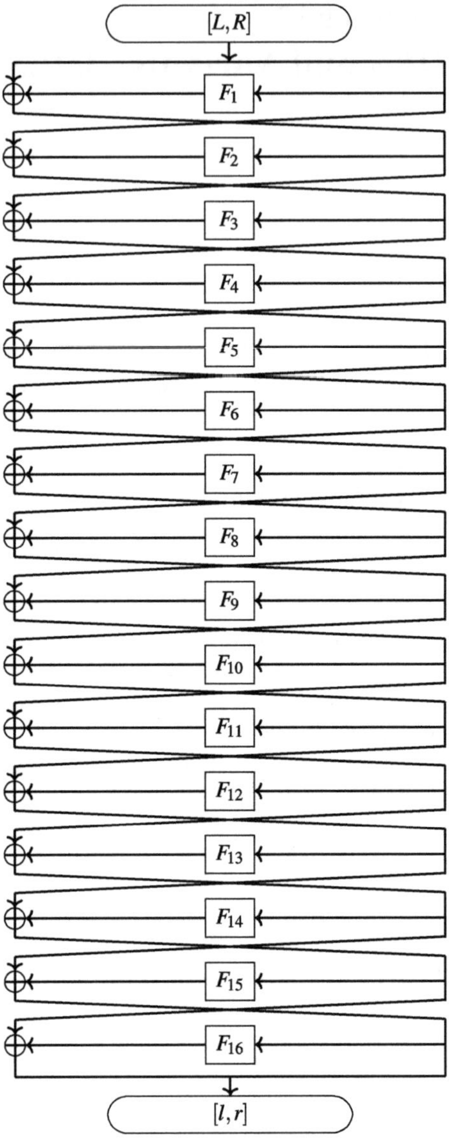

Fig. 12.1 16 rounds of *DES*

12.1.2 Design of the Functions F_i

Figure 12.2 shows how the F_i, $1 \leq i \leq 16$ functions are computed.

We have to perform the following computations:

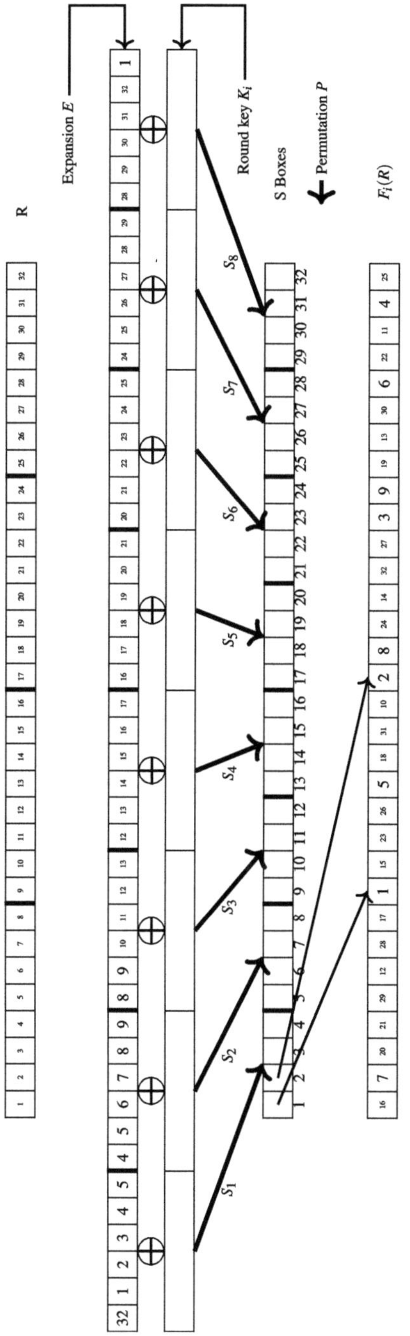

Fig. 12.2 Computation of $F_i(R)$

12.1.2.1 The Expansion Function E

Half of the bits of R are duplicated (see Fig. 12.2) in order to obtain 48 bits. For example, bits 1, 4, 5, 8, 9 are duplicated but this is not the case for bits 2 and 3. Notice that this expansion function E is "linear," which means that each output bit $Y_1, \ldots, Y_{48}$ can be computed with a polynomial of degree 1 involving the input bits $R_1, \ldots R_{32}$. For example, we have $Y_1 = R_{32}, Y_2 = R_1, Y_3 = R_2$. This expansion function is public and each output bit depends on only one input bit.

12.1.2.2 XOR with Subkeys and Key Schedule

The 48-bit output of the expansion function is then XORed (bitwise addition) with the subkey derived from the 56-bit master key K. The key schedule for the 16 subkeys (one at each round) is described in Table 12.1. Notice that each bit of a subkey is exactly equal to a bit of K. Moreover, although we have a 56-bit key, we consider that we have a 64-bit key where bits placed at positions 0, 8, 16, 24, 32, 40, 48, and 56 are not used.

12.1.2.3 S-Boxes

The 48 bits obtained after the XOR with a subkey are separated into eight blocks of 6 bits $A_1, A_2, A_3, A_4, A_5, A_6, A_7, A_8$, and each of these blocks goes through 8 S-boxes which are given in Table 12.2. Each S-box maps 6 bits to 4 bits as follows. We now explain how the S-box S_j maps $A_j = a_1^j a_2^j a_3^j a_4^j a_5^j a_6^j$. The 2 bits $a_1^j a_6^j$ determine the binary representation of a number r, $0 \le r \le 3$ which gives the row of the array S_j. Similarly the 4 bits $a_2^j a_3^j a_4^j a_5^j$ determine the binary representation of a number c, $0 \le c \le 15$ which gives the column of the array S_j. Then $S_j(A_j)$ is the entry $S_j(r, c)$ written in binary as bit string of length 4. For example, suppose that with S_1, we want to compute the value $S_1(101100)$. Then $a_1^1 a_6^1 = 10$. This gives the row $r = 2$. Then $a_2^1 a_3^1 a_4^1 a_5^1 = 0110$ and we get $c = 6$. Thus $S_1(101100)$ is the binary representation of $S_1(2, 6) = 2$, i.e., $S_1(101100) = 0010$.

12.1.2.4 The Permutation *P*

The last applied transformation is a permutation P that will modify the position of the 32 bits (see Fig. 12.2). We have $P(a_1, a_2, \ldots, a_{32}) = (b_1, b_2, \ldots, b_{32})$ with $b_1 = a_{17}, b_2 = a_7, \ldots, b_{32} = a_{25}$. Notice that this permutation P is linear, public, and each bit b_i depends on only 1 bit a_j. However, this permutation plays an important role for the security of DES since thanks to this permutation, the 4 output bits of a S-box will be sent to four different S-boxes at the next round. Due to this, an

Table 12.1 Key schedule

Round 1											
10	51	34	60	49	17	33	57	2	9	19	42
3	35	26	25	44	58	59	1	36	27	18	41
22	28	39	54	37	4	47	30	5	53	23	29
61	21	38	63	15	20	45	14	13	62	55	31

Round 2											
2	43	26	52	41	9	25	49	59	1	11	14
60	27	18	17	36	50	51	58	57	19	10	33
14	20	31	46	29	63	39	22	28	45	15	21
53	13	30	55	7	12	37	6	5	54	47	23

Round 3											
51	27	10	36	25	58	9	33	43	50	60	18
44	11	2	1	49	34	35	42	41	3	59	17
61	4	15	30	13	47	23	6	12	29	62	5
37	28	14	39	54	63	21	53	20	38	31	7

Round 4											
35	11	59	49	9	42	58	17	27	34	44	2
57	60	51	50	33	18	19	26	25	52	43	1
45	55	62	14	28	31	7	53	63	13	46	20
21	12	61	23	38	47	5	37	4	22	15	54

Round 5											
19	60	43	33	58	26	42	1	11	18	57	51
41	44	35	34	17	2	3	10	9	36	27	50
29	39	46	61	12	15	54	37	47	28	30	4
5	63	45	7	22	31	20	21	55	6	62	38

Round 6											
3	44	27	17	42	10	26	50	60	2	41	35
25	57	19	18	1	51	52	59	58	49	11	34
13	23	30	45	63	62	38	21	31	12	14	55
20	47	29	54	6	15	4	5	39	53	46	22

Round 7											
52	57	11	1	26	59	10	34	44	51	25	19
9	41	3	2	50	35	36	43	42	33	60	18
28	7	14	29	47	46	22	5	15	63	61	39
4	31	13	38	53	62	55	20	23	37	30	6

Round 8											
36	41	60	50	10	43	59	18	57	35	9	3
58	25	52	51	34	19	49	27	26	17	44	2
12	54	61	13	31	30	6	20	62	47	45	23
55	15	28	22	37	46	39	4	7	21	14	53

Round 9											
57	33	52	42	2	35	51	10	49	27	1	60
50	17	44	43	26	11	41	19	18	9	36	59
4	46	53	5	23	22	61	12	54	39	37	15
47	7	20	14	29	38	31	63	62	13	6	45

Round 10											
41	17	36	26	51	19	35	59	33	11	50	44
34	1	57	27	10	60	25	3	2	58	49	43
55	30	37	20	7	6	45	63	38	23	21	62
31	54	4	61	13	22	15	47	46	28	53	29

Round 11											
25	1	49	10	35	3	19	43	17	60	34	57
18	50	41	11	59	44	9	52	51	42	33	27
39	14	21	4	54	53	29	47	22	7	5	46
15	38	55	45	28	6	62	31	30	12	37	13

Round 12											
9	50	33	59	19	52	3	27	1	44	18	41
2	34	25	60	43	57	58	36	35	26	17	11
23	61	5	55	38	37	13	31	6	54	20	30
62	22	39	29	12	53	46	15	14	63	21	28

Round 13											
58	34	17	43	3	36	52	11	50	57	2	25
51	18	9	44	27	41	42	49	19	10	1	60
7	45	20	39	22	21	28	15	53	38	4	14
46	6	23	13	63	37	30	62	61	47	5	12

Round 14											
42	18	1	27	52	49	36	60	34	41	51	9
35	2	58	57	11	25	26	33	3	59	50	44
54	29	4	23	6	5	12	62	37	22	55	61
30	53	7	28	47	21	14	46	45	31	20	63

Round 15											
26	2	50	11	36	33	49	44	18	25	35	58
19	51	42	41	60	9	10	17	52	43	34	57
38	13	55	7	53	20	63	46	21	6	39	45
14	37	54	12	31	5	61	30	29	15	4	47

Round 16											
18	59	42	3	57	25	41	36	10	17	27	50
11	43	34	33	52	1	2	9	44	35	26	49
30	5	47	62	45	12	55	38	13	61	31	37
6	29	46	4	23	28	53	22	21	7	63	39

Table 12.2 DES S-Boxes

S-Box 1

14	4	13	1	2	15	11	8	3	10	6	12	5	9	0	7
0	15	7	4	14	2	13	1	10	6	12	11	9	5	3	8
4	1	14	8	13	6	2	11	15	12	9	7	3	10	5	0
15	12	8	2	4	9	1	7	5	11	3	14	10	0	6	13

S-Box 2

15	1	8	14	6	11	3	4	9	7	2	13	12	0	5	10
3	13	4	7	15	2	8	14	12	0	1	10	6	9	11	5
0	14	7	11	10	4	13	1	5	8	12	6	9	3	2	15
13	8	10	1	3	15	4	2	11	6	7	12	0	5	14	9

S-Box 3

10	0	9	14	6	3	15	5	1	13	12	7	11	4	2	8
13	7	0	9	3	4	6	10	2	8	5	14	12	11	15	1
13	6	4	0	8	15	3	0	11	1	2	12	5	10	14	7
1	10	13	0	6	9	8	7	4	15	14	3	11	5	2	12

S-Box 4

7	13	14	3	0	6	9	10	1	2	8	5	11	12	4	15
13	8	11	5	6	15	0	3	4	7	2	12	1	10	14	9
10	6	9	0	12	11	7	13	15	1	3	14	5	2	8	4
3	15	0	6	10	1	13	8	9	4	5	11	12	7	2	14

S-Box 5

2	12	4	1	7	10	11	6	8	5	3	15	13	0	14	9
14	11	2	12	4	7	13	1	5	0	15	10	3	9	8	6
4	2	1	11	10	13	7	8	15	9	12	5	6	3	0	14
11	8	12	7	1	14	2	13	6	15	0	9	10	4	5	3

S-Box 6

12	1	10	15	9	2	6	8	0	13	3	4	14	7	5	11
10	15	4	2	7	12	9	5	6	1	13	14	0	11	3	8
9	14	15	5	2	8	12	3	7	0	4	10	1	13	11	6
4	3	2	12	9	5	15	10	11	14	1	7	6	0	8	13

S-Box 7

4	11	2	14	15	0	8	13	3	12	9	7	5	10	6	1
13	0	11	7	4	9	1	10	14	3	5	12	2	15	8	6
1	4	11	13	12	3	7	14	10	15	6	8	0	5	9	2
6	11	13	8	1	4	10	7	9	5	0	15	14	2	3	12

S-Box 8

13	2	8	4	6	15	11	1	10	9	3	14	5	0	12	7
1	15	13	8	10	3	7	4	12	5	6	11	0	14	9	2
7	11	4	1	9	12	14	2	0	6	10	13	15	3	5	8
2	1	14	7	4	10	8	13	15	12	9	0	3	5	6	11

"avalanche effect" is expected to occur: a small change between two inputs should generate a large change after several rounds.

12.2 Simple *DES*

12.2.1 Presentation

DES (data encryption standard) is the famous secret-key algorithm published in 1976, that we have just seen above. It uses only a 56-bit key and encrypts blocks of 64 bits. *DES* features are summarized in Table 12.3.

We now describe best known attacks on simple *DES*.

12.2.2 Brute Force Attack

This is a KPA with $q = 1$ plaintext/ciphertext pair, 2^{56} computations and very little memory (we just have to be able to compute a DES, to be able to increase the key by 1, and to be able to check if the result is a given value). It is a generic attack that applies on any algorithm with a key that has only 56 bits. This attack was unrealistic in 1976 when *DES* was published. However, now computers are much more powerful than in 1976 and now this attack can be done (and was done, cf below). In fact it is at present the simplest to mount attack on simple *DES*.

This attack was done for the first time on July 15, 1998. Distributed.net and the Electronic Frontier Foundation broke a DES key by using brute force in 56 h. On January 1999, they broke another DES key challenge in 22 h. For this, they used a machine built in collaboration with the society Cryptography Research. The machine (nicknamed "Deep Crack") costs less than $ 250 000.

We will see below the three best known attacks on the DES with a complexity smaller than the exhaustive search of the key (however the exhaustive search of the key is still the simplest way to find the key since it requires only one cleartext/ciphertext). These attacks are very interesting since they can be used on many other schemes than the DES. Interestingly the weakest S-boxes are not the same for these three attacks:

Table 12.3 Simple *DES* features

Name	Simple *DES*
Input and output block size	64 bits
Key size	56 bits
Scheme type	16 round Feistel scheme
Computation time	$\leq 10^{-3}$ s on a smart card

they are boxes S_1-S_2-S_3 for differential cryptanalysis (and the change of the other S-boxes will almost not change the complexity of differential cryptanalysis of the DES), the box S_5 for linear cryptanalysis (however linear cryptanalysis do not only use S_5) and the boxes S_7-S_8 for Davies attack (and the change of the other S-boxes will almost not change the complexity of Davies attack). Differential cryptanalysis is a chosen plaintext attack while linear and Davies attack are just known plaintext attacks. However differential cryptanalysis can be implemented on DES with different keys (with almost the same total complexity if we send about 2^24 messages for each keys) while for linear and Davies attack we need to have messages from the same key in order to have about the same complexity.

12.2.3 Differential Cryptanalysis of DES

We present here a short explanation about the differential cryptanalysis of DES. More explanations can be found in the original paper (Biham and Shamir 1991).

On the full 16-round DES differential cryptanalysis will give a CPA (non-interactive chosen plaintext attack) that will recover the secret 56-bit Key K of the DES after about 2^{47} computations if can ask for the encryption of about 2^{47} chosen messages. So differential cryptanalysis of the DES seems to be less efficient than linear cryptanalysis of the DES, because with linear cryptanalysis we need about 2^{43} computations (instead of 2^{47}) and the encryption of about 2^{43} message (instead of 2^{47}). Moreover linear cryptanalysis is a KPA (known plaintext attack) while differential cryptanalysis is a CPA (chosen plaintext attack). This is why in the first edition of this book we did not give many information about differential cryptanalysis of the DES.

However, it is important to notice that in many real situations differential cryptanalysis may be easier to implement than linear cryptanalysis. This is because in linear cryptanalysis of the DES you need to have access to 2^{43} cleartext/ciphertext pairs encrypted with the same secret key K to recover K. While in differential cryptanalysis of the DES you need to have access to 2^{47} ciphertext of chosen plaintexts, but they do not need to be encrypted with the same key K. Each time that you will find a (cleartext/ciphertext) pair with the good differentials, then you will recover the corresponding secret key. This is why we will give more information on (the very famous) differential cryptanalysis of the DES in this second edition. (Due to the increases of computing power of modern computers, the easiest attack on the DES is however now generally just to try all the possible keys, i.e., to perform 2^{56} computations on a single cleartext/ciphertext pair).

The basic idea of differential cryptanalysis is to find values A and B such that the probability that $F(x \oplus A) \oplus F(x) = B$ will be much higher than for average values A and B. So we will select pairs of cleartexts $(x, x \oplus A)$ and ask for their corresponding ciphertexts (y, z). If $y \oplus z = B$, then a precise analysis of these two messages will give information on the secret key. Let us see what this idea gives on the full 16-round DES.

Theorem 12.1 *Let* (L, R) *be the input of a DES round function (for example round* k*). Let* f_k *be the functions such that if* (L, R) *is the input of this round the value after this round becomes* $(R, L \oplus f_k(R))$*. Let* ψ *be this specific value of 32 bits:* $\psi = 0001\ \ 1001\ \ 0110$ *and the next 20 bits of* ψ *are 0. Then the probability to have* $f_k(R) = f_k(R \oplus \psi)$ *is about* $1/234$.

Proof After the expansion E the differential becomes

$$\psi' = 000011\ \ 110010\ \ 101100$$

and then only 0 values.

The differential 000011 will go to S-box S_1. Therefore the differential of the output of S_1 will be 0000 with probability $14/64$. (See below the details on S-box S_1.)

The differential 110010 will go to S-box S_2. Therefore the differential of the output of S_2 will be 0000 with probability $8/64$. (See below the details on S-box S_2.)

The differential 101100 will go to S-box S_3. Therefore the differential of the output of S_3 will be 0000 with probability $10/64$. (See below the details on S-box S_3.)

The differential 000000 will go to S-boxes S_4, S_5, S_6, S_7, and S_8. Therefore the differential of the output of S_4, S_5, S_6, S_7, and S_8 is 0000 with probability 1 (because the same input always gives the same output).

We see that the probability to have $f_k(R) = f_k(R \oplus \psi)$ is about $(14/64) \cdot (8/64) \cdot (10/64)$, i.e., about $1/234$ as claimed. □

Theorem 12.2 *Let L, R, L, and R be values of 32 bits. Let* ψ *be this value of 32 bits:* $\psi = 0001\ \ 1001\ \ 0110$ *and the next 20 bits of* ψ *are 0. Let* F *be 2 rounds of DES. Let* $(S, T) = F(L, R)$ *and* $(S, T) = F(L, R)$*. Then if* $L = L and R = R \oplus \psi$*, the probability that* $S = S$ *and* $T = T \oplus \psi$ *is about* $1/234$.

Proof If the input is (L, R): the value after 1 round is: $(R, L \oplus f_k(R))$, the value after 2 rounds is: $(L \oplus f_k(R), R \oplus f_{k+1}(L \oplus f_k(R)))$. From Theorem 12.1 above we know that the probability to have $f_k(R) = f_k(R \oplus \psi)$ is about $1/234$. Now if $f_k(R) = f_{k+1}(R)$ then after 2 rounds we see immediately that we have $S = S$ (since $L = L$) and $T = T \oplus \psi$ (since $R = R \oplus \psi$).

12.2.3.1 Differential Cryptanalysis of 8-Round *DES*

Before analyzing the full 16-round DES let us see what we could obtain after Theorem 12.2 if we had only 8 rounds. We would have to use Theorem 12.2 four times (to have 8 rounds), so from an initial differential of ψ we would obtain a final differential of also ψ with a probability about $(1/234)^4$, i.e., about 2^{31}.

So we will ask for the encryption of about 2^{31} pairs of messages (L, R), $(L, R \oplus A)$ in order to find one of these pairs with an output differential of ψ. When this occurs, then we concentrate the analysis on this single pair (all the other cleartext/ciphertext pairs will not be used).

There is a very good probability that we have obtained such a differential not due to random values but thanks to the fact that for the 4 rounds involves we had $f_k(R) = f_k(R \oplus \psi)$. (Because for random outputs the probability to have the differential ψ at the end is $1/2^{64}$ while for the four equations $f_k(R) = f_k(R \oplus \psi)$ it is about $1/2^{31}$).

However a detailed analysis of the good DES subkeys that give these four equations will reduce the number of possible DES key by a factor of about 2^{31}. Thus this will give an attack in 2^{31} instead of 2^{56} for 8-round DES. (Even better complexities can also be obtained on 8-round DES, for example, by using some ideas that we will now introduce for the full DES.)

12.2.3.2 Differential Cryptanalysis of the Full 16-Round DES

If we want to attack the full 16-round DES as we did above for 8 rounds, then we would need about 234^8 message, i.e., about 2^{63} messages and 2^{63} computations. This is worse than the complexity 2^{56} of exhaustive search. This certainly comes from the fact that the conceptors of the DES (including Don Coppermith) new about differential cryptanalysis and that they have chosen specific S-boxes to avoid such attacks. (It is possible to show that for random S-boxes much more efficient differential cryptanalysis exist with a very high probability).

Nevertheless Eli Biham and Adi Shamir discovered a way to improve their differential cryptanalysis on the full DES in order to obtain a complexity of 2^{47} messages and computations (instead of 2^{63}). We will quickly present here the main ideas for this improvement (see the paper of Biham and Shamir for more details). Essentially the idea will be that instead of eight equations of the form $f_k(R) = f_k(R \oplus A)$ we will have to use only six of them, the first one and the last will be, in a way, free.

12.2.3.3 Ideas to Remove the Last Equation $f_k(R) = f_k(R \oplus \psi)$

If we do not have the last equation $f_{15}(R) = f_{15}(R \oplus \psi)$ (i.e., if we do not have a differential of 0 after f_{15}) then let w be the differential after f_{15}.

This differential w will have at least 20 bits at 0 with known positions, because the bits different from 0 of w can only be bits of the output of S-boxes S_1, S_2, or S_3 permuted by the permutation P.

The differential of the final outputs (after 16 rounds) will be $(f_{16}(w) \oplus \psi, w)$. With high probability at least one of the S-boxes used in the last round will have a differential at 0, therefore we will have at the end a test on 24 bits (20 bits from w, and 4 bits from $f16(w) \oplus \psi$). This test is less efficient than the previous test on 64 bits (if we had the last equation $f_k(R) = f_k(R \oplus \psi)$ nevertheless it will be enough. The probability that a wrong pair (cleartext/ciphertext) succeeds this test is $1/2^{24}$ and we will have 2^{47} pairs, therefore some of them will pass this test. However for these pairs, we have the possibility to test them by performing about 2^{24} computations without degrading the total complexity significatively, and this is in fact enough to test if these solutions are valid or not.

12.2.3.4 Ideas to Remove the First Equation $f_k(R) = f_k(R \oplus \psi)$

We will not assume now that $f_1(R) = f_1(R \oplus \psi)$.

Let then v be the differential after f_1.

This differential v will have at most 12 bits different from 0, and we know the position of these 12 bits (they are the output of S-boxes S_1, S_2, or S_3 permuted by the permutation P).

Now the idea is that instead of sending pairs of messages with a differential of ψ, we will send families of 2^{13} messages of the form $(L \oplus v, R)$ or $((L \oplus v, R \oplus \psi)$. In these families we have all the possibilities for the 12 bits of v (therefore 2^{12} possibilities here), and R or $R \oplus \psi$ (therefore two possibilities here).

Moreover, we will have about 2^{35} families like this (i.e., 2^{34} different values for (L, R)).

From the 2^{13} messages of a family, we can generate 2^{24} pairs of message of the form $((L \oplus v_1, R), ((L \oplus v_2, R \oplus \psi))$.

For all of these pairs we have a probability $1/2^{12}$ that $f_1(R) = f_1(R \oplus \psi) \oplus v_1 \oplus v_2$.

If this occurs, then the differential after round 1 is $(\psi, 0)$ exactly as before (but we did not need here to have $f_1(R) = f_1(R \oplus \psi)$ for that now).

Now if we combine the ideas to first and the last equations $f_k(R) = f_k(R \oplus \psi)$, we will need only six equations like this. The probability that all these six equations are satisfied is $1/234^6 = 2^{-47.2}$ and the probability that $f_1(R) = f_1(R \oplus \psi) \oplus v_1 \oplus v_2$ is $1/2^{12}$. However and we have potentially $2^{24} \cdot 2^{35}$ possible differentials (because we have 2^{35} families). So at the end we will obtain about one solution, as wanted.

Moreover it is possible to find directly the members of a family that has the right differential at the end without testing all the 2^{24} pairs that we can obtain from this family. For this we will store the input at the address of the 24 bits where we do the test explained above when we remove the last equation. Thus we can detect immediately the collisions (i.e., a differential of 0 on these 24 bits).

Finally the total complexity for the full 16-round DES with this differential cryptanalysis is then of 2^{47} chosen messages and 2^{47} computations (Tables 12.4, 12.5, and 12.6).

Table 12.4 Detailed presentation of DES S-box S_1 (in bits and in base 10). (i) denotes one of the 14 values (i.e., 7 pairs) such that: $S_1(x \oplus 000011) = S_1(x)$. (These values will be useful in differential cryptanalysis of the DES)

Input		Output	
000000	0	14	1110
000001	1	0	0000
000010	2	4	0100
000011	3	15	1111
000100	4	13	1101
000101	5	7	0111
000110	6	1	0001
000111	7	4	0100
001000 (1)	8	2	0010
001001	9	14	1110
001010	10	15	1111
001011 (1)	11	2	0010
001100	12	11	1011
001101	13	13	1101
001110	14	8	1000
001111	15	1	0001
010000	16	3	0011
010001 (2)	17	10	1010
010010 (2)	18	10	1010
010011	19	6	0110
010100	20	6	0110
010101 (3)	21	12	1100
010110 (3)	22	12	1100
010111	23	11	1011
011000 (4)	24	5	0101
011001 (5)	25	9	1001
011010 (5)	26	9	1001
011011 (4)	27	5	0101
011100	28	0	0000
011101	29	3	0011
011110	30	7	0111
011111	31	8	1000
100000	32	4	0100
100001	33	15	1111
100010	34	1	0001

(continued)

Table 12.4 (continued)

Input		Output	
100011	35	12	1100
100100	36	14	1110
100101 (6)	37	8	1000
100110 (6)	38	8	1000
100111	39	2	0010
101000	40	13	1101
101001	41	4	0100
101010	42	6	0110
101011	43	9	1001
101100	44	2	0010
101101	45	11	0001
101110	46	11	1011
101111	47	7	0111
110000	48	15	1111
110001	49	5	0101
110010	50	12	1100
110011	51	11	1011
110100	52	9	1001
110101	53	3	0011
110110	54	7	0111
110111	55	14	1110
111000	56	3	0011
111001 (7)	57	10	1010
111010 (7)	58	10	1010
111011	59	0	0000
111100	60	5	0101
111101	61	6	0110
111110	62	0	0000
111111	63	13	1101

12.2.4 Linear Cryptanalysis

This attack was described by Matsui (1993). It is a KPA with 2^{43} plaintext–ciphertext pairs, 2^{43} computations and very little memory. This very famous paper by Matsui, together with Gilbert's paper on FEAL (Tardy-Corfdir and Gilbert 1998), initiated the technique of linear cryptanalysis.

We now present a short explanation about this attack.

Table 12.5 Detailed presentation of DES S-box S_2 (in bits and in base 10). (i) denotes one of the 8 values (i.e., 4 pairs) such that: $S_2(x \oplus 110010) = S_2(x)$ (These values will be useful in differential cryptanalysis of the DES)

Input		Output	
000000	0	15	1111
000001	1	3	0011
000010	2	1	0001
000011	3	13	1101
000100	4	8	1000
000101	5	4	0100
000110	6	14	1110
000111 (1)	7	7	0111
001000	8	6	0110
001001	9	15	1111
001010	10	11	1011
001011	11	2	0010
001100	12	3	0011
001101	13	8	1000
001110	14	4	0100
001111 (2)	15	14	1110
010000	16	9	1001
010001	17	12	1100
010010	18	7	0111
010011	19	0	0000
010100	20	2	0010
010101 (3)	21	1	0001
010110	22	13	1101
010111 (4)	23	10	1010
011000	24	12	1100
011001	25	6	0110
011010	26	0	0000
011011	27	9	1001
011100	28	5	0101
011101	29	11	1011
011110	30	10	1010
011111	31	5	0101
100000	32	0	0000
100001	33	13	1101
100010	34	14	1110

(continued)

Table 12.5 (continued)

Input		Output	
100011	35	8	1000
100100	36	7	0111
100101 (4)	37	10	1010
100110	38	11	1011
100111 (3)	39	1	0001
101000	40	10	1010
101001	41	3	0011
101010	42	4	0100
101011	43	15	1111
101100	44	13	1101
101101	45	4	0100
101110	46	1	0001
101111	47	2	0010
110000	48	5	0101
110001	49	11	1011
110010	50	8	1000
110011	51	6	0110
110100	52	12	1100
110101 (1)	53	7	0111
110110	54	6	0110
110111	55	12	1100
111000	56	9	1001
111001	57	0	0000
111010	58	3	0011
111011	59	5	0101
111100	60	2	0010
111101 (2)	61	14	1110
111110	62	15	1111
111111	63	9	1001

12.2.4.1 Equation A

Let $x_0, x_1, \ldots, x_5$ be the 6 input bits of an S-box, and let $y_0, y_1, \ldots, y_5$ be the 4 output bits.

We choose here the order from the right to the left (this is the order chosen in Matsui's paper), i.e., from low bits to high bits.

Table 12.6 Detailed presentation of DES S-box S_3 (in bits and in base 10). (i) denotes one of the 10 values (i.e., 5 pairs) such that: $S_2(x \oplus 101100) = S_3(x)$ (These values will be useful in differential cryptanalysis of the DES)

Input		Output	
000000	0	10	1010
000001	1	13	1101
000010 (1)	2	0	0000
000011 (2)	3	7	0111
000100	4	9	1001
000101	5	0	0000
000110	6	14	1110
000111 (3)	7	9	1001
001000	8	6	0110
001001	9	3	0011
001010	10	3	0011
001011	11	4	0100
001100	12	15	1111
001101	13	6	0110
001110	14	5	0101
001111 (4)	15	10	1010
010000	16	1	0001
010001 (5)	17	2	0010
010010	18	13	1101
010011	19	8	1000
010100	20	12	1100
010101	21	5	0101
010110	22	7	0111
010111	23	14	1110
011000	24	11	1011
011001	25	12	1100
011010	26	4	0100
011011	27	11	1011
011100	28	2	0010
011101	29	15	1111
011110	30	8	1000
011111	31	1	0001
100000	32	13	1101
100001	33	1	0001
100010	34	6	0110

(continued)

Table 12.6 (continued)

Input		Output	
100011 (4)	35	10	1010
100100	36	4	0100
100101	37	13	1101
100110	38	9	1001
100111	39	0	0000
101000	40	8	1000
101001	41	6	0110
101010	42	15	1111
101011 (3)	43	9	1001
101100	44	3	0011
101101	45	8	1000
101110 (1)	46	0	0000
101111 (2)	47	7	0111
110000	48	11	1011
110001	49	4	0100
110010	50	1	0001
110011	51	15	1111
110100	52	2	0010
110101	53	14	1110
110110	54	12	1100
110111	55	3	0011
111000	56	5	0101
111001	57	11	1011
111010	58	10	1010
111011	59	5	0101
111100	60	14	1110
111101 (5)	61	2	0010
111110	62	7	0111
111111	63	12	1100

By looking at S-box S_5, we can notice that, for exactly 14 of the 64 possible inputs $(x_0, x_1, \ldots, x_5)$, we have

$$x_4 = y_0 \oplus y_1 \oplus y_2 \oplus y_3. \tag{12.1}$$

So this Eq. (12.1) is expected to be true with probability about $12/64 = 0.1875$.

(In average we can expect 32 values among the 64 to satisfy such an equation. It is possible to show that in fact the deviation of $32 - 12 = 20$ that we have here is the largest linear deviation of an S-box of the *DES*.)

Taking into account the E expansion and the P permutation of the *DES*, Eq. (12.1) can be written like this:

$$\text{(Equation A):} \quad x[15] \oplus F(X, K)[7, 18, 24, 29] = K[22],$$

and this equation A will hold with probability 12/64.

This is because bit x_4 of S_5 is input bit number 15 when we start from the left (and from 0), that this bit is XORed with the bit number 22 of the subkey (again we start counting from the left and from 0), and the four output bits of S_5 are in positions 7, 18, 24, and 29 (again when we start counting from the left and from 0).

12.2.4.2 Equations B, C, D, E

Similarly by looking the S-boxes we can notice that we have

$$\text{(Equation B):} X[27, 28, 30, 31] \oplus F(X, K)[15] = K[42, 43, 45, 46] \quad \text{with probability } 22/64,$$

$$\text{(Equation C):} X[29] \oplus F(X, K)[15] = K[44] \quad \text{with probability } 30/64,$$

$$\text{(Equation D):} X[15] \oplus F(X, K)[7, 18, 24] = K[22] \quad \text{with probability } 42/64,$$

$$\text{(Equation E):} X[12, 16] \oplus F(X, K)[7, 18, 24] = K[19, 23] \quad \text{with probability } 16/64.$$

12.2.4.3 Equation for the 16-Round Full *DES*

Let us denote "–" the fact that no approximation is needed in a round. Then by using the equations A, B, C, D, E seen above in that order for the rounds: E-DCA-ACD-DCA-A we will obtain this equation after 15 rounds of *DES*:

$$\begin{aligned} &P_H[7, 18, 24] \oplus P_L[12, 16] \oplus C_H[7, 18, 24, 29] \oplus C_L[15] \\ &= K_1[19, 23] \oplus L_3 \oplus L_7 \oplus L_{11} \oplus K_{15}[22]. \end{aligned} \tag{12.2}$$

Here P_H denotes the 32-bit high values of the plaintext, P_L denotes the 32-bit low values of the plaintext, C_H denotes the 32-bit high values of the ciphertext, and C_L denotes the 32-bit low values of the ciphertext.

By using the pilling-up lemma (see below), the probability for this equation (after 15 rounds) is about $\frac{1}{2} + 1.19 \cdot 2^{-22}$.

(The rounds are not perfectly independent, however computer simulations show that the probability obtained in reality are not far from this, and in fact are even slightly better).

Now by intruding the last round we obtain from Eq. (12.2):

$$\begin{aligned}&P_H[7, 18, 24] \oplus P_L[12, 16] \oplus C_H[15] \oplus C_L[7, 18, 24, 29] \oplus F_{16}(C_L, K_{16})[15]\\ =&K_1[19, 23] \oplus K_3[22] \oplus K_4[44] \oplus K_5[22] \oplus K_7[22] \oplus K_8[44] \oplus K_9[22] \oplus K_{11}[22]\\ &\oplus K_{12}[44] \oplus K_{13}[22] \oplus K15[22].\end{aligned} \quad (12.3)$$

It is this Eq. (12.3) that we will use it the linear cryptanalysis of the full *DES*, with its correcting mirror equation by using the fact that the inverse of a Feistel scheme is another Feistel scheme by introducing the key rounds in inverse order.

In the term $F_{16}(C_L, K_{16})[15]$ of Eq. (12.3), 6 bits of key are used. We will perform exhaustive search on these 6 bits. For the right values of these 6 key bits Eq. (12.3) will then have a probability of about $\frac{1}{2} + 1.19 \cdot 2^{-22}$ (since we then obtain the probability of Eq. (12.2)). However for wrong values this probability is much closer to 1/2.

Therefore, thanks to the theorem below (extraction of information from noise Theorem), if we have access to about 2^{43} cleartext/ciphertext couples, we will be able to find here these 6 bits of key plus $K_1[19, 23] \oplus K_3[22] \oplus K_4[44] \oplus K_5[22] \oplus K_7[22] \oplus K_8[44] \oplus K_9[22] \oplus K_{11}[22] \oplus K_{12}[44] \oplus K_{13}[22] \oplus K_{15}[22]$, i.e., plus 1 bit of information.

Similarly from the mirror equation of Eq. (12.3) we also obtain 7 bits.

Finally we have obtained 14 key bits, and the remaining 42 bits of the secret key can be obtained by exhaustive search.

The total complexity of this KPA attack is therefore in about 2^{43} computations and 2^{43} cleartext/ciphertext couples.

Remark 12.1 In Matsui's paper, a variant of this attack is also given to obtain a ciphertext-only attack (with a slightly worse complexity).

12.2.4.4 Piling-Up Lemma

Theorem 12.3 ("Piling-up Lemma") *Let* $(X_i)_{1\le i\le n}$ *be independent random variables such that* X_i *has value* 1 *with probability* p_i*, and* X_i *has value* 0 *with probability* $1 - p_i$*. Then the probability that* $X_1 \oplus X_2 \cdots \oplus X_n = 0$ *is exactly*

$$\frac{1}{2} + 2^{n-1} \prod_{i=1}^{n} \left(p_i - \frac{1}{2}\right).$$

Proof Let us start with $n = 2$. One has

$$X_1 \oplus X_2 = 0 \iff [(X_1 = 0) \text{ and } (X_2 = 0)] \text{ or } [(X_1 = 1) \text{ and } (X_2 = 1)].$$

Therefore, the probability to have $X_1 \oplus X_2 = 0$ is

$$p_1p_2 + (1-p_1)(1-p_2) = 1 - p_1 - p_2 + 2p_1p_2 = \frac{1}{2} + 2\left(p_1 - \frac{1}{2}\right)\left(p_2 - \frac{1}{2}\right)$$

as claimed.

Now we will prove the pilling-up lemma by induction on n.

Let $Y = X_1 \oplus X_2 \cdots \oplus X_n$. One has

$$\begin{aligned} & X_1 \oplus X_2 \cdots \oplus X_{n+1} = 0 \\ & \quad \Longleftrightarrow Y \oplus X_{n+1} = 0 \\ & \quad \Longleftrightarrow \left[(Y = 0) \text{ and } (X_{n+1} = 0)\right] \text{ or } \left[(Y = 1) \text{ and } (X_{n+1} = 1)\right]. \end{aligned}$$

Therefore, by using the hypothesis that the pilling-up lemma is true for the value n, and by using the result seen above for $n = 2$, we have

$$\begin{aligned} \Pr\left[X_1 \oplus X_2 \cdots \oplus X_{n+1} = 0\right] &= \Pr\left[Y \oplus X_{n+1} = 0\right] \\ &= \frac{1}{2} + 2 \cdot 2^{n-1} \prod_{i=1}^{n} \left(p_i - \frac{1}{2}\right) \cdot \left(p_{n+1} - \frac{1}{2}\right) \\ &= \frac{1}{2} + 2^n \prod_{i=1}^{n+1} \left(p_i - \frac{1}{2}\right), \end{aligned}$$

and this means that the pilling-up lemma is true for $n + 1$. □

12.2.4.5 How to Distinguish Information from Random Noise?

Theorem 12.4 (Extraction of information from noise) *Let $(X_i)_{1 \le i \le n}$ be independent random variables. We are in one of these two situations:*

1. *(Case 1) Each value X_i has value 1 with probability $1/2$, and value 0 with probability $1/2$.*
2. *(Case 2) Or each value X_i takes value 1 with probability $1/2 + \varepsilon$, and value 0 with probability $1/2 - \varepsilon$.*

We want to distinguish between the two cases by computing $X = X_1 \oplus X_2 \cdots \oplus X_n$. Then when n becomes larger than about $1/4\varepsilon^2$ we will be able to distinguish these two cases with a good probability.

Remark 12.2 We will use this theorem here for linear cryptanalysis (with a value ε of about 2^{-22} we will need n to be about 2^{43}) but this theorem is also useful, for example, for rigged dice, or for polls (in each case to distinguish information from random noise with a good probability).

Proof ***(of Theorem 12.4)*** Let $q = p = 1/2$.

Since the variables X_i are independent, X follows a binomial distribution. In the first case, the mean value of X is $np = n/2$, the variance of X is $npq = n/4$, and the standard deviation of X is therefore $\sqrt{n}/2$. Similarly in the second case, the mean value of X is $n(1/2 + \varepsilon) = n/2 + n\varepsilon$, the variance of X is $n(1/2 + \varepsilon)(1/2 - \varepsilon)$, and the standard deviation of X is the square root of this value. We are able to distinguish between these two cases when the difference of these two mean values becomes larger than the standard deviation, i.e., when $n\varepsilon$ becomes larger than $\sqrt{n}/2$, i.e., when n becomes larger than $1/4\varepsilon^2$ as claimed. □

12.2.5 Biham-Type Attack (Biham 2025)

We suppose here that the same plaintext block A is encrypted under distinct keys. With simple *DES* only a few seconds are needed to find a small percentage of keys that will be used (and not a specific key). The attack is based on a very simple process. First we generate a table containing the encryption of A under 2^{28} keys (i.e., about one key every 2^{28}). Then, as soon as A is encrypted under one of these keys, it is possible to detect the key just by observing the table. This is a generic attack that applies to any algorithm with a 56-bit key, when the same plaintext block A often is encrypted under distinct keys.

12.2.6 Davies' Attacks on the DES

We present here a short explanation about Davies' attack on *DES*. More explanations can be found in Davies and Murphy (1995) and Biham and Biryukov (1995). The general idea of this attack is to use the fact that the distribution of the 8-bit output of a pair of S-boxes is non-uniform. This comes from the duplication of the 2 bits in the expansion function E, and from the values of the S-boxes. From the S-box values (given in Table 12.2), it is easy to obtain Tables 12.7 and 12.8. Now from these two tables it is easy to obtain Table 12.9.

Let (R_{-1}, R_0) be the plaintext. The encryption of each round is given as follows:

$$R_{i+1} = R_{i-1} \oplus f(R_i, K_{i+1}), \qquad i = 0, \ldots, 15.$$

The ciphertext is (R_{16}, R_{15}) with:

$$\begin{aligned} R_0 \oplus R_{16} = & f(R_1, K_2) \oplus f(R_3, K_4) \oplus f(R_5, K_6) \oplus f(R_7, K_8) \oplus f(R_9, K_{10}) \\ & \oplus f(R_{11}, K_{12}) \oplus f(R_{13}, K_{14}) \oplus f(R_{15}, K_{16}), \end{aligned} \tag{12.4}$$

Table 12.7 Number of inputs x of S_7 such that $S_7(x) = A$ and the 2 last input bits of S_7 are equal to C

A \ C	00	01	10	11
0	1	1	1	1
1	1	1	2	0
2	1	1	1	1
3	1	1	1	1
4	1	1	1	1
5	1	1	1	1
6	2	0	1	1
7	1	1	0	2
8	1	1	1	1
9	2	0	1	1
10	1	1	1	1
11	1	1	1	1
12	1	1	0	2
13	0	2	2	0
14	0	2	2	0
15	1	1	0	2

Table 12.8 Number of inputs x of S_8 such that $S_8(x) = B$ and the first 2 input bits of S_8 are equal to C

B \ C	00	01	10	11
0	0	0	2	2
1	2	2	0	0
2	1	2	1	0
3	1	0	1	2
4	2	2	0	0
5	0	0	2	2
6	1	0	1	2
7	1	2	1	0
8	2	1	0	1
9	0	1	2	1
10	1	1	1	1
11	1	1	1	1
12	0	1	2	1
13	2	1	0	1
14	0	2	2	0
15	2	0	0	2

Table 12.9 Number of inputs x such that $S_7(x) = A$, $S_8(x) = B$ and the value C of the two last input bits of S_7 is equal to the value of the first two input bits of S_8. Tables with a different parity P_1 of these four bits can immediatly be obtained from this Table and Formula (12.6)

$B : S_8$ / $A : S_7$	0	1	2	3	4	5	6	7	8	9	10	11	12	13	14	15
0	4	4	4	4	4	4	4	4	4	4	4	4	4	4	4	4
1	4	4	5	3	4	4	3	5	3	5	4	4	5	3	6	2
2	4	4	4	4	4	4	4	4	4	4	4	4	4	4	4	4
3	4	4	4	4	4	4	4	4	4	4	4	4	4	4	4	4
4	4	4	4	4	4	4	4	4	4	4	4	4	4	4	4	4
5	4	4	4	4	4	4	4	4	4	4	4	4	4	4	4	4
6	4	4	3	5	4	4	5	3	5	3	4	4	3	5	2	6
7	4	4	3	5	4	4	5	3	5	3	4	4	3	5	2	6
8	4	4	4	4	4	4	4	4	4	4	4	4	4	4	4	4
9	4	4	3	5	4	4	5	3	5	3	4	4	3	5	2	6
10	4	4	4	4	4	4	4	4	4	4	4	4	4	4	4	4
11	4	4	4	4	4	4	4	4	4	4	4	4	4	4	4	4
12	4	4	3	5	4	4	5	3	5	3	4	4	3	5	2	6
13	4	4	6	2	4	4	2	6	2	6	4	4	6	2	8	0
14	4	4	6	2	4	4	2	6	2	6	4	4	6	2	8	0
15	4	4	3	5	4	4	5	3	5	3	4	4	3	5	2	6

and

$$R_0 \oplus R_{16} = f(R_0, K_1) \oplus f(R_2, K_3) \cdots \oplus f(R_{14}, K_{15}). \tag{12.5}$$

Davies and Murphy (1995) found that the distribution of the output of a pair of S-boxes depends on the XOR (i.e., the parity) of the 4 key bits which are mixed with the duplicate bits. Thus we can write after 1 round for the distribution

$$D_1(x, y, P_1) = 4 + (-1)^{P_1} d_1(x, y), \tag{12.6}$$

where x is the output of the left S-box of the pair, y is the output of the right S-box, and P_1 is the parity of the 4 key bits involved.

Moreover, Davies found that after $2n$ rounds (i.e., n XORs in Eqs. (12.4) or (12.5)) we have a similar expression for the distribution of the output of such a pair of S-boxes:

$$D_n(x, y, P_n) = 2^{10n-8} + (-1)^{P_n} d_n(x, y), \tag{12.7}$$

where P_n is the parity of the $4n$ subkey bits mixed with the data bits. $D_n(x, y, P_n)$, and therefore the values $d_n(x, y)$ can be computed by the recurrent equations:

$$D_n(x, y, 0) = \sum_{\substack{x_1 \oplus x_2 = x \\ y_1 \oplus y_2 = y}} D_{n-1}(x_1, y_1, 0) D_1(x_2, y_2, 0), \qquad n = 2, 3, \cdots, 8$$

$$D_n(x, y, 1) = 2^{10n-7} - D_n(x, y, 0).$$

How the attack works. Now that the $d_n(x, y)$ are known, we want to find the parity P_n by using Eq. (12.7) and m pairs of plaintext/ciphertext. For this, after $2n$ rounds, Davies shows that we can compute this indicator value:

$$I = \sum_{i=0}^{255} m_i d_n(x, y),$$

where m_i is the number of times that the output i occurs. If $I > 0$, then $P_n = 0$, and if $I < 0$, then $P_n = 1$. Davies showed that this attack will succeed with a good probability when

$$m \geq \frac{2^{20n-6}}{\sum_{x,y} d_n(x, y)^2}.$$

Remark 12.3 This is a KPA attack, and as for all KPA attacks it involves input values (the parity P_n and output values (the values m_i).

Proof We follow here the proof given in Davies and Murphy (1995). Let $i = (x, y)$ be the output, $i = 0, \ldots, 255$. From Eq. (12.7), the probability $p(i, P_n)$ of the output i conditioned by P_n is given by

$$\begin{aligned} p(i, P_n) &= 2^{-8} + 2^{-10n}(-1)^{P_n} d_n(i) \\ &= 2^{-8} + (-1)^{P_n} d_i', \end{aligned}$$

where $d_i' = 2^{-10n} d_n(i)$. The problem is to decide whether $P_n = 0$ or $P_n = 1$.

Let X denote the output of m inputs (m is the number of plaintext/ciphertext pairs that we have in KPA) to the pair of neighboring S-boxes. Suppose we have a realization of X, $x = (x_1, \ldots, x_m)$, then let m_i be the number of times output i occurs, so we have

$$\Pr[X = x] = \prod_{i=0}^{255} p(i, P_k)^{m_i}.$$

We wish to test whether $P_n = 0$ or $P_n = 1$. The likelihood ratio statistic λ is given by

$$\lambda = \prod_{i=0}^{255} \left(\frac{p(i, 0)}{p(i, 1)} \right)^{m_i},$$

and $\lambda > 1$ means $\log \lambda > 0$.

Now,

$$\log \lambda = \sum_{i=0}^{255} m_i \log\left(\frac{p(i,0)}{p(i,1)}\right)$$
$$= \sum_{i=0}^{255} m_i \log\left(\frac{2^{-8}+d_i'}{2^{-8}-d_i'}\right).$$

Let $w_i = \log\left(\frac{2^{-8}+d_i'}{2^{-8}-d_i'}\right)$. Since the d_i' values are small compared with 2^{-8}, we have

$$w_i = \log\left(1 + \frac{2d_i'}{2^{-8}-d_i'}\right) \simeq \frac{2d_i'}{2^{-8}}.$$

Thus, we can use the sign of

$$I' = \sum_{i=0}^{255} m_i d_i',$$

and since $d_i' = 2^{-10n} d_n(i)$, the sign of I' is also the sign of

$$I = \sum_{i=0}^{255} m_i d_n(i),$$

as claimed.

Now

$$E(I') = \sum_{i=0}^{255} d_i' E(m_i)$$
$$\simeq \sum_{i=0}^{255} d_i' m(2^{-8} + (-1)^{P_n} d_i') \qquad = m(-1)^{P_n} \sum_{i=0}^{255} d_i'^2 = (-1)^{P_n} mT,$$

with $T = \sum_{i=0}^{255} d_i'^2$.

We also have

$$\mathrm{Var}(I') \simeq \sum_{i=0}^{255} d_i'^2 \mathrm{Var}(m_i)$$
$$\simeq m2^{-8} \sum_{i=0}^{255} d_i'^2 = m \cdot 2^{-8} \cdot T.$$

The test is expected to succeed with a good probability when $\mathrm{E}(I)$ is larger than twice the standard deviation of I, i.e., when

$$mT \geq 2\sqrt{m2^{-8}T},$$

i.e., when

$$mT \geq 2^{-6},$$

or

$$m \geq \frac{2^{-6}}{\sum_{i=0}^{255} d_i'^2}.$$

Now since $d_i' = 2^{-10n} d_n(i)$ and $i = (x, y)$, we obtain

$$m \geq \frac{2^{20n-6}}{\sum_{(x,y)} d_n(x, y)^2}$$

as claimed. □

Table 12.10 gives this value of m from the chosen pair of S-boxes and the number of rounds. We see that the pair S_7/S_8 is the best pair. However so far we have found only 1 bit (the involved parity) for the key and the complexity is still $\geq 2^{56}$ for the full *DES*.

In 1992, Henri Gilbert and in 1993 Eli Biham and Alex Biryukov showed how to improve Davies' attack. The idea is to peel up 1 round of *DES* by guessing all the possible values of the key bits of the pair of S-boxes. Then, we can compute the distribution that results for each value of the key bits entering the pair of S-boxes in the last round after XORing the plaintext and ciphertext bits with the output of the S-boxes.

The total complexity of this KPA attack is now of about 2^{50} plaintext/ciphertext pairs and 2^{50} computations, i.e., less than the 2^{56} computations of exhaustive search.

Table 12.10 The complexities of Davies' attack

Rounds	Distribution	S_1/S_2	S_2/S_3	S_3/S_4	S_4/S_5	S_5/S_6	S_6/S_7	S_7/S_8	S_8/S_1
2, 3	D_1	$2^{8.4}$	$2^{6.1}$	$2^{8.8}$	$2^{6.7}$	$2^{7.4}$	$2^{7.1}$	$2^{6.2}$	$2^{7.7}$
4, 5	D_2	$2^{16.3}$	$2^{15.7}$	$2^{20.4}$	$2^{16.7}$	$2^{17.6}$	$2^{16.8}$	2^{145}	$2^{18.5}$
6, 7	D_3	$2^{25.2}$	$2^{24.9}$	$2^{31.4}$	$2^{26.0}$	$2^{27.0}$	$2^{25.4}$	$2^{21.8}$	$2^{28.6}$
8, 9	D_4	$2^{33.6}$	$2^{33.9}$	$2^{42.3}$	$2^{35.1}$	$2^{36.1}$	$2^{33.7}$	$2^{28.9}$	$2^{38.5}$
10, 11	D_5	$2^{41.8}$	$2^{42.8}$	$2^{53.1}$	$2^{44.1}$	$2^{45.0}$	$2^{41.8}$	$2^{35.9}$	$2^{48.2}$
12, 13	D_6	$2^{49.9}$	$2^{51.6}$	$2^{64.0}$	$2^{52.9}$	$2^{53.9}$	$2^{49.9}$	$2^{42.8}$	$2^{57.9}$
14, 15	D_7	$2^{57.9}$	$2^{60.5}$	$2^{74.8}$	$2^{61.8}$	$2^{62.8}$	$2^{57.9}$	$2^{49.7}$	$2^{67.6}$
16	D_8	$2^{66.0}$	$2^{69.3}$	$2^{85.6}$	$2^{70.6}$	$2^{71.6}$	$2^{66.0}$	$2^{56.6}$	$2^{77.3}$

12.2.7 Conclusion on Simple DES

Three different attacks are known with a complexity smaller than exhaustive search (i.e. $< 2^{56}$). It is interesting to notice that the best implementations of these attacks involve different S-boxes: differential cryptanalysis involves S_1, S_2, S_3, linear cryptanalysis involves S_5 and Davies' attack involves S_7, S_8. Currently, the simple *DES* key is clearly too weak for security. Thus simple *DES* should not be used anymore. Indeed, with specific computers it is possible to retrieve the key by exhaustive search (brute force attack) in only several minutes or hours. The key size is now clearly too small. This is why several variants of *DES* have been proposed. The most frequently used are 3*DES* with two keys, 3*DES* with three keys and $DES - X$. In the next section, we compare these different algorithms. Notice that 3*DES* with two keys is still currently used in many banking solutions.

12.3 3*DES* with Two Keys

3*DES* is a variant of *DES* published in 1998 that aimed to solve the problem of the too small key (56 bits for simple *DES*). At present, a length of at least 80 bits is recommended.(Some cryptographers recommend even larger key size: 100 or 128 bits for very high security, or for some applications that may last 30 years.) 3*DES* is designed with either two keys or three keys. Since *DES* was widely used, 3*DES* has been very successful and is still used in many applications (especially banking solutions), even if *AES* (which is not a Feistel scheme) is now the most used secret-key algorithm.

12.3.1 Presentation

3*DES* with two keys is defined by

$$3DES_{k_1,k_2}(X) = DES_{k_1}(DES_{k_2}^{-1}(DES_{k_1}(X))).$$

Notice that using DES^{-1} instead of *DES* in the middle of the encryption scheme does not affect security. If the keys are chosen such that $k_1 = k_2$ then we get simple *DES* and this could be interesting in some applications to move from simple *DES* to 3*DES*. Notice also that using k_1, k_2 and again k_1 comes from the fact that the designer of 3*DES* wanted to avoid man-in-the-middle attacks. However, as we will see below, this is partially satisfied since some attacks on 3*DES* with two keys can be considered as a generalization of man-in-the-middle attacks.

3*DES* with two key features are summarized in Table 12.11.

We now describe best known attacks on 3*DES* with two keys.

Table 12.11 *3DES* with two key features

Name	*3DES* with two keys
Input and output block size	64 bits
Key size	112 bits = 2× 56 bits
Scheme type	48-round Feistel scheme
Computation time	Three times slower than *DES*

12.3.2 Brute Force Attack

This is a KPA with $q = 2$ plaintext/ciphertext pair, 2^{112} computations and very little memory. Moreover, this attack is parallelizable and needs very little memory. However, 2^{112} computations are currently unrealistic. Therefore, this attack is currently unrealistic.

12.3.3 Merkle–Hellman Attack (Merkle and Hellman 1981)

This is an NCPA with $q = 2^{56}$ messages, 2^{56} computations, and 2^{56} in memory. This attack is very efficient in terms of computations. However this attack requires to choose and know $1/256$ of the number of plaintext/ciphertext pairs. This is a huge number of plaintext/ciphertext pairs. We now describe this attack (Fig. 12.3).

$$C = DES_{K_1}(DES_{K_2}^{-1}(DES_{K_1}(P))) = Enc(P).$$

Aim: Here we want to obtain the keys K_1 and K_2 after performing about 2^{56} computations in NCPA with $q \simeq 2^{56}$ messages. The attacks (cf. Merkle and Hellman 1981) proceed as follows:

- **Step 1**. We create a table T such that, when we give a value B to the table, we obtain the keys K_2 such that $B = DES_{K_2}^{-1}(0)$. This means that table $T : B \to K_2$. For this, for all keys K_2, we compute $B = DES_{K_2}^{-1}(0)$ and store K_2 at the address B (or at the address $\lceil p \rceil$, where $\lceil p \rceil$ has only 56 bits instead of 64 bits). It is possible to manage the collisions via pointers in order to need only about 2^{56} in memory. This step 1 needs 2^{56} in time, and 2^{56} in memory.
- **Step 2**: For each key K_1, we compute $P = DES_{K_1}^{-1}(0)$, and we ask for $C = Enc(P)$. Then, we compute $B = DES_{K_1}^{-1}(C)$, and we look in Table T of Step 1 if we have a

Fig. 12.3 Merkle–Hellman attack

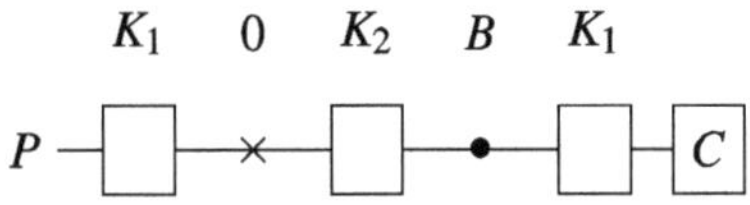

key K_2 such that $B = DES_{K_2}^{-1}(0)$. Then we check if (K_1, K_2) is a valid solution (for example, by checking if these values are compatible on another plaintext/ciphertext pair that we have).
For Step 2, we need 2^{56} computations and very little memory (outside Table T). Finally, we have obtained the real keys (K_1, K_2) in NCPA with $q \simeq 2^{56}$ messages, with about 2^{57} computations and about 2^{56} in memory. We can notice that the central idea of the attack is that thanks to NCPA we can "connect" P and C, i.e., "connect" the two blocks with key K_1, and thanks to this we can obtain here a kind of generalization of meet-in-the-middle attack.

12.3.4 Van Oorschot and Wiener Attack (Van Oorschot and Wiener 1991)

This is a KPA with q messages, $\frac{2^{120}}{q}$ computations, and 2^{56} in memory. For example, with $q = 2^{42}$ plaintext/ciphertext pairs, we obtain a KPA with 2^{78} computations and 2^{56} in memory. This attack requires more computations than Merkle–Hellman attack. However, it is a KPA instead of a NCPA, and we need to know fewer messages. This attack is described below (Fig. 12.4).

Aim: Here we want to generalize Merkle–Hellman attack in order to obtain a KPA instead of an NCPA. Let q be the number of plaintext/ciphertext pairs that we have for a KPA. The attack proceeds like this (cf. Van Oorschot and Wiener (1991)). Clearly, the value 0 in Merkle–Hellman attack above can be replaced by any value α. For each fixed value α, the probability that Merkle–Hellman attack succeeds in finding (K_1, K_2) is about $\frac{q}{2^{64}}$ with 2^{57} computations and in 2^{56} in memory. Let $\lambda = \frac{2^{64}}{q}$. If we perform Merkle–Hellman attack about λ times, with λ values α, and λ tables $T_\alpha : B \to K_2$, we will have a high probability to find (K_1, K_2). However we will need $\lambda.2^{57}$ computations, and still 2^{56} in memory (since we can re-use the same memory for Tables T with various α). Therefore, we have here a KPA with q plaintext/ciphertext pairs, about $\frac{2^{121}}{q}$ computations, and 2^{56} in memory. For example, with $q = 2^{32}$, we find (K_1, K_2) after 2^{89} computations, and with $q = 2^{42}$, we find (K_1, K_2) after 2^{79} computations.

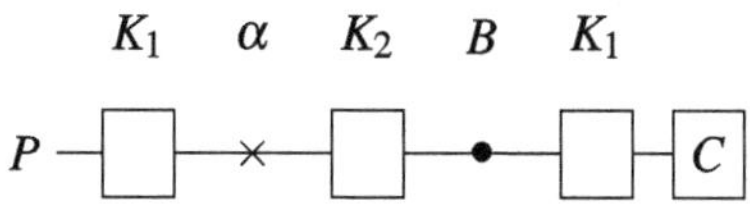

Fig. 12.4 Oorschot–Wiener Attack

12.3.5 Mitchell Attack (Mitchell 2016)

In Mitchell (2016), pointed out in 2016 that in Oorschot and Wiener attack described above, we do not need that all the q plaintext/ciphertext that we have for an attack come from the same (K_1, K_2). Thanks to this remark, the attack can be much more practical, and it does not matter if the keys (K_1, K_2) are often changed. However, when the attack is successful, only one of the key will be found, and can only be used to decrypt other materials encrypted using that key. (The attack also requires that for each (K_1, K_2) we have at least two, and preferably three plaintext/ciphertext pairs so that candidates for (K_1, K_2) can be checked).

In Mitchell (2016), also pointed out that we can use the *DES* complementation property in order to double the number of plaintext/ciphertext pairs available to conduct the attack, since every plaintext/ciphertext pair for the key K will give us another pair for the key $\bar{K}$, i.e., all bits are flipped: "0" becomes "1" and vice versa: $\overline{e_K(P)} = e_{\bar{K}}(\bar{P})$. Therefore, we have a KPA with q plaintext/ciphertext pairs, that do not necessary comes from the same (K_1, K_2), with about $\frac{2^{120}}{q}$ computations and 2^{56} on memory. For example, if $q = 2^{32}$, we find one of the keys (K_1, K_2) after 2^{88} computations, and with $q = 2^{42}$, we find one of the keys (K_1, K_2) after 2^{78} computations.

12.3.6 Codebook Attack

In this attack, we need about 2^{64} plaintex/ciphertext pairs (almost the whole codebook), 2^{64} computations and 2^{64} in memory and although this attack does not allow to retrieve the key, it is possible to decrypt almost all messages, since we know the corresponding plaintexts.

12.3.7 Attack with Partial Decryption

When we encrypt more than 2^{32} blocks (each of 64 bits) with the same key, then by the birthday paradox, the probability to have two blocks with the same input, and consequently the same output is high. This is satisfied for most encryption modes currently used, for example, the CBC mode. So an attacker will be able to get the XOR of two plaintext in a cipher-only attack with 2^{32} computations (a few seconds) when he observes 2^{32} blocks encrypted with the same key. Generally this allows him (by KPA) to retrieve the two plaintext block if he knows one, or if the messages have a given structure due to the language, for example.

More generally, if more than $A.2^{32}$ blocks are encrypted with the same key, where A is an integer greater than 1, by the birthday paradox, an adversary can find (by KPA) with a good probability about A^2 plaintext blocks, or the XOR of about A^2 plaintext blocks in a cipher-only attack. Notice that when $A = 2^{32}$, we get a Codebook attack. This attack does not allow to retrieve the key but gives the possibility to retrieve some plaintexts from time to time.

To avoid this simple attack, it is recommended not to encrypt 2^{32} messages under the same key. It is even recommended to stay significantly apart from this value: for example, if 2^{30} messages are encrypted the probability that this attack succeeds is about $\frac{1}{16}$, and if 2^{20} messages are encrypted the probability is then about $\frac{1}{2^{24}}$. Another possibility, that we will not detail here, is to use very specific encryption mode, for example, tweakable block that introduces diversification.

12.3.8 Biham-Type Attack (Biham 2025)

We suppose here that the same plaintext block A is encrypted under distinct keys. Then it will be possible to find one key every 2^{28} keys, for example, after 2^{84} computations and by using 2^{84} in memory. It is also possible to find one key every 2^{56} keys, for example, after 2^{56} computations and by using 2^{56} in memory. This attack is unrealistic here since a huge memory is needed. However, it is possible to lower the memory by a factor of t, but then the number of computations is multiplied by t.

This attack is based on a very simple process. First we generate a table containing the encryption of A under 2^{84} keys (or 2^{56} keys), i.e., about one key every 2^{28}, (or one key every 2^{56}). Then, as soon as A is encrypted under one of these keys, it is possible to detect the key just by observing the table. This is a generic attack that applies to any algorithm with a 112-bit key, when the same plaintext block A is often encrypted under distinct keys.

12.3.9 Related-Key Attack

Vaudenay (2025) discovered a related-key attack on 3*DES* with two keys whose characteristics are close to Merkle–Hellman attack: $q = 2^{56}$ for NCPA, with 2^{56} computations and 2^{56} in memory. Thus this attack is not easier to implement than Merkle–Hellman's attack.

12.3.10 Related-Key Distinguisher

In *DES* and in 3*DES* as well, if we replace all bits equal to 1 by 0 and all bits equal to 0 by 1 in the message, but also in the key and in the ciphertext, then we get back to the original ciphertext (complementarity property). This allows to distinguish a black box of 3*DES* from an ideal cipher black box on 64 bits thanks to a related-key attack with $q = 2$ messages and two computations. However, this does not provide here the value of the 3*DES* key, but just a related-key distinguisher. This is why generally this attack is not seen as important for security.

12.3.11 Conclusion on 3DES with Two Keys

3*DES* security is in 2^{56} computations (like *DES*) and not in 2^{112} as expected. However, from a practical point of view, most of the previous attacks do not apply to smart cards since these cards will usually never encrypt more than 2^{20} values under the same key. This is why 3*DES* with two keys is still used for credit card and this could be the case for several years. However, it is better to abandon aging algorithm whose major flaw is the 64-bit block size. On the contrary, the fact to have a Feistel cipher with 48 rounds is better than 16 rounds, since a greater number of rounds increases security against many cryptanalysis (such as linear or differential cryptanalysis for example).

12.4 3*DES* with Three Keys

12.4.1 Presentation

3*DES* with three keys is defined by

$$3DES_{k_1,k_2,k_3}(X) = DES_{k_3}(DES_{k_2}(DES_{k_1}(X))).$$

3*DES* with three key features are summarized in Table 12.12.

We now describe the best known attacks on 3*DES* with three keys.

Table 12.12 3*DES* with three keys features

Name	3*DES* with three keys
Input and output block size	64 bits
Key size	168 bits = 3× 56 bits
Scheme type	48-round Feistel scheme
Computation time	Three times slower than *DES*

12.4.2 Man-in-the-Middle Attack and Refinements by Lucks

The basic man-in-the-middle attack is a KPA with $q = 3$ plaintex/ciphertext pairs, 2^{112} computations and 2^{56} in memory. Lucks (1998) presented several refinements of this attack. For example, he gave a KPA with $q = 2^{45}$, 2^{109} computations and 2^{56} memory, or another KPA with $q = 2^{32}$, 2^{113} simple computations together with 2^{90} DES computations and 2^{88} in memory.

12.4.3 Codebook Attack

The attack is similar to the attack given for 3*DES* with two keys. Only the block size is taken into account. Here, we need 2^{64} plaintex/ciphertext pairs (almost the whole codebook), 2^{64} computations and 2^{64} memory and although this attack does not allow to retrieve the key, it is still possible to decrypt almost all messages.

12.4.4 Attack with Partial Decryption

Again the attack is similar to the attack given for 3*DES* with two keys. Only the block size is taken into account. If we encrypt more than 2^{32} of 64 bits under the same key, then by the birthday paradox, an attacker will find with high probability, with a KPA, one plaintext 64-bit block, or the XOR of two plaintext block in a cipher-only attack. More generally, if more than $A.2^{32}$ blocks are encrypted with the same key, where A is an integer greater than 1, always by the birthday paradox, an adversary can find with a good probability about A^2 plaintext blocks in a KPA, or the XOR of about A^2 plaintext blocks in a cipher-only attack. Notice that when $A = 2^{32}$, we get a Codebook attack.

12.4.5 Biham-Type Attack (Biham 2025)

We suppose here that the same plaintext block A is encrypted under distinct keys. Then it will be possible to find one key every 2^{28} keys, for example, after 2^{84} computations and by using 2^{84} memory (2^{84} computations and not 2^{112} or 2^{109} like in a man-in-the-middle attack; this was not obviously a priori). This attack is unrealistic here since a huge memory is needed. However, it is possible to lower the memory by a factor of t, but then the number of computations is multiplied by t.

This attack is based on a very simple process: we combine the previous simple Biham attack together with a man-in-the-middle attack. More precisely, let z be the value obtained after A has been encrypted under the two first *DES*, i.e., under the keys k_1 and k_2. Then we generate a correlation table between z and the keys (k_1, k_2), where we have tested and memorized 2^{84} values for (k_1, k_2), i.e., about one key every 2^{28} for (k_1, k_2). Then, as soon as A is encrypted under one of these keys, it is possible to detect the key just by performing 2^{56} computations for the exhaustive search of k_3 and by observing the table. This is a generic attack that applies to any algorithm composed three times with three times 56 bits for the key, when the same plaintext block A is often encrypted under distinct keys.

12.4.6 Related-Key Attack

Kelsey et al. (1998) found a related-key attack on 3*DES* with three keys needing three chosen plaintexts and 2^{56} computations. Surprisingly, this attack is more efficient on 3*DES* with three keys than on 3*DES* with two keys.

12.4.7 Related-Key Distinguisher

In *DES* and in 3*DES* as well, if we replace all bits equal to 1 by 0 and all bits equal to 0 by 1 in the message, but also in the key and in the ciphertext, then we get back to the original ciphertext (complementarity property). This allows to distinguish a black box of 3*DES* from an ideal cipher black box on 64 bits thanks to a related-key attack with $q = 2$ messages and two computations. However, this does not provide here the value of the 3*DES* key, but just a related-key distinguisher. This is why generally this attack is not seen as important for security.

12.4.8 Conclusion on 3DES with Three Keys

One of the main issues for 3*DES* with three keys comes from the block size, which is only 64 bits. For banking solutions, most of the attacks presented here are unrealistic due to the fact that the card will encrypt a limited number of messages with the same key (typically less than 10^6).

Related-key attacks need only 2^{56} computations that are surprisingly less than for 3*DES* with two keys. However, 3*DES* with three keys remains generally better than 3*DES* with two keys but not as much as expected. For example, Biham's attack needs here 2^{84} computations in order to find one every 2^{84} keys and the codebook attack stays in 2^{64}.

Overall, 3*DES* with three keys can be still considered as a strong algorithm for smart cards.

12.5 $DES-X$

12.5.1 Presentation

$DES-X$ was proposed by Ron Rivest in 1984. It is defined by

$$DES-X(M) = k_3 \oplus DES_{k_1}(M \oplus k_2)$$

where k_1, k_2, k_3 represent the three parts of the secret key.

$DES-X$ with two key features are given in Table 12.13.

Remark 12.4 This construction is a particular case of Even-Mansour construction which is currently very popular in Cryptography Conferences (see Dunkelman 2012 or Lampe 2012). We now describe best known attacks on $DES-X$.

12.5.2 Codebook Attack

This attack is similar to that performed on $3DES$, only the block size is involved here. In this attack, we need 2^{64} plaintext/ciphertext pairs (i.e., almost all of them), 2^{64} computations and 2^{64} memory although this attack does not allow to retrieve the key, it is still possible to decrypt almost all messages.

12.5.3 Linear Cryptanalysis (Rogaway 1996)

This is a KPA with $q = 2^{60}$ plaintext/ciphertext pairs, 2^{60} computations and the need memory is small. Notice here that we need to have about $\frac{1}{16}$ of the total space of plaintext/ciphertex pairs. This implies that here the codebook attack can be performed, even without the key (but we can find the key with linear cryptanalysis, even if we do not really need it to decrypt).

Table 12.13 $DES-X$ features

Name	$DES-X$
Input and output block size	64 bits
Key size	184 bits = (64 + 56 + 64) bits
Scheme type	16-round Feistel scheme
Computation time	About same time as simple *DES*

12.5.4 Daemen's Attack

In 1991, Daemen (1991) presented a NCPA on $DES - X$ able to retrieve the key with $q = 2^{32}$ messages and 2^{88} computations.

12.5.5 Attack with Partial Decryption

This attack is similar to that performed on 3*DES*, only the block size is involved here. If we encrypt more than 2^{32} of 64 bits under the same key, then by the birthday paradox, an attacker will find with high probability in KPA one 64-bit block plaintext, or the XOR of two plaintext block in a cipher-only attack. More generally, if more than $A.2^{32}$ blocks are encrypted with the same key, where A is an integer greater than 1, always by the birthday paradox, an adversary can find with a good probability about A^2 plaintext blocks in a KPA, or the XOR of about A^2 plaintext blocks in a cipher-only attack. Notice that when $A = 2^{32}$, we get a Codebook attack.

12.5.6 Biham-Type Attack

We suppose here that the same plaintext block A is encrypted under distinct keys. Then it will be possible to find one key every 2^{28} keys, for example, in 2^{92} computations and 2^{92} memory. This attack is a simple variant of the same attack on 3*DES* with three keys. This attack is unrealistic and rather less good than the previous ones.

12.5.7 Related-Key Attack

Biryukov (2009) presented a related-key attack that allowed to retrieve the $DES - X$ keys with 2^{32} messages with related keys, 2^{88} computations and 2^{32} memory ("Advanced Slide Attack"). The characteristics are close to those of Daemen's attack but here we have a related-key attack instead of an NCPA.

By using the complementarity property of DES (exchanging 0 and 1 in the plaintext and the key), it is possible to describe a related-key attack much more efficient: about two encryptions to determine each key bit, i.e., only 368 encryptions and computations (see Kelsey et al. 1997).

12.5.8 Related-Key Distinguisher

In *DES* like in 3*DES*, if we replace all the 0 by 1 and all the 1 by 0 in the plaintext, and also in the key and the ciphertext, then we get the original ciphertext (complementarity property).

This allows to distinguish a $DES - X$ black box from an ideal cipher black box on 64 bits with a related-key attack and with only $q = 2$ messages and two computations.

However here we do not obtain the value of the $DES - X$ key, but only a related-key distinguisher. This is why generally this attack is not seen as important for security.

12.5.9 Conclusion on DES − X

The attack by Daemen shows that it is better to use 3*DES* with three keys than $DES - X$. At present, $DES - X$ offers an effective security, when, like with credit card, the number of messages encrypted with the same key stay small (typically less than 10^6). This is a remarkable feature since $DES - X$ construction is very simple and fast.

Problems

12.1 Show that the 16-round functions used in the DES design are not pseudo-random, i.e., that it is easy to distinguish them from truly random functions from 32 bits to 32 bits.

12.2 Imagine that, from DES, we design an algorithm A as follows:

1. The secret key K of A is 256 bits instead of 56 bits.
2. The number of rounds of A is 32 instead of 16.
3. The eight public S_i boxes of DES from 6 bits to 4 bits are replaced by eight affine transformations of the form $X \mapsto A_i X + B_i$, $1 \leq i \leq 8$ where the B_i are secret 4-bit vectors generated from K and the A_i are secret 6×4 matrices generated by K as well. The input X is a 6-bit vector.

Finally, with all these changes, is A more secure than DES?

12.3 Is the query complexity of Davies' attack better in CPA?

12.4 Is the query complexity of the linear cryptanalysis of DES better in CPA?

References

Biham, E., and A. Biryukov. 1995. An improvement of Davies' attack on DES. In *Advances in cryptology—EUROCRYPT'94. EUROCRYPT 1994*, vol. 950, ed. A. De Santis. Lecture notes in computer science. Berlin: Springer.

Biham, E. 2025. How to forge DES encrypted messages in 2^{28} steps. Available at http://www.cs.technion.ac.il/users/wwwb/cgi-bin/tr-get.cgi/1996/CS/CS0884.ps.gz

Biham, E., and S. Shamir. 1991. Differential cryptanalysis of DES-like cryptosystems. *Journal of Cryptology* 4: 3–72.

Biryukov, A. 2009. Related-key cyptanalysis of the full AES-192 and AES-256. In *Advances in Cryptology—ASIACRYPT '09*, vol. 5912, ed. M. Matsui. Lecture notes in computer science, 1–18. Berlin: Springer.

Daemen, J. 1991. Limitations of the even-mansour construction. In *Advances in cryptology—ASIACRYPT '91*, vol. 739, ed. H. Imai, R. L. Rivest, and T. M. Sako. Lecture notes in computer science, 495–498. Berlin: Springer.

Davies, D., and S. Murphy. 1995. Pairs and triplets of DES S-boxes. *Journal of Cryptology* 8: 1–25.

Dunkelman, O., N. Keller, and A. Shamir. 2012. Minimalism in cryptography: the even-mansour scheme revisited. In *Advances in cryptology—EUROCRYPT '12*, vol. 7237, ed. D. Pointcheval and T. Johansson. Lecture notes in computer science, 336–354. Berlin: Springer.

Kelsey, J., B. Schneier, and D. Wagner. 1998. Key-Schedule cryptanalysis of IDEAn G-DES, GOST, SAFER and triple-DES. In *Advances in cryptology—CRYPTO '96*, vol. 1109, ed. N. Koblitz. Lecture notes in computer science, 237–251. Berlin: Springer.

Kelsey, J., B. Schneier, and D. Wagner. 1997. Related-Key Cryptanalysis of 3-WAY, Biham-DES, CAST, DES-X, New DES, RC2, nad TEA. In *Information and Communications Security – ICICS '97*, vol. 1334, ed. T. Okamoto and S. Qing. Lecture Notes in Computer Science, 233–246. Berlin: Springer.

Lampe, R., J. Patarin, and Y. Seurin. 2012. An Asymptotically tight security analysis if the iterated even-Mansour Cipher. In *Advances in cryptology—ASIACRYPT '12*, vol. 7658, ed. X. Wang and K. Sako. Lecture notes in computer science, 278–295. Berlin: Springer.

Lucks, S. 1998. Attacking triple encryption. In *Fast software encrytion—FSE '98*, vol. 1372, ed. S. Vaudenay. Lecture notes in computer science, 239–253. Berlin: Springer.

Matsui, M. 1993. Linear Cryptanalysis method for DES Cipher. In *Advances in cryptology—EUROCRYPT '93*, vol. 775, ed. T. Helleseth. Lecture notes in computer science, 386–387. Berlin: Springer.

Merkle, R. C., and M. E. Hellman. 1981. On the security of multiple encryption. *Communications of the ACM* 24 (7): 465–467.

Mitchell, C. J. 2016. On the Security of 2-key triple DES. Available at ArXiv: 1602.06229v1 [CS.CR].

Rogaway, P. S. 1996. The security of DES-X. *Crypto Bytes* 2 (2).

Tardy-Corfdir, A., and H. Gilbert. 1998. A Known plaintext attack of FEAL-4 and FEAL-6. In *Advances in cryptology—CRYPTO '91*, vol. 567, ed. J. Feigenbaum. Lecture notes in computer science, 172–181. Berlin: Springer.

Van Oorschot, P. C., and M. J. Wiener. 1991. A Known plaintext attack on two-key triple DES. In *Advances in cryptology—EUROCRYPT '90*, vol. 473, ed. I. Bjerre Damgård. Lecture notes in computer science, 318–325. Berlin: Springer.

Vaudenay, S. 2025. Related-key attack against triple encryption based on fixed points. In *Security and cryptography—SECRYPT11 '11*, 59–67.

Chapter 13
GOST, SIMON, BEAR-LION, CAST-256, CLEFIA

Abstract In this chapter, we present concrete ciphers based on the constructions studied previously. We provide examples of balanced, unbalanced, and generalized Feistel ciphers. For each of them, we give the description and a survey of attacks performed on these ciphers.

13.1 Ciphers Based on Balanced Feistel Constructions

13.1.1 GOST

13.1.1.1 Description of GOST

The GOST block cipher was developed in 1989 by the Soviet Union in order to be the Russian encryption standard. It was finally published in 1994 but even today the description of the S-boxes is not always public. In the GOST standard, the S-boxes are not specified. A different choice can be made for each application. For example, one of the S-boxes used in the Central Bank of the Russian Federation is known (Schneier 1994). The design of GOST is based on a 32-round balanced Feistel structure with 64-bit block and 256-bit key size. The round function consists of key addition, eight 4×4-bit S-boxes $S_1, S_2, \ldots, S_8$ and a 11-bit left rotation as shown in Fig. 13.1.

The key schedule is very simple: the 256-bit master key K is divided eight 32-bit words, i.e., $K = (k_1, \ldots, k_8)$. Each subkey k_i is used as a round key in each round function. The key schedule is given in Table 13.1.

13.1.1.2 Cryptanalysis of GOST

The structure of GOST is similar to that of DES. However, GOST uses larger parameters than DES. Thus DES was theoretically and practically broken 20 years ago, but most attacks on GOST before 2011 were single-key attacks on reduced-round versions of GOST. In Seki and Kaneko (2001), Seki and Kaneko proposed a

J. Patarin et al., *Guide to Feistel Ciphers*, Information Security and Cryptography,
https://doi.org/10.1007/978-3-031-99850-8_13

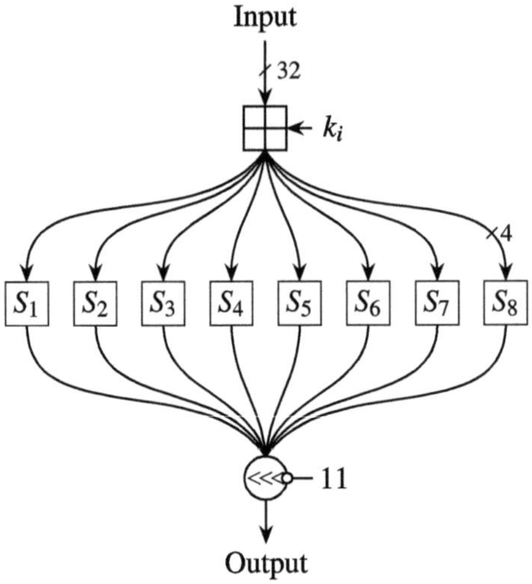

Fig. 13.1 GOST. Function of round *i*

Table 13.1 Key schedule of GOST

Round	1	2	3	4	5	6	7	8	9	10	11	12	13	14	15	16
Subkey	k_1	k_2	k_3	k_4	k_6	k_6	k_7	k_8	k_1	k_2	k_3	k_4	k_6	k_6	k_7	k_8
Round	17	18	19	20	21	22	23	24	25	26	27	28	29	30	31	32
Subkey	k_1	k_2	k_3	k_4	k_6	k_6	k_7	k_8	k_8	k_7	k_6	k_5	k_4	k_2	k_2	k_1

differential attack on 13 rounds. A related-key differential attack on the full GOST was given by Ko et al. (2015). However, these attacks are valid only on the GOST that uses the S-boxes of the Central Bank of the Russian Federation. Fleischmann et al. (2009) provided a related-key boomerang attack on the full GOST which works for any S-boxes. Biham et al. showed slide attacks on 24 and 30 round reduced (Biham et al. 2007). Their attacks make use of self-similarities among the round functions. Then Kara proposed a reflection attacks on 30 rounds (Kara 2008). Also some slide and reflection attacks were mounted for classes of weak keys (Biham et al. 2007; Kara 2008). The first single-key attack on the fill 32-round GOST was given by Isobe (2011). In Courtois (2011), Courtois proposed an attack on full GOST that used an algebraic approach. Later, Courtois and Misztal (2001) gave a differential attacks on full GOST. Several aspects of previous attacks were improved by Dinur et al. (2012). They used either the reflection property or a new fixed point property to reduce the problem of attacking full GOST into an attack on 8-round GOST with two known input/output pairs. Then in Courtois (2016), Courtois provided a multi-stage advanced differential attack on full GOST which is the fastest single-key attack known so far.

The results for key recovery attacks are summarized in Table 13.2 and for single-key attacks on the 8-round GOST used for full GOST attacks in Table 13.3.

Table 13.2 Key recovery attack on GOST

Key setting	Attack type	r	Data	Time complexity	References
Single key	Differential	13	2^{51} CP	–	Seki and Kaneko (2001)
	Slide	24	2^{63} ACP	2^{63}	Biham et al. (2007)
	Slide	30	2^{63} ACP	$2^{253.7}$	Biham et al. (2007)
	Reflection	30	2^{32} KP	2^{224}	Kara (2008)
	Reflection Meet-in-the-Middle	32	2^{32} KP	2^{224}	Isobe (2011)
Single key	Slide (2^{128} weak keys)	32	2^{63} ACP	2^{63}	Biham et al. (2007)
(weal key)	Reflection (2^{224} weak keys)	32	2^{32} CP	2^{192}	Kara (2008)
Related key	Differential	21	2^{56} CP	–	Seki and Kaneko (2001)
	Differential	32	2^{35} CP	2^{244}	Ko et al. (2015)
	Bommerang	32	$2^{7.5}$ CP	2^{248}	Fleischmann et al. (2009)

Table 13.3 Single-key attacks on 8-round GOST used for full GOST attacks

8-round attacks	Self-similarity	Data	Time	Memory	References
	Property	KP			
–	Reflection	2^{32}	2^{224}	2^{64}	Isobe (2011)
Algebraic	Other	2^{64}	2^{248}	2^{64}	Courtois (2011)
–	Differential (not based on self-similarity)	2^{64}	2^{226}	2^{64}	Courtois and Misztal (2001)
2DMITM	Fixed point	2^{64}	2^{192}	2^{36}	Dinur et al. (2012)
Law-memory	Fixed point	2^{64}	2^{204}	2^{19}	Dinur et al. (2012)
2DMITM	Reflection	2^{32}	2^{224}	2^{36}	Dinur et al. (2012)
Law-memory	Reflection	2^{32}	2^{236}	2^{19}	Dinur et al. (2012)
–	Multi-stage advanced differential	2^{62}	2^{179}	2^{70}	Courtois (2016)

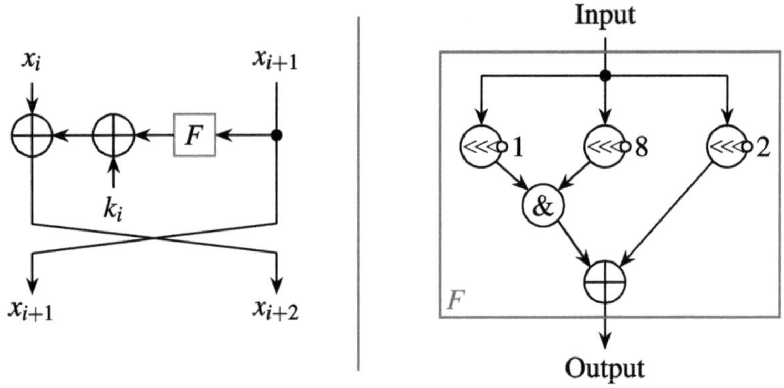

Fig. 13.2 Round i of SIMON and round function

13.1.2 *SIMON*

13.1.2.1 Description of SIMON

Simon is a balanced Feistel cipher published by the NSA in June 2013 (Beaulieu et al. 2013). SIMON consists of a family of ciphers with different sizes of blocks and keys (Fig. 13.2).

13.1.2.2 Cryptanalysis of SIMON

There are some papers with differential analysis (Biryukov et al. 2015; Abed et al. 2015), a paper with linear analysis (Wang 2014), and several other papers (Abdelraheem et al. 2015; Ahmadian et al. 2015; Kölbl et al. 2015). A recent paper of Raddum uses the algorithm ElimLin of Courtois et al. (2012) in order to attack 16 rounds of SIMON (Raddum 2015).

13.2 Ciphers Based on Expanding and/or Feistel Constructions

13.2.1 *BEAR-LION*

The BEAR and LION block ciphers were invented by Ross Anderson and Eli Biham in 1996 (Anderson and Biham 1996), by combining a stream cipher and a cryptographic hash function. The algorithms use a very large variable block size, on the order of 213 to 223 bits or more. Both are 3-round generalized Feistel ciphers, using the hash function and the stream cipher as round functions. BEAR uses the hash function twice with independent keys, and the stream cipher once. LION uses the stream cipher twice and the hash function once. The inventors proved that an attack on either BEAR or LION that recovers the key would break both the stream cipher and the hash. We suppose that the block size is m and the block size of the hash function is k. The message M is divided into blocks: $M = [L, R]$ with $|L| = k$ and $|R| = m - k$.

13.2.1.1 BEAR

BEAR performs encryption and decryption by using two applications of a keyed hash function and one application of a stream cipher. A BEAR key is a pair of subkeys $K = (K_1, K_2)$ such that $|K_1| > k$ and $|K_2| > k$. Then encryption and decryption are done as follows:

Encryption	Decryption
$L = L \oplus H_{K_1}(R)$	$L = L \oplus H_{K_2}(R)$
$R = R \oplus S(L)$	$R = R \oplus S(L)$
$L = L \oplus H_{K_2}(R)$	$L = L \oplus H_{K_1}(R)$

The keyed hash function H_K satisfies

- It is based on an unkeyed hash function H, in which we append/or prepend the key to the message.
- It is one-way and collision-free, i.e., given Y, it is hard to find X such that $Y = H(X)$, X and $X' \neq X$, such that $H(X) = H(X')$.
- It is pseudo-random, i.e., if $H(X_i)$ is given for any set of inputs, it is hard to predict any bit of $H(X')$ for any new input X'.

The stream cipher $S(M)$ is such that:

- It is pseudo-random.
- It resists key recovery attacks.
- It resists expansion attacks.

13.2.1.2 LION

The construction of LION is similar to that of BEAR except that it uses one application of a hash function and two applications of a stream cipher. Again the key is a pair of subkeys $K = (K_1, K_2)$ such that $|K_1| > k$ and $|K_2| > k$. Then encryption and decryption are done as follows:

Encryption	Decryption
$R = R \oplus S(L \oplus K_1)$	$R = R \oplus S(L \oplus K_2)$
$L = L \oplus H(R)$	$L = L \oplus H(R)$
$R = R \oplus S(L \oplus K_2)$	$R = R \oplus S(L \oplus K_1)$

The hash function H is again one-way and collision-free.
The stream cipher $S(M)$ is such that:

- It is pseudo-random.
- It resists key recovery attacks.
- It resists expansion attacks.

13.2.1.3 Security of BEAR and LION Ciphers

The security of BEAR and LION ciphers is based on the properties of the hash function and the stream cipher. With these properties, Anderson and Biham, the designers of both ciphers, proved that BEAR and LION were immune to an efficient known plaintext key recovery attack that can use as input only one plaintext/ciphertext pair (Anderson and Biham 1996). Later, Morin (1996) presented a meet-in-the-Middle attack that reduces the complexity of a brute force key search on both ciphers. In Maines et al. (2011), Maines et al. showed that BEAR and LION are immune to any efficient known plaintext key recovery attack that can use an input any number of plaintext/ciphertext pairs. They also use slightly weaker hypothesis on the hash function and the stream cipher.

13.2.2 Other Examples of Unbalanced Feistel Ciphers

Other structures make use of unbalanced Feistel ciphers. For example, the SMS4 cipher (Diffie and Ledin 2008), a 4-branch unbalanced Feistel cipher with contracting functions. SMS4 is the underlying block cipher used in the WAPI (WLAN authentication and privacy infrastructure) standard for securing wireless LANs in China. SMS4 has a 128-bit block size, a 128-bit user key, and a total of 32 rounds. Until now, there exist attacks up to 23 rounds (Su et al. 2011).

Unbalanced Feistel ciphers with expanding functions were used in the design of the CRUNCH hash function (Goubin et al. 2008) which was submitted to the SHA-3 competition. This hash function has not been broken so far.

13.3 Ciphers Based on Generalized Feistel Constructions

13.3.1 CAST-256

13.3.1.1 Description of CAST-256

CAST-256 is a symmetric-key block cipher published in June 1998. It was submitted as a candidate for the Advanced Encryption Standard (AES). However, it was not among the five AES finalists. It is an extension of an earlier cipher, CAST-128; both were designed according to the "CAST" design methodology invented by Carlisle Adams and Stafford Tavares. Howard Heys and Michael Wiener also contributed to the design. CAST-256 uses the same elements as CAST-128, including S-boxes, but is adapted for a block size of 128 bits—twice the size of its 64-bit predecessor. Key sizes can be 128, 160, 192, 224, or 256 bits. CAST-256 is composed of 48 rounds, sometimes described as 12 "quad-rounds," arranged in a type-1 generalized Feistel network with four branches. The structure is described in Fig. 13.3.

We now explain the "quad-rounds" that use two types of round functions, the *forward quad-round* $Q_i(.)$ and the *reverse quad-round* $\bar{Q}_i(.)$.

We denote 128-bit block as $\beta = [A, B, C, D]$ where A, B, C, D are each 32 bits in length. The forward quad-round $\beta \leftarrow Q_i(\beta)$ is defined as the following 4 rounds:

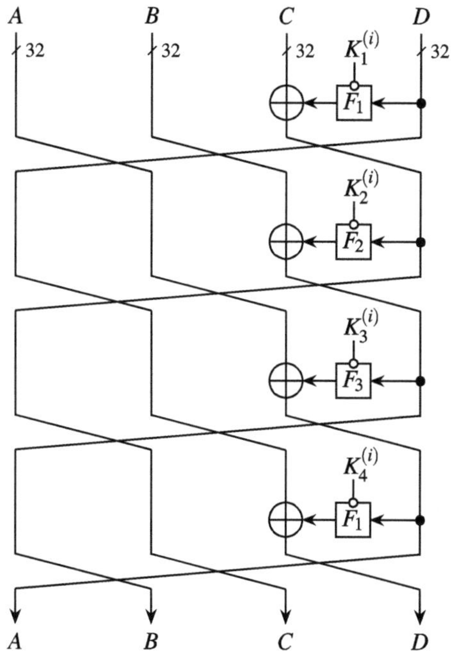

Fig. 13.3 Forward quad-round of CAST-256

$$\begin{aligned}
C &= C \oplus F_1(D, K_{r1}^{(i)}, K_{m1}^{(i)}) \\
B &= B \oplus F_2(C, K_{r2}^{(i)}, K_{m2}^{(i)}) \\
A &= A \oplus F_3(B, K_{r3}^{(i)}, K_{m3}^{(i)}) \\
D &= D \oplus F_1(A, K_{r4}^{(i)}, K_{m4}^{(i)}).
\end{aligned}$$

The reverse quad-round $\beta \leftarrow \bar{Q}_i(\beta)$ is defined as the following 4 rounds:

$$\begin{aligned}
D &= D \oplus F_1(A, K_{r4}^{(i)}, K_{m4}^{(i)}) \\
A &= A \oplus F_3(B, K_{r3}^{(i)}, K_{m3}^{(i)}) \\
B &= B \oplus F_2(C, K_{r2}^{(i)}, K_{m2}^{(i)}) \\
C &= C \oplus F_1(D, K_{r1}^{(i)}, K_{m1}^{(i)}),
\end{aligned}$$

where $K_r^{(i)} = \left\{K_{r1}^{(i)}, K_{r2}^{(i)}, K_{r3}^{(i)}, K_{r4}^{(i)}\right\}$ is the set of rotation keys for the ith quad-round, and $K_m^{(i)} = \left\{K_{m1}^{(i)}, K_{m2}^{(i)}, K_{m3}^{(i)}, K_{m4}^{(i)}\right\}$ is the set of masking keys for the ith quad-round. CAST-256 consists of six forward quad-rounds and six reverse quad-rounds.

13.3.1.2 Cryptanalysis of CAST-256

Many kinds of attacks were mounted against CAST-256. A distinguishing attack on 12 rounds was given by Nakahara and Rasmussen (2007): they found 12-round linear approximations. In Wang et al. (2009), Wang et al. constructed a 21-round linear approximation that allowed a key recovery on 24-round CAST-256. A boomerang attack was provided by Wagner (1999) on 16 rounds. Under a weak-key assumption that covers 2^{-35} of the keys, Seki and Kaneko (2009) gave a differential attack on 36 rounds. Bogdanov et al. (2012) presented an attack on 28 rounds by using multidimensional zero-correlation linear cryptanalysis. Zhao et al. (2014) provided linear attacks on 32 rounds that allows partial key recovery.

The results are summarized in Table 13.4.

13.3.2 CLEFIA

13.3.2.1 Description of CLEFIA

CLEFIA (Beaulieu et al. 2013) is a 128-bit block cipher with the key length of 128, 192, and 256 bits developed by Sony in 2007. It employs a type-2 Feistel structure with four data lines, and the width of each data line is 32 bits. Additionally, there are key whitening parts at the beginning and the end of the cipher.

Table 13.4 Summary of attacks on CAST-256

Attack type	r	Key size	Chosen plaintexts	Time complexity (encryptions)	References
Distinguishing	12	128	$2^{101.0}$	$2^{101.0}$	Nakahara and Rasmussen (2007)
Boomerang	16	128	$2^{49.3}$	–	Wagner (1999)
Differential	36	256	$2^{123.0}$	$2^{182.00}$	Seki and Kaneko (2009)
Linear	24	192	$2^{124.1}$	$2^{156.52}$	Wang et al. (2009)
Multidimensional ZC	28	256	$2^{98.8}$	$2^{246.90}$	Bogdanov et al. (2012)
Linear	32	256	$2^{126.8}$	$2^{251.00}$	Zhao et al. (2014)

We introduce the following notation:

- The 128-bit plaintext is denoted by $[I_1, I_2, I_3, I_4]$.
- The 128-bit ciphertext is denoted by $[C_1, C_2, C_3, C_4]$.
- C^r is the 128-bit output after r rounds and C_i^r is the i-th word of C^r.
- The 2-round functions are denoted by F_0 and F_1.
- The transposition of a vector a is denoted by a^T.

Figure 13.4 shows the encryption process of r-round CLEFIA.

The encryption process is as follows. Let $WK_0, WK_1, WK_2, WK_3 \in \{0, 1\}^{32}$ be whitening keys and $RK_i \in \{0, 1\}^{32}$ $(0 \leq i \leq 2r)$ be round subkeys produced by the key schedule. For a 128-bit plaintext $[I_1, I_2, I_3, I_4]$, the ciphertext $[C_1, C_2, C_3, C_4]$ is computed as follows:

1. $C_1^0 = I_1, C_2^0 = I_2 \oplus WK_0, C_3^0 = I_3, C_4^0 = I_4 \oplus WK_1$.
2. For $i = 1$ to $r - 1$,
   $C_1^i = C_2^{i-1} \oplus F_0(C_1^{i-1}, RK_{2i-2}), C_2^i = C_3^{i-1}$
   $C_3^i = C_4^{i-1} \oplus F_1(C_3^{i-1}, RK_{2i-1}), C_4^i = C_1^{i-1}$.
3. $C_1^r = C_1^{r-1}, C_2^r = C_2^{r-1} \oplus F_0(C_1^{r-1}, RK_{2r-2}) \oplus WK_2$
   $C_3^r = C_3^{r-1}, C_4^r = C_4^{r-1} \oplus F_1(C_3^{r-1}, RK_{2r-1}) \oplus WK_3$.

The number of rounds r can be 18, 22, and 26 for CLEFIA-128, CLEFIA-192, and CLEFIA-256, respectively. We now described the round functions F_0 and F_1.

Let us denote the 32-bit output of the round function F_0, F_1 as $T = [T_{i,1}, T_{i,2}, T_{i,3}, T_{i,4}], i = 0, 1$ with $T_{i,j} \in \{0, 1\}^{32}$. Then $F_0(C_1^{i-1}, RK_{2i-2}), (1 \leq i \leq r)$ is computed as follows:

1. $U_i = C_1^{i-1} \oplus RK_{2i-2}$.
2. Let $U_i = [U_{i,1}, U_{i,2}, U_{i,3}, U_{i,4}]$ with $U_{i,j} \in \{0, 1\}^8$. Then $V_i = [V_{i,1}, V_{i,2}, V_{i,3}, V_{i,4}]$ is defined by $V_{i,1} = S_0(U_{i,1}), V_{i,2} = S_1(U_{i,2}), V_{i,3} = S_0(U_{i,3}), V_{i,4} = S_1(U_{i,4})$.
3. $[T_{i,1}, T_{i,2}, T_{i,3}, T_{i,4}]^T = M_0([V_{i,1}, V_{i,2}, V_{i,3}, V_{i,4}]^T)$.

Fig. 13.4 r rounds of CLEFIA

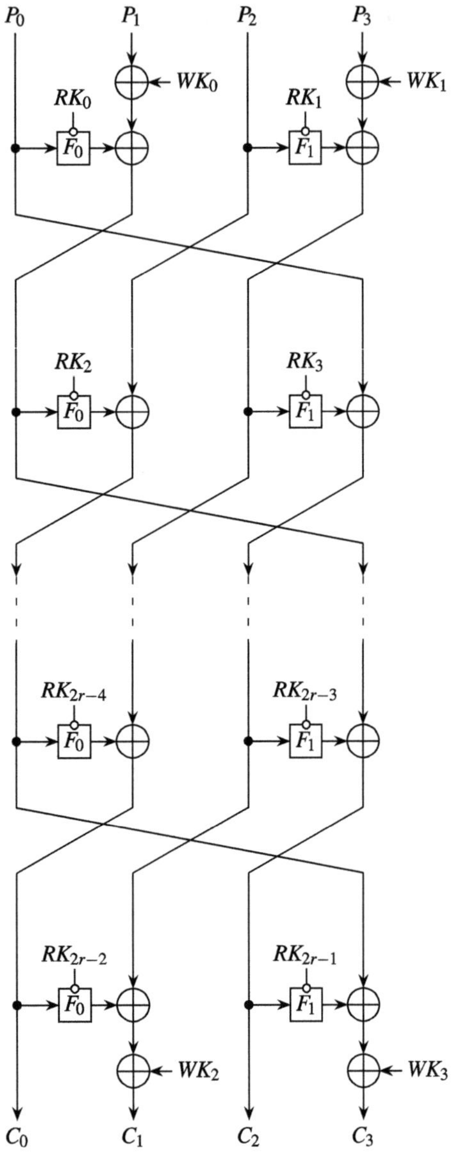

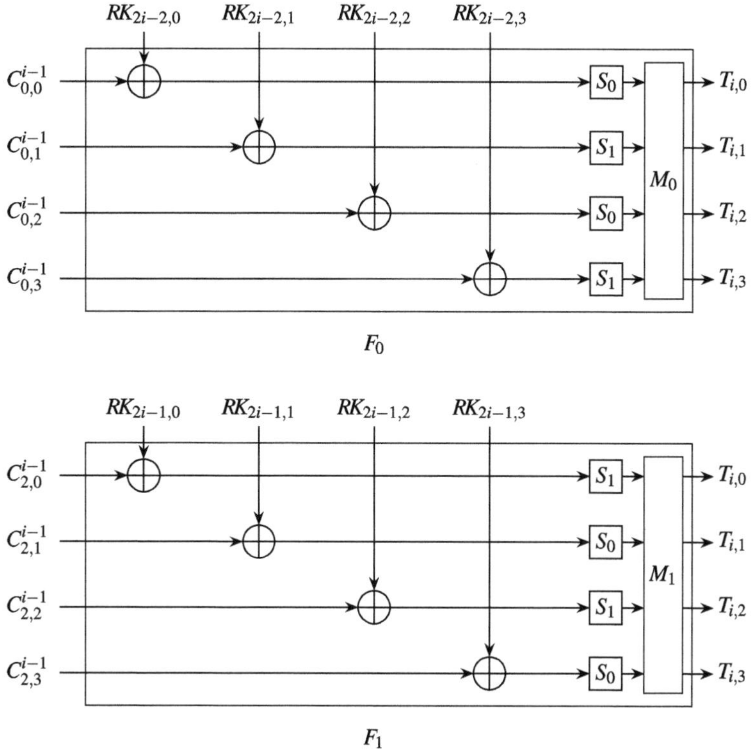

Fig. 13.5 CLEFIA. F-Functions

Here, S_0 and S_1 are two nonlinear 8-bit S-boxes, and M_0 is a 4×4 Hadamard- type matrix. $F_1(C_3^{i-1}, RK_{2i-1})$, $(1 \leq i \leq r)$ is similar to F_0 by replacing S_0 with S_1, S_1 with S_0, and M_0 with another 4×4 Hadamard-type matrix M_1. The functions F_0 and F_1 are represented in Fig. 13.5.

The round subkeys and whitening keys are independent of each other. The description of the matrices M_0 and M_1 is given in Tsunoo et al. (2008). The key schedule is explained in Chen et al. (2007).

13.3.2.2 Cryptanalysis of CLEFIA

Attacks on CLEFIA are NCPA based on impossible differentials on 9 rounds and then are extended on key recovery attacks on more rounds. For 10 and 11 rounds with whitening keys and 12 rounds without whitening keys, the 9-round differential is given by $[0, 0, 0, \alpha] \not\rightarrow [0, 0, 0, \alpha]$, with four 32-bit words (Shirai et al. 2007; Sony Corporation 2007) and different sizes for the key. In Tsunoo et al. (2008), the authors proposed another 9-round impossible differential that allowed to get key

recovery attacks on 12, 13, and 14 rounds according to the key length. The 9-round differential is given by
$[0, 0, 0, \alpha_{in}] \not\rightarrow [0, 0, 0, \alpha_{out}]$
$\alpha_{in} = [0, 0, 0, X]$, and $\alpha_{in} = [0, 0, Y, 0]$˙
In $[0, 0, 0, \alpha_{in}]$ (resp., $[0, 0, 0, \alpha_{out}]$, there are four 32-bit words and in $\alpha_{in} = [0, 0, 0, X]$, (resp., $[0, 0, Y, 0]$) there are four 8-bit words.

Truncated differential attacks were studied by Li et al. (2015). In Teczan (2010), Teczan presented improbable differential attacks on 13, 14, and 15 rounds of, respectively, CLEFIA-128, CLEFIA-192, and CLEFIA-256. However, the validity of this model was discussed by Blondeau (2013), where the author showed that the computation of the probability was not correct. Integral and zero-correlation distinguishers were introduced by Bogdanov et al. (2012). They showed that an integral attack implies the existence of a zero-correlation distinguisgher and that under some conditions the inverse is also true. In Bogdanov et al. (2014), provided multidimensional zero-correlation distinguisher. Finally, in Yi and Chen (2016), studied multiple zero-correlation linear attacks.

The summary of the attacks on CLEFIA is listed in Table 13.5, whereas usual r is the number of rounds.

Table 13.5 Summary of attacks on CLEFIA

Attack type	r	Key size	Chosen plaintexts	Time complexity (encryptions)	Reference
Impossible	13	192	$2^{119.8}$	2^{146}	Tsunoo et al. (2008)
Truncated differential	14	192	2^{100}	2^{135}	Li et al. (2015)
Integral	13	192	2^{113}	$2^{180.5}$	Li et al. (2011)
Integral	14	192	2^{128}	$2^{166.7}$	Yi and Chen (2016)
Multidimensional ZC	14	192	$2^{127.5}$	$2^{180.2}$	Bogdanov et al. (2014)
Multiple ZC	14	192	$2^{124.5}$	$2^{173.9}$	Yi and Chen (2016)
Impossible	14	256	$2^{120.3}$	2^{212}	Tsunoo et al. (2008)
Truncated differential	15	256	2^{100}	2^{203}	Li et al. (2015)
Integral	14	256	2^{113}	$2^{244.5}$	Li et al. (2011)
Integral	15	256	2^{128}	$2^{230.7}$	Yi and Chen (2016)
Multidimensional ZC	15	256	$2^{127.5}$	$2^{244.2}$	Bogdanov et al. (2014)
Multiple ZC	15	256	$2^{124.5}$	$2^{237.9}$	Yi and Chen (2016)

References

Abdelraheem, M. A., J. Alizadeh, H. Alkhzaimi, M. R. Aref, N. Bagheri, and P. Gauravaram. 2015. Improved linear cryptanalysis of reduced-round SIMON-32 and SIMON-48. In *Progress in cryptology? INDOCRYPT 2015*, vol. 9462, ed. A. Biryukov and V. Goya. Lecture notes in computer science, 153–179. Berlin: Springer.

Abed, F., E. List, S. Lucks, and J. Wenzel. 2015. Differential cryptanalysis of round-reduced SIMON and SPECK. In *Fast software encryption—FSE 2014*, ed. C. Cid and C. Rechberger, 546–570. Berlin: Springer.

Ahmadian, Z., S. Rasoolzadeh, M. Salmasizadeh, and M. R. Aref. 2015. Automated dynamic cube attack on block ciphers: cryptanalysis of SIMON and KATAN. In *Cryptology ePrint Archive, Report 2015/26136.*

Anderson, R., and E. Biham. 1996. Two practical and provably secure block ciphers: BEAR and LION. In *Fast software encryption—FSE '96*, vol. 1039, ed. D. Golmann. Lecture notes in computer science, 113–120. Springer.

Beaulieu, R., D. Shors, J. Smith, S. Treatman-Clark, B. Weeks, and L. Wingers. 2013. The SIMON and SPECK families of lightweight block ciphers. In *Cryptology ePrint Archive, Report 2013/404.*

Biham, E., O. Dunkelman, and N. Keller. 2007. Improved slide attacks. In *Fast software encryption—FSE '07*, vol. 4593. Lecture notes in computer science, 153–166. Berlin: Springer.

Biryukov, A., A. Roy, and V. Velichkov. 2015. Differential analysis of block ciphers SIMON and SPECK. In *Fast software encryption—FSE 2014*, ed. C. Cid and C. Rechberger, 546–570. Berlin: Springer.

Blondeau, C. 2013. Improbable differential from impossible differential: on the validity of the model. In *Progress in cryptology—INDOCRYPT '13*, vol. 6498, ed. G. Paul and G. Vaudenay. Lecture notes in computer science, 149–160. Berlin: Springer.

Bogdanov, A., H. Geng, M. Wang, L. Wen, and B. Collard. 2014. Zero-correlation linear cryptanalysis with FFT and improved attacks on ISO standard camellia and CLEFIA. In *Selected areas in cryptography—SAC '13*, vol. 8282, ed. T. Lange, K. Lauter, and P. Lisonek. Lecture notes in computer science, 306–324. Berlin: Springer.

Bogdanov, A., G. Leander, K. Nyberg, and M. Wang. 2012. Integral and multidimensional linear distinguishers with correlation zero. In *Advances in cyptology—ASIACRYPT 2012*, vol. 7658, ed. X. Wang and K. Sako. Lecture notes in computer science, 244–261. Berlin: Springer.

Chen, H., W. L. Wu, and D. G. Feng. 2007. Differential fault analysis on CLEFIA. In *Information and communications security—ICICS 2007*, vol. 4861, ed. S. Qing, H. Imai, and G. Wang. Lecture notes in computer science, 284–295. Berlin: Springer.

Courtois, N. T., and M. Misztal. 2011. Differential cryptanaysis of GOST. In *Cryptology ePrint Archive: Report 2011/312.*

Courtois, N. T., P. Sepehrdad, P. Sušil, and S. Vaudenay. 2012. *ElimLin algorithm revisited*, 306–325. Berlin: Springer.

Courtois, N. T. 2011. Security evaluation if GOST 28147-89 in View of national standardization. In *Cryptology ePrint Archive: Report 2011/211.*

Courtois, N. T. 2016. An improved differential attacks on full GOST. In *The new codebreakers*, vol. 9100, ed. P. Y. A. Ryan, D. Naccache, and J. J. Quisquater. Lecture notes in computer science, 282–303. Berlin: Springer.

Diffie, W., and G. Ledin. 2008. (translators): SMS4 encryption algorithm for wireless networks. In *Cryptology ePrint Archive: Report 2008/329.*

Dinur, I., O. Dunkelman, and A. Shamir. 2012. Improved attacks on full GOST cipher. In *Software encryption—FSE '12*, vol. 7549, ed. A. Canteaut. Lecture notes in computer science, 9–28. Berlin: Springer.

Fleischmann, E., M. Gorski, J. Hüehne, and S. Lucks. 2009. Key recovery attack on full GOST block cipher with negligible time and memory. In *WEWoRC '09*, vol. 6429. Lecture notes in computer science. Berlin: Springer.

Goubin, L., M. Ivasco, W. Jalby, O. Ly, V. Nachef, J. Patarin, J. Treger, and E. Volte. 2008. *CRUNCH*. Submission to NIST.

Isobe, T. 2011. A single-key attack on the full GOST cipher. In *Fast software encryption—FSE '11*, vol. 6733, ed. A. Joux. Lecture notes in computer science, 290–305. Berlin: Springer.

Kara, O. 2008. Reflection cryptanalysis of some ciphers. In *Progress in cryptology—INDOCRYPT '08*, vol. 5365, ed. D. R. Chowdhury, V. Rijmen, and A. Das. Lecture notes in computer science, 294–307. Berlin: Springer.

Ko, Y., S. H. Hong, W. I. Lee, S. J. Lee, and J. S. Kang. 2015. Related key differential attacks on 27 rounds of XTEA and full-round GOST. In *Fast software encryption—FSE '04*, vol. 3017, ed. B. K. Roy and W. Meier. Lecture notes in computer science, 299–316. Berlin: Springer.

Kölbl, S., G. Leander, and T. Tiessen. 2015. Observations on the SIMON block cipher family. In *Cryptology ePrint Archive: Report 2015/145.*

Li, L., K. Jia, X. Wang, and X. Dong. 2015. Meet-in-the-middle technique for truncated differential and its applications to CLEFIA. In *Fast software encryption—FSE '15*, vol. 9054. Lecture notes in computer science, 48–70. Berlin: Springer.

Li, Y., W. Wu, and L. Zhang. 2011. Improved integral attacks on reduced-round CLEFIA block cipher. In *WISA '11*, vol. 7115, ed. S. Jung and M. Yung. Lecture notes in computer science, 28–39. Berlin: Springer.

Maines, L., M. Piva, A. Rimoldi, and M. Sala. 2011. On the provable security of BEAR and LION schemes. *AAECC* 22 (3): 413–423.

Morin, P. 1996. Zero-correlation linear cryptanalysis with FFT and improved attacks on ISO standard camellia and CLEFIA. In *Selected areas in cryptography—SAC '96*. Lecture notes in computer science, 30–37. Berlin: Springer.

Nakahara, J. J., and M. Rasmussen. 2007. Linear analysis of reduced-round CAST-128 and CAST-256. In *SBSEG '07*, 45–55.

Raddum, H. 2015. *Algebraic analysis of the Simon Block Cipher family*, 157–169. Cham: Springer International Publishing.

Schneier, B. 1994. *Applied Cryptography*, 2nd ed. Protocols: Algorithms, and Source Code. C. Wiley, New-York.

Seki, K., and T. Kaneko. 2001. Differential cryptanalysis on reduced rounds of GOST. In *Selected areas in cryptography—SAC '00*, vol. 2012, ed. D. R. Stinson and S. E. Tavares. Lecture notes in computer science, 315–323. Berlin: Springer.

Seki, K., and T. Kaneko. 2009. Differential cryptanalysis of CAST-256 reduced to nine quad-rounds. *IEICE Transactions on Fundamentals of Electronics Communications and Computer Sciences* E844 (4): 913–918.

Shirai, T., K. Shibutani, S. Moriai, and T. Iwata. 2007. The 128-bit block cipher CLEFIA (Estended Abstract). In *Fast software encryption—FSE '07*, vol. 4593, ed. A. Biryukov. Lecture notes in computer science, 181–195. Berlin: Springer.

Sony Corporation: The 128-bit Block Cipher CLEFIA, Security and Performance Evaluations, Revision 1.0, June 1, 2007. Available at http://www.sony.co.jp/Products/clefia/technical/data/clefia-eval-1.0.pdf

Su, B. Z., W. L. WU, and W. T. Zhang. 2011. Security of the SMS block cipher against differential cryptanalysis. *Journal of Computer Science and Technology* 26 (1): 130–138.

Teczan, C. H. 2010. The improbable differential attack: cryptanalysis of reduced-round CLEFIA. In *Progress in cryptology—INDOCRYPT '10*, vol. 6498, ed. G. Gong and K. C. Gupta. Lecture notes in computer science, 197–209. Berlin: Springer.

Tsunoo, Y., E. Tsujihara, M. Shigeri, T. Saito, T. Suzaki, and H. Kubo. 2008. Impossible differential cryptanalysis of CLEFIA. In *Fast software encryption—FSE '08*, vol. 5086, ed. K. Nyberg. Lecture notes in computer science, 398–411. Berlin: Springer.

Wagner, D. 1999. The boomerang attack. In *Fast software encryption—FSE '99*, vol. 1636, ed. L. Knudsen. Lecture notes in computer science, 156–170. Berlin: Springer.

Wang, Q., Z. Liu, K. Varıcı, Y. Sasaki, V. Rijmen, and Y. Todo. 2014. Cryptanalysis of reduced-round SIMON32 and SIMON48. In *Progress in cryptology–INDOCRYPT 2014*, vol. 8885, ed. W. Meier and D. Mukhopadhyay. Lecture notes in computer science, 143–160. Berlin: Springer.
Wang, M., X. Wang, and C. Hu. 2009. New linear crypanalytic results of reduced-round of CAST-128 and CAST-256. In *Selected areas in cryptography—SAC '08*, vol. 5381, ed. R. Avanzi, L. Keliher, and F. Sica. Lecture notes in computer science, 429–441. Berlin: Springer.
Wen, L., M. Wang, A. Bogdanov, and H. Chen. 2014. General application of FFT in cryptanalysis and improved attacks on CAST-256. In *Progress in cryptology—INDOCRYPT '14*, vol. 8885, ed. W. Meier and D. Mukhopadhyay. Lecture notes in computer science, 161–176. Berlin: Springer.
Yi, W., and S. Chen. 2016. Improved integral and zero-correlation linear cyptanalysis of reduced-round CLEFIA Block cipher. In *Cryptology ePrint Archive: Report 2016/149.*
Zhao, J. Y., M. Q. Wang, and L. Wen. 2014. Improved linear cryptanalysis of CAST-256. *Journal of Computer Science and Technology* 29 (16): 1134–1139.

Part IV
Advanced Security Results

Chapter 14
Proof Beyond the Birthday Bound with the Coupling Technique

Abstract The coupling technique originates from the theory of Markov chains, and allows to upper bound the rate at which a Markov chain approaches its stationary distribution. Recasting the encryption process of a tuple of plaintexts through an iterative block cipher as a Markov chain, it is possible to use this tool to upper bound the advantage of a distinguisher against various kinds of Feistel schemes.

14.1 Feistel Networks as Shuffles

Block ciphers have a lot in common with card shuffles. Both aim at "mixing" a set of elements (n-bit strings for modern block ciphers, or a set of N cards for shuffles) by iterating (variations of) the same elementary transformation. While a block cipher should be indistinguishable from a random permutation in face of a computationally bounded adversary (which, in particular, might not have the entire codebook at his disposal), the traditional goal in card shuffling is much stronger: the permutation of the set of N cards generated by the shuffle should be statistically close to a uniformly random permutation of the set (in other words, it should be indistinguishable from a random permutation even by a computationally unbounded distinguisher which knows *all* the final positions of the cards).

Consider, for example, the Thorp shuffle, which works as follows. Assume N is even. The deck of N cards is split into two halves; for $i = 1, \ldots, N/2$, one drops the bottom card from either the right or the left half with probability $1/2$, and then the bottom card from the other half. Let us call this process a *round* of the shuffle. One round of the Thorp shuffles hence requires $N/2$ bits of randomness. However, note that tracking the location of a card after each round is a "local" process that does not require to look at other cards; such a shuffle is sometimes called *oblivious*, a desirable property if we were to use it to implement a block cipher.

When $N = 2^n$ is a power of two, the Thorp shuffle can be recast as a "maximally unbalanced" contracting Feistel cipher, i.e., a contracting Feistel cipher with domain $\{0, 1\}^n$ based on random round functions from $n - 1$ bits to 1 bit. Indeed, writing an n-bit string $x \in \{0, 1\}^n$ as $x = b\|x'$, where b and x are, respectively, the leftmost bit of x and the $n - 1$ rightmost bit of x, 1 round of a maximally unbalanced contracting

J. Patarin et al., *Guide to Feistel Ciphers*, Information Security and Cryptography,
https://doi.org/10.1007/978-3-031-99850-8_14

Feistel with round function f sends x to $x'\|b \oplus f(x')$. We see that each pair of strings $0\|x'$ and $1\|x'$ gets mapped to "consecutive" positions $x'\|0$ and $x'\|1$, in a way that depends on the random bit $f(x')$. This is exactly 1 round of the Thorp shuffle, the round function encoding the $N/2 = 2^{n-1}$ bits of randomness required for such a round.

Note that a shuffle defines a Markov chain on the state of the set of cards (in other words, on the set of permutations of the set of cards). Under mild conditions on the elementary transformation of the shuffle, the Markov chain is *ergodic* (i.e., there exists an integer t such as after t step, any final state is reachable from any starting state within $t$steps) and its stationary distribution is uniform, i.e., starting from any ordering of the cards, the distribution of the cards approaches the uniform. The obvious question is, *how fast*? This question can be tackled with the coupling technique, as we explain in the next section.

14.2 Definition and History of the Coupling Technique

The coupling technique is a major tool from the theory of Markov chains that allows to conveniently upper bound the so-called mixing time of a chain, i.e., the number of steps it takes for the chain, starting from any distribution, to be at statistical distance at most ε from its stationary distribution. For this, one considers running in a joint process (the *coupling*) two chains (C_1, C_2) with the same transitions probabilities, where the initial state of C_1 is distributed arbitrarily, and the initial state of C_2 has the stationary distribution of the chain (e.g., uniform). The crucial point is to "design" this joint process so as to ensure that the two chains will become equal quickly. Indeed, the statistical distance between the distributions of the two chains is upper bounded by the probability that the two states are different, as we explain now. This is in fact a consequence of a much more general result that applies to any two distributions, not just Markov chains.

Let μ and ν be two probability distributions on some non-empty finite set Ω. A *coupling* of μ and ν is a probability distribution λ on $\Omega \times \Omega$ whose marginal distributions are μ and ν, i.e.:

$$\begin{cases} \forall x \in \Omega, \sum_{y\in\Omega} \lambda(x, y) = \mu(x) \\ \forall y \in \Omega, \sum_{x\in\Omega} \lambda(x, y) = \nu(y). \end{cases}$$

The fundamental lemma of the coupling technique is the following one.

Lemma 14.1 *Let μ and ν be two probability distributions on some non-empty finite set Ω and let λ be a coupling of μ and ν. Let (X, Y) be a pair of random variables following distribution λ. Then*

$$\|\mu - \nu\| \le \Pr[X \ne Y]. \tag{14.1}$$

Proof Let λ be a coupling of μ and ν, and (X, Y) be a pair of random variables with distribution λ. By definition, we have that for any $z \in \Omega$, $\lambda(z, z) \leq \min\{\mu(z), \nu(z)\}$. Moreover, $\Pr[X = Y] = \sum_{z\in\Omega} \lambda(z, z)$. Hence we have

$$\Pr[X = Y] \leq \sum_{z\in\Omega} \min\{\mu(z), \nu(z)\}.$$

Therefore,

$$\begin{aligned}
\Pr[X \neq Y] &\geq 1 - \sum_{z\in\Omega} \min\{\mu(z), \nu(z)\} \\
&= \sum_{z\in\Omega} (\mu(z) - \min\{\mu(z), \nu(z)\}) \\
&= \sum_{\substack{z\in\Omega \\ \mu(z)\geq\nu(z)}} (\mu(z) - \nu(z)) \\
&= \max_{S\subset\Omega}\{\mu(S) - \nu(S)\} \\
&= \|\mu - \nu\|.
\end{aligned}$$

This proves the lemma. □

Given two distribution μ and ν, many couplings can be defined, yielding various upper bounds on the statistical distance $||\mu - \nu\|$. It is known there always exists an *optimal* coupling, i.e., a coupling which achieves equality in (14.1).

14.3 Application to Feistel Ciphers

Let E be a block cipher with key space $\mathcal{K}$ and message space $\mathcal{M}$, and let $(\mathcal{M})_q$ be the set of tuples of q distinct elements of $\mathcal{M}$. Let D be a non-adaptive chosen plaintext (NCPA) distinguisher against E. Recall that such a distinguisher makes q distinct queries $\mathbf{x} = (x_1, \ldots, x_q) \in (\mathcal{M})_q$ and receives q answers $(y_1, \ldots, y_q)$, based on which it outputs a single bit b, and that the best advantage it can achieve is upper bounded by

$$\mathscr{A}_E^{\mathrm{NCPA}}(q) = \max_{\mathbf{x}\in(\mathcal{M})_q} \|\mu_{\mathbf{x}} - \mu^*\|, \tag{14.2}$$

where $\mu_{\mathbf{x}}$ is the distribution of $(E_K(x_1), \ldots, E_K(x_q))$ over the draw of K uniformly at random in $\mathcal{K}$ and μ^* is the uniform distribution over $(\mathcal{M})_q$. Hence, we need to upper bound the statistical distance between the distribution of the outputs of the cipher on any tuple of inputs $\mathbf{x}$ and the uniform distribution over $(\mathcal{M})_q$. For this, we can remark that the uniform distribution over $(\mathcal{M})_q$ is exactly the distribution of the outputs of the cipher when given inputs $(X_1, \ldots, X_q)$ that are themselves uniformly distributed over $(\mathcal{M})_q$ (and independent from the random key K) and use the following hybrid argument in order to reason "query-by-query."

Lemma 14.2 *Let $\mathbf{x} = (x_1, \ldots, x_q) \in (\mathcal{M})_q$. For any $\ell \in \{0, \ldots, q-1\}$, let ν_ℓ be the distribution of $(E_K(x_1), \ldots, E_K(x_\ell), E_K(x_{\ell+1}))$ over the draw of K uniformly at random in $\mathcal{K}$, and let ν_ℓ^* be the distribution of $(E_K(x_1), \ldots, E_K(x_\ell), E_K(X_{\ell+1}))$, over the draw of K uniformly at random in $\mathcal{K}$ and of $X_{\ell+1}$ uniformly at random in $\mathcal{M} \setminus \{x_1, \ldots, x_\ell\}$. Then*

$$\|\mu_{\mathbf{x}} - \mu^*\| \leq \sum_{\ell=0}^{q-1} \|\nu_\ell - \nu_\ell^*\|.$$

Proof For a distribution ν on $(\mathcal{M})_q$, let

$$\nu(x_1, \ldots, x_\ell) \stackrel{\text{def}}{=} \Pr\left[X_1 = x_1, \ldots, X_\ell = x_\ell\right]$$
$$\nu(x_{\ell+1}|x_1, \ldots, x_\ell) \stackrel{\text{def}}{=} \Pr\left[X_{\ell+1} = x_{\ell+1}|X_1 = x_1, \ldots, X_\ell = x_\ell\right],$$

where $(X_1, \ldots, X_q)$ follows distribution ν. We define a coupling $(\mathbf{Y}, \mathbf{Z})$, where $\mathbf{Y} = (Y_1, \ldots, Y_q)$ has distribution $\mu_{\mathbf{x}}$ and $\mathbf{Z} = (Z_1, \ldots, Z_q)$ has distribution μ^*, as follows. First, we draw (Y_1, Z_1) according to the optimal coupling of Y_1 and Z_1. Then, for $\ell = 1, \ldots, q-1$, we proceed as follows: if $(Y_1, \ldots, Y_\ell) = (Z_1, \ldots, Z_\ell) = (y_1, \ldots, y_\ell)$, we draw $(Y_{\ell+1}, Z_{\ell+1})$ according to the optimal coupling of $\mu_{\mathbf{x}}(\cdot|y_1, \ldots, y_\ell)$ and $\mu^*(\cdot|y_1, \ldots, y_\ell)$. Otherwise, if $(Y_1, \ldots, Y_\ell) \neq (Z_1, \ldots, Z_\ell)$, we couple $(Y_{\ell+1}, Z_{\ell+1})$ arbitrarily. Then, we have

$$\begin{aligned}
\|\mu_{\mathbf{x}} - \mu^*\| &\leq \Pr\left[\mathbf{Y} \neq \mathbf{Z}\right] \qquad (14.3)\\
&\leq \sum_{\ell=0}^{q-1} \Pr\left[(Y_1, \ldots, Y_\ell) = (Z_1, \ldots, Z_\ell) \wedge Y_{\ell+1} \neq Z_{\ell+1}\right]\\
&\leq \sum_{\ell=0}^{q-1} \sum_{(y_1, \ldots, y_\ell)} \Pr\left[(Y_1, \ldots, Y_\ell) = (Z_1, \ldots, Z_\ell) = (y_1, \ldots, y_\ell)\right]\\
&\qquad \times \Pr\left[Y_{\ell+1} \neq Z_{\ell+1}|(Y_1, \ldots, Y_\ell) = (Z_1, \ldots, Z_\ell) = (y_1, \ldots, y_\ell)\right]\\
&= \sum_{\ell=0}^{q-1} \sum_{(y_1, \ldots, y_\ell)} \Pr\left[(Y_1, \ldots, Y_\ell) = (Z_1, \ldots, Z_\ell) = (y_1, \ldots, y_\ell)\right]\\
&\qquad \times \|\mu_{\mathbf{x}}(\cdot|y_1, \ldots, y_\ell) - \mu^*(\cdot|y_1, \ldots, y_\ell)\| \qquad (14.4)\\
&\leq \sum_{\ell=0}^{q-1} \sum_{(y_1, \ldots, y_\ell)} \Pr\left[(Y_1, \ldots, Y_\ell) = (y_1, \ldots, y_\ell)\right]\\
&\qquad \times \|\mu_{\mathbf{x}}(\cdot|y_1, \ldots, y_\ell) - \mu^*(\cdot|y_1, \ldots, y_\ell)\|
\end{aligned}$$

$$
\begin{aligned}
&= \sum_{\ell=0}^{q-1} \sum_{(y_1,\ldots,y_\ell)} \Pr\left[(Y_1,\ldots,Y_\ell) = (y_1,\ldots,y_\ell)\right] \\
&\qquad\qquad \times \frac{1}{2} \sum_{y_{\ell+1}} |\mu_{\mathbf{x}}(y_{\ell+1}|y_1,\ldots,y_\ell) - \mu^*(y_{\ell+1}|y_1,\ldots,y_\ell)| \\
&= \sum_{\ell=0}^{q-1} \frac{1}{2} \sum_{(y_1,\ldots,y_{\ell+1})} |\nu_\ell(y_1,\ldots,y_{\ell+1}) - \nu_\ell^*(y_1,\ldots,y_{\ell+1})| \qquad (14.5) \\
&= \sum_{\ell=0}^{q-1} \|\nu_\ell - \nu_\ell^*\|,
\end{aligned}
$$

where (14.3) comes from Lemma 14.1, (14.4) comes from the fact that $(Y_{\ell+1}, Z_{\ell+1})$ follows the optimal coupling of $\mu_{\mathbf{x}}(\cdot|y_1,\ldots,y_\ell)$ and $\mu^*(\cdot|y_1,\ldots,y_\ell)$ when $(Y_1,\ldots,Y_\ell) = (Z_1,\ldots,Z_\ell) = (y_1,\ldots,y_\ell)$, and (14.5) comes from the definition of ν_ℓ and ν_ℓ^*. □

We are now left with the easier task of upper bounding $\|\nu_\ell - \nu_\ell^*\|$ for each $\ell \in \{0,\ldots,q-1\}$. For this, we will use a coupling. While the reasoning up to now was completely generic, the specifics of the coupling will of course depend on the structure of the state transition function of the cipher. We will illustrate the technique with the example of balanced Feistel ciphers with independent and perfectly random round functions. Consider the r-round Feistel cipher Ψ^r with random n-bit round functions, and recall that the key space for this cipher is $(\mathscr{F}_n)^r$, a key being a tuple $(f_1,\ldots,f_r) \in (\mathscr{F}_n)^r$. In the following, we denote, respectively, $x_{i,0}$ and $x_{i,1}$ the left and right n-bit halves of x_i and for $j \in \{1,\ldots,r\}$ we let $x_{i,j+1}$ be recursively defined as

$$x_{i,j+1} = x_{i,j-1} \oplus f_j(x_{i,j}),$$

the ciphertext corresponding to x_i begin $x_{i,r}\|x_{i,r+1}$.

Our goal is now to define a coupling of ν_ℓ and ν_ℓ^* for the specific case at hand where $E = \Psi^r$. Recall that for this we must define a joint probability distribution on $(\mathscr{M})_{\ell+1} \times (\mathscr{M})_{\ell+1}$ (note that in this specific case, $\mathscr{M} = \{0,1\}^{2n}$). For this, we will consider two Feistel ciphers in parallel, one with inputs $(x_1,\ldots,x_\ell,x_{\ell+1})$ the second one with inputs $(x_1,\ldots,x_\ell,X_{\ell+1})$, where $X_{\ell+1}$ is uniformly random in $\mathscr{M} \setminus \{x_1,\ldots,x_\ell\}$, and we will define the round functions $(f_1',\ldots,f_r')$ in the second Feistel cipher based on what happens in the first Feistel cipher with round functions $(f_1,\ldots,f_r)$. More precisely, we prove the following lemma.

Lemma 14.3 *Let n, r, ℓ, ν_ℓ, and ν_ℓ^* be as above. Then*

$$\|\nu_\ell - \nu_\ell^*\| \le \left(\frac{4\ell}{2^n}\right)^{r'},$$

where $r' = \lfloor (r-1)/2 \rfloor$.

Proof In order to define the coupling between the two distributions ν_ℓ and ν_ℓ^*, we imagine two Feistel ciphers run in parallel, one with inputs $(x_1, \ldots, x_\ell, x_{\ell+1})$, the other with inputs $(x_1, \ldots, x_\ell, X_{\ell+1})$, where $X_{\ell+1}$ is uniformly random in $\mathcal{M} \setminus \{x_1, \ldots, x_\ell\}$, and we will "choose" the round functions $(f_1', \ldots, f_r')$ in the second Feistel cipher depending on what happens in the first one. First, we set $f_f' = f_f$ on every input encountered when enciphering $x_1, \ldots, x_\ell$, i.e., for $i = 1, \ldots, \ell$ and $j = 1, \ldots, r$, we simply let $f_j'(x_{i,j}) = f_j(x_{i,j})$. Note that this implies that the intermediate states and the outputs corresponding to inputs $x_1, \ldots, x_\ell$ are the same for both Feistel schemes. It remains to couple the last inputs $x_{\ell+1}$ and $X_{\ell+1}$. Let us denote, respectively, $X_{\ell+1,0}$ and $X_{\ell+1,1}$ the left and right n-bit halves of $X_{\ell+1}$. For $j = 1, \ldots, r$, we proceed as follows:

(i) if $X_{\ell+1,j} \in \{x_{1,j}, \ldots, x_{\ell,j}\}$, then $f_j'(X_{\ell+1,j})$ has already been set to $f_j(x_{i,j})$ for the index i such that $X_{\ell+1,j} = x_{i,j}$;
(ii) if $X_{\ell+1,j} \notin \{x_{1,j}, \ldots, x_{\ell,j}\}$ and $x_{\ell+1,j} \in \{x_{1,j}, \ldots, x_{\ell,j}\}$, we draw $f_j'(X_{\ell+1,j})$ uniformly at random in $\{0, 1\}^n$;
(iii) if $X_{\ell+1,j} \notin \{x_{1,j}, \ldots, x_{\ell,j}\}$ and $x_{\ell+1,j} \notin \{x_{1,j}, \ldots, x_{\ell,j}\}$, then we define

$$f_j'(X_{\ell+1,j}) = f(x_{\ell+1,j}) \oplus x_{\ell+1,j-1} \oplus X_{\ell+1,j-1}.$$

Note that in case (iii) we have $X_{\ell+1,j+1} = x_{\ell+1,j+1}$. The process we just described defines a joint probability distribution on $(\mathcal{M})_q \times (\mathcal{M})_q$. To show that it is a coupling of ν_ℓ and ν_ℓ^*, we must prove that it has the expected marginal distributions. This is clear for the outputs of the first Feistel cipher. For the outputs of the second Feistel cipher, it is easy to see that functions $(f_1', \ldots, f_r')$ are uniformly distributed. This is clear when $f_j'(X_{\ell+1,j})$ is chosen according to rule (i) or (ii). When it is chosen according to rule (iii), note that $f(x_{\ell+1,j})$ is uniformly random since $x_{\ell+1,j} \notin \{x_{1,j}, \ldots, x_{\ell,j}\}$, so that $f_j'(X_{\ell+1,j})$ is uniformly random as well.

In order to apply Lemma 14.1, we need to upper bound the probability that the outputs of the two Feistel ciphers are different. Let us denote the two outputs corresponding to the $(\ell + 1)$th input

$$Y = x_{\ell+1,r} \| x_{\ell+1,r+1}$$
$$Z = X_{\ell+1,r} \| X_{\ell+1,r+1}.$$

Note that, as already mentioned, the ℓ first outputs are always equal by construction. Hence, we only have to upper bound the probability that the outputs Y and Z for the $(\ell + 1)$-th input are different. For this, note that by the very way the coupling was defined, $Y = Z$ as soon as there exists two consecutive rounds j and $j + 1$ such that $X_{\ell+1,j} = x_{\ell+1,j}$ and $X_{\ell+1,j+1} = x_{\ell+1,j+1}$. (Indeed, once this is true, subsequent round function values will always be set according to rule (i) or (iii), and it is easy to check that this will ensure that subsequent values of the state will be equal.) Moreover, we already observed that setting f_j' according to rule (iii) immediately implies that

$X_{\ell+1,j+1} = x_{\ell+1,j+1}$. Hence, $Y = Z$ as soon as there exists two consecutive rounds $j-1$ and j such that

$$\begin{cases} x_{\ell+1,j-1} \notin \{x_{1,j}, \dots, x_{\ell,j-1}\} \\ X_{\ell+1,j-1} \notin \{x_{1,j}, \dots, x_{\ell,j-1}\} \\ x_{\ell+1,j} \notin \{x_{1,j}, \dots, x_{\ell,j}\} \\ X_{\ell+1,j} \notin \{x_{1,j}, \dots, x_{\ell,j}\}. \end{cases} \tag{14.6}$$

Note that $\mathbf{x}$ is freely chosen by the adversary, so we cannot hope to couple at the first round (indeed, the adversary is free to set $x_{\ell+1,1} = x_{i,1}$ for some $i < \ell+1$). However, starting from $j = 2$, the probability that $x_{\ell+1,j} \in \{x_{1,j}, \dots, x_{\ell,j}\}$ is at most $\ell/2^n$. Indeed, for each $i \in \{1, \dots, \ell\}$, either $x_{\ell+1,j-1} = x_{i,j-1}$, in which case the probability that $x_{\ell+1,j} = x_{i,j}$ is zero (since otherwise this would contradict the assumption that $x_{\ell+1} \neq x_i$), or $x_{\ell+1,j-1} \neq x_{i,j-1}$, in which case the probability that

$$x_{\ell+1,j} = x_{i,j} \Leftrightarrow x_{\ell+1,j-2} \oplus f'_{j-1}(x_{\ell+1,j-1}) = x_{i,j-2} \oplus f'_{j-1}(x_{i,j-1})$$

is exactly 2^{-n} since f'_{j-1} is uniformly random. By a union bound, the probability that conditions (14.6) are not fulfilled for $j = 3$ (and that the coupling fails at rounds 2 and 3) is at most $4\ell/2^n$. If the coupling fails at rounds 2 and 3, we can start again at rounds 4 and 5, etc. Hence, in total we have $r' = \lfloor (r-1)/2 \rfloor$ trials, so that the probability that $Y \neq Z$ is at most

$$\left(\frac{4\ell}{2^n}\right)^{r'}.$$

This concludes the proof. □

From Lemma 14.3, it is now easy to upper bound the advantage of any NCPA distinguisher against an r-round balanced Feistel cipher.

Theorem 14.1 *Let n, r, and q be integers and Ψ^r be the r-round balanced Feistel scheme with random n-bit round functions. Then*

$$\mathscr{A}^{\mathrm{NCPA}}_{\Psi^r}(q) \leq \frac{2^{r'}}{r'+1} \times \frac{q^{r'+1}}{2^{r'n}},$$

where $r' = \lfloor (r-1)/2 \rfloor$.

Proof Combining Eq. (14.2) and Lemmas 14.2 and 14.3, we obtain

$$\begin{aligned}\mathscr{A}^{\mathrm{NCPA}}_{\Psi^r}(q) &\leq \sum_{\ell=0}^{q-1} \left(\frac{4\ell}{2^n}\right)^{r'} \\ &\leq \left(\frac{2}{2^n}\right)^{r'} \int_{\ell=0}^{q} \ell^{r'} d\ell \\ &\leq \frac{2^{r'}}{r'+1} \times \frac{q^{r'+1}}{2^{r'n}},\end{aligned}$$

which proves the result. □

In other words, the r-round balanced Feistel cipher with random round functions is secure up to roughly $2^{r'n/(r'+1)}$ queries, which approaches 2^n as r' (and hence r) increases.

A drawback of the coupling technique is that it only allows to consider non-adaptive adversaries. In order to lift the results to adaptive adversaries, one has to appeal to the following *security amplification* result.

Lemma 14.4 *Let E and F be two block ciphers with the same message space $\mathscr{M}$ and respective key spaces $\mathscr{K}_E$ and $\mathscr{K}_F$. Let $F^{-1} \circ E$ be the block cipher with key space $\mathscr{K}_E \times \mathscr{K}_F$ and message space $\mathscr{M}$ which maps a plaintext $M \in \mathscr{M}$ under a key (K_E, K_F) to $F^{-1}(K_F, E(K_E, M))$. Then*

$$\mathscr{A}^{\mathrm{CCA}}_{F^{-1}\circ E}(q) \leq \mathscr{A}^{\mathrm{NCPA}}_{E}(q) + \mathscr{A}^{\mathrm{NCPA}}_{F}(q).$$

Hence, given a block cipher E which resists NCPA attacks, we can obtain a block cipher which resists CCA attacks by simply cascading E and its inverse with two independent keys. Using this result, we obtain the following for the CCA security of the r-round Feistel cipher.

Theorem 14.2 *Let n, r, and q be integers and Ψ^{2r-1} be the $2r-1$-round balanced Feistel scheme with random n-bit round functions. Then*

$$\mathscr{A}^{\mathrm{CCA}}_{\Psi^{2r-1}}(q) \leq \frac{2^{r'}}{r'+1} \times \frac{q^{r'+1}}{2^{r'n}},$$

where $r' = \lfloor (r-1)/2 \rfloor$.

Proof Let σ be the function which swaps the two n-bit halves of a $2n$-bit string. Then a $(2r-1)$-round Feistel scheme with round functions $f_1, \ldots, f_{2r-1}$ can be written as $\sigma \circ F^{-1} \circ E$, where E and F are r-round Feistel schemes. This can be seen by writing the middle round function f_r as the XOR of two independent round functions $f_r' \oplus f_r''$ (clearly, this does not change the distribution of the outputs of the system): then E is the Feistel scheme with round functions $f_1, \ldots, f_{r-1}, f_r'$, while F is the Feistel scheme with round functions $f_{2r-1}, \ldots, f_{r+1}, f_r''$. The result then follows from Lemma 14.4 and Theorem 14.1 (clearly composing with σ does not change the advantage). □

14.4 Further Reading

The Thorp shuffle was first studied in Thorp (1973). The coupling technique was introduced by Doeblin in order to upper bound the mixing time XXX Lemma 14.1 is originally due to Aldous (1983). The first use of couplings in cryptography is due to Mironov (2002), who used it to analyze the RC4 stream cipher. It was first applied to (maximally unbalanced) Feistel ciphers by Morris et al. (2009). This was generalized to other types of generalized Feistel ciphers by Hoang and Rogaway (2010). Subsequently, the coupling technique was used to analyze the iterated Even-Mansour cipher (Lampe et al. 2012), tweakable block cipher constructions (Lampe and Seurin 2013), and Feistel schemes where round functions are of the form $x \mapsto F(x \oplus k)$ where F is a random oracle and k the secret key (Lampe and Seurin 2014). A proof of the amplification Lemma 14.4 can be found in Maurer et al. (2007) or Cogliati et al. (2014).

References

Aldous, D. J. 1983. Random walks on finite groups and rapidly mixing Markov chains. In *Séminaire de Probabilités XVII*. Volume 986 of Lecture notes in mathematics, 243–297. Berlin: Springer.

Cogliati, B., J. Patarin, and Y. Seurin. 2014. Security amplification for the composition of block ciphers: simpler proofs and new results. In *Selected areas in cryptography—SAC 2014*, ed. A. Joux and A. M. Youssef. Volume 8781 of LNCS, 129–146. Berlin: Springer.

Hoang, V. T., and P. Rogaway. 2010. On generalized feistel networks. In *Advances in cryptology—CRYPTO 2010*, ed. T. Rabin. Volume 6223 of LNCS, 613–630. Berlin: Springer.

Lampe, R., J. Patarin, and Y. Seurin. 2012. An asymptotically tight security analysis of the iterated even-mansour cipher. In *Advances in cryptology—ASIACRYPT 2012*, ed. X. Wang and K. Sako. Volume 7658 of LNCS, 278–295. Berlin: Springer.

Lampe, R., and Y. Seurin. 2013. Tweakable blockciphers with asymptotically optimal security. In *Fast software encryption—FSE 2013*, ed. S. Moriai. Volume 8424 of LNCS, 133–151. Berlin: Springer.

Lampe, R., and Y. Seurin. 2014. Security analysis of key-alternating feistel ciphers. In *Fast software encryption—FSE 2014*, ed. C. Cid and C. Rechberger. Volume 8540 of LNCS, 243–264. Berlin: Springer.

Maurer, U. M., K. Pietrzak, and R. Renner. 2007. Indistinguishability amplification. In *Advances in cryptology—CRYPTO 2007*, ed. A. Menezes. Volume 4622 of LNCS, 130–149. Berlin: Springer. Full version available at http://eprint.iacr.org/2006/456.

Mironov, I. 2002. (Not So) Random Shuffles of RC4. In *Advances in cryptology—CRYPTO 2002*, ed. M. Yung. Volume 2442 of LNCS, 304–319. Berlin: Springer.

Morris, B., P. Rogaway, and T. Stegers. 2009. How to encipher messages on a small domain. In *Advances in cryptology—CRYPTO 2009*, ed. S. Halevi. Volume 5677 of LNCS, 286–302. Berlin: Springer.

Thorp, E. O. 1973. Nonrandom shuffling with applications to the game of faro. *Journal of the American Statistical Association* 68 (344): 842–847.

Chapter 15
Introduction to Mirror Theory

Abstract "Mirror Theory" is the theory that evaluates the number of solutions of affine systems of equalities (=) and non-equalities ($\neq$) in finite groups. It is deeply related to the security and attacks of many generic cryptographic secret-key schemes, like random Feistel schemes (balanced or unbalanced), Misty schemes, XOR of two pseudo-random bijections to generate a pseudo-random function, etc. In this chapter, we will assume that the groups are abelian. Most of the time in cryptography the group is $((\mathbb{Z}/2\mathbb{Z})^n, \oplus)$ and this chapter concentrates on these cases. We present here general definitions, some theorems, and many examples and computer simulations.

15.1 Definitions

Definition 15.1 (*Mirror System T, and $H(T)$*) A "Mirror system" T is a set of affine equations (=) or affine non-equalities ($\neq$) in a finite group G. We assume that the group G is abelian. Very often in cryptography G will be $G = \big((\mathbb{Z}/2\mathbb{Z})^n, \oplus\big)$. Then T is a set of equations (and, respectively, of non-equalities) of the form: $X_1 \oplus X_2 \ldots \oplus X_k = c$ (respectively, $\neq c$), where c is a constant in G. Let m be the number of equalities in T, and v be the number of variables $X_1, \ldots, X_v$ in T.

Each equality of G has a linear part and a constant. We denote by $c_1, \ldots, c_m$ these m constants (c_i can be 0 or not). We denote by $H(T)$, or by $H(c_1, \ldots, c_m)$, or simply by H, the number of solutions $(X_1, \ldots, X_v)$ of T.

Remark 15.1 In more general abelian groups, we can have variables X_i or X_i^{-1} in these equalities or non-equalities, and the same variable can appear more than one time (for example, $X_1 * X_1 * X_1 * X_2 * X_2 = c$), i.e., we can have some coefficients in the affine equations. However in the cryptographic applications if we change $G = ((\mathbb{Z}/2\mathbb{Z})^n, \oplus)$ for another abelian group, we will generally have no coefficient (except 0 or 1) in the variables of the mirror systems that we want to study. Therefore, the definitions, analysis, and results obtained for $G = ((\mathbb{Z}/2\mathbb{Z})^n, \oplus)$ will be generally very similar for other abelian groups in most cryptographic applications.

J. Patarin et al., *Guide to Feistel Ciphers*, Information Security and Cryptography,
https://doi.org/10.1007/978-3-031-99850-8_15

Definition 15.2 First case: When $G = ((\mathbb{Z}/2\mathbb{Z})^n, \oplus)$.

Let T be a mirror system. We will say that an equation E can be obtained (or deduced) from T "by linearity" when we can obtain E by XORing some equalities of T.

Second case: More general abelian groups $(G, *)$.

Here the definition is more complex since from one equation we also have its inverse equation (for example, from $X_1 * X_2 = c$, we can also use $X_1^{-1} * X_2^{-1} = c^{-1}$), and since it is useful to integrate some simplification rules on the coefficients, but also because we have to be careful about a coefficient making a variable to 0. All these points can be solved, but we give no details here since in this chapter we concentrate only on First Case. (Moreover as said in Remark 15.1 above for cryptographic applications we generally have mirror systems with no coefficient.)

Definition 15.3 Let X_1 be a variable of T. A "minimal equation" for X_1 is an equation B such that:

- we can deduce B from T by linearity;
- B has the variable X_1;
- all the other equations with the variables X_1 that we can deduce by linearity from T have at least more or the same number of variables than B.

Definition 15.4 (*block of variables*, $\xi(A)$, $\xi_{\max}$, depth(A)) We will say that two variables X_i and X_j are "in the same block" when we can deduce this from these rules:

- If $(i = j)$ then X_i and X_j are in the same block.
- If there is a minimal equation for X_i with the variable X_j then X_i and X_j are in the same block.
- If there is a variable X_k such that (X_i and X_k are in the same block) and (X_j and X_k are in the same block) then X_i and X_j are in the same block.

When A is a block of variables $\xi(A)$ denotes the number of variables in A. $\xi_{\max}$ denotes the maximum value $\xi(A)$ for a block A. depth(A) denotes the minimum number of variables that we have to fix in order to fix by linearity all the variables of A.

For example if T is: $x_1 \oplus x_3 = x_2 \oplus x_4$, then in T we have only one block A, with $\xi(A) = 4$, and depth$(A) = 3$. However, if we add the equation $x_1 = x_2 \oplus a$, then now T has two blocks: $x_1 = x_2 \oplus a$ and $x_3 = x_4 \oplus a$, with $\xi = 2$ and depth $= 1$ in these two blocks.

From Definition 15.4, we see that "being in the same block" is (as expected) an equivalence relation.

Example 15.1 ("*$\xi = 3$ and $\xi = 2$ system*") Let T be this system of four equations on $(\mathbb{Z}/2\mathbb{Z})^n$, with pairwise distinct variables $P_1, P_2, P_3, P_4, P_5, P_6, P_7$:

$$P_2 = P_1 \oplus c_1$$

$$P_4 = P_3 \oplus c_2$$

$$\begin{cases} P_6 = P_5 \oplus c_3 \\ P_7 = P_5 \oplus c_4 \end{cases}$$

c_1, c_2, c_3, c_4 are the constants.

We have here three blocks of equations, two blocks with $\xi = 2$, and one block with $\xi = 3$, so $\xi_{\max} = 3$.

Definition 15.5 (*Regular Systems*) We will say that T is a "regular system" if it is a mirror system that satisfies these properties S1 and S2:

S1: By linearity from the equalities of T we cannot obtain $X_i =$ a constant, or $X_i \neq$ a constant (where X_i is one of the variables of T), i.e., we always have by linearity at least two variables in $=$ or $\neq$.
S2: Let $X'_1, \ldots, X'_a$ be the variables of a block B, and $X_1, \ldots, X_q$ the other variables of T (not in B). We say that we have property S2 when: for each block B of T, when $X_1, \ldots, X_q$ are fixed, the number of $X'_1, \ldots, X'_a$ that satisfy the non-equalities ($\neq$) do not depend on $X_1, \ldots, X_q$.

Example 15.2 In Example 15.1 above system T is a "regular system." For example when P_1, P_2, P_3, P_4 are fixed, for (P_5, P_6, P_7) we have exactly $(2^n - 4)(2^n - 5)(2^n - 6)$ solutions that satisfy all the non-equalities ($\neq$).

However with this system T':

$$X_1 \oplus X_2 = c_1,$$

$$X_3 \oplus X_4 = c_2,$$

$$X_5 \oplus X_6 = c_3,$$

$$X_5 \neq X_1, \; X_5 \neq X_3,$$

T' is not a "regular system" since when X_1, X_2, X_3, X_4 are fixed, the number of solutions (X_5, X_6) that satisfy $X_5 \neq X_1$ and $X_5 \neq X_3$ depend on the fact that $X_1 = X_3$ or not.

Definition 15.6 ("*Standard form*") A system T is in "standard form" when all the non-equalities ($\neq$) of T are of the form $X_i \neq X_j$ (with $i \neq j$). By introducing new variables X_k it is always possible to write a system T in standard form.

Definition 15.7 (Weight(T)) Generally in cryptographic applications we will have a constant on each equation. Then, Weight(T) is the number of $(X_1, \ldots, X_v)$ that satisfy only the non-equalities ($\neq$) of T (i.e., we give up here the equalities). Now if we have equations with no constants, then Weight(T) will be the nomber of solutions of these equations and of the non-equalities. In both cases, we have

$$\text{Weight}(T) = \sum_{c_1,\ldots,c_m} H(T).$$

In standard form, Weight(T) is always easy to compute.

Example 15.3 In Example 15.1, Weight$(T) = 2^n(2^n - 1)(2^n - 2)(2^n - 3)(2^n - 4)(2^n - 5)(2^n - 6)$.

Definition 15.8 (*Block conditions and* Space(T)) The "block conditions" are equalities or non-equalities on the constants c_i that we can deduce by linearity by using the equalities and non-equalities of T (when we consider $c_1, \ldots, c_m$ as variables).

Example 15.4 In Example 15.1, the "Block conditions" are

$$c_1 \neq 0, c_2 \neq 0, c_3 \neq 0, c_4 \neq 0, c_3 \neq c_4.$$

We say that a constant c_i is "compatible with T by linearity" if c_i satisfies all the block conditions.

When T is in standard form, Space(T) is the number of $(c_1, \ldots, c_m)$ that are compatible with T by linearity. Space(T) is also easy to compute.

Example 15.5 In Example 15.1

$$\text{Space}(T) = (2^n - 1)^3(2^n - 2).$$

Remark 15.2 As said above, any system can be written in standard form. However the resulting system will generally not have the same weight, or space.

Definition 15.9 ($\tilde{H}$ *and* $\tilde{M}$) We will denote

$$\tilde{H} = \frac{\text{Weight}(T)}{\text{Space}(T)}$$

$\tilde{H}$ is the mean value of H when $(c_1, \ldots, c_m)$ are randomly chosen compatible by linearity with T.

$$\tilde{H} = \sum_{c_1,\ldots,c_m \text{ compatible by linearity}} \frac{H(T)}{\text{Number of } c_1, \ldots, c_m \text{ compatible by linearity}}.$$

We will denote

$$\tilde{M} = \frac{\text{Weight}(T)}{|G|^m}.$$

$\tilde{M}$ is the mean value of H when $(c_1, \ldots, c_m)$ are randomly chosen in G^m.

$$\tilde{M} = \sum_{(c_1,\dots c_m)\in G^m} \frac{H(T)}{|G|^m}.$$

Definition 15.10 (*Tame and Wild systems*) We say that T is "Tame" (on $(c_1, \dots, c_m)$) when $H \simeq \tilde{H}$, i.e., $H \simeq \frac{\text{Weight}(T)}{\text{Space}(T)}$ (here $c_1, \dots, c_m$ are fixed).

We say that T is "Wild" when T is not Tame.

For a system T (now $c_1, \dots, c_m$ are not fixed), the "wild" coefficient is defined as $\text{W}(T) = \frac{E(|H-\tilde{H}|)}{\tilde{H}}$ where E is the mean value function.

We say that T is "Tame on average" when for randomly chosen $(c_1, \dots, c_m)$ compatible by linearity with T there is a high probability that T is Tame, i.e., when $\text{W}(T) \ll 1$.

We say that T is "Always Tame" or "Tame in the worst case" when for all constants $(c_1, \dots, c_m)$ compatible with T by linearity

$$H(c_1, \dots, c_m) \gtrsim \frac{\text{Weight}(T)}{\text{Space}(T)}.$$

We say that H is "homogeneous" when for all constants $(c_1, \dots, c_m)$ compatible with T by linearity

$$H(c_1, \dots, c_m) \simeq \frac{\text{Weight}(T)}{\text{Space}(T)}.$$

Remark 15.3 The use of the fuzzy term ($\simeq$) in the definition of "Tame" can appear at first surprising. However, as we will see, "Tame" will be closely related to "Secure" in most generic cryptographic applications. In the same way that we use "Advantage" to evaluate precisely security, we can use the wild coefficient (defined without the fuzzy term) to evaluate Tame.

Remark 15.4 Very often systems T will have a very small number of $(c_1, \dots, c_m)$ with $H(c_1, \dots, c_m)$ much larger than $\frac{\text{Weight}(T)}{\text{Space}(T)}$. This will generally not be a problem (as long as this number of $(c_1, \dots, c_m)$ is small) and this is why in the definition of "Always Tame" we used $\gtrsim$ instead of $\simeq$. Homogeneous systems seldom appear and are at present much less important than Always Tame systems in systems that are used in cryptography.

We say that H is "σ Tame" when for constants $(c_1, \dots, c_m)$ randomly chosen compatible by linearity with T the standard deviation $\sigma(H)$ of H satisfies: $\sigma(H) \ll \frac{\text{Weight}(T)}{\text{Space}(T)}$ ($a \ll b$ means as usual that a is small compared with b).

"Mirror Theory" is the theory that evaluates the number of solutions H of mirror systems T. A particularly important aim in Mirror Theory is to evaluate when T is Tame. As we will see this is closely related to the security of many generic cryptographic designs, where "Tame" will be associated with "secure" (with a proof of security).

Remark 15.5 The nickname "Mirror" comes from the fact that we will have a huge number of induction formulas between these systems, and also a huge number of induction formulas with the systems related to $\sigma(H)$.

15.2 First Properties

Theorem 15.1 *For all mirror systems T,*

$$\begin{array}{ccccc} & \nearrow & \text{Always Tame} & \searrow & \\ \text{homogeneous} & & & & \text{Tame in average} \\ & \searrow & \sigma\ \text{Tame} & \nearrow & \end{array}$$

Proof · Homogeneous $\Longrightarrow$ Always Tame: Comes immediately from the definitions.

· Homogeneous $\Longrightarrow$ σ Tame: If $\forall c_1, \ldots, c_m, |H - E(H)| \leq \varepsilon$, then $\sigma(H) \leq \varepsilon$.

· Always Tame $\Longrightarrow$ Tame in average: If $\forall c_1, \ldots, c_m$, $H \geq E(II) - \varepsilon$, then $E(|H - E(H)|) \leq 2\varepsilon$.

· σ Tame $\Longrightarrow$ Tame in average: $E(|H - E(H)|) \leq \sigma(H)$ (see Cauchy–Schwarz or Jensen's inequality since $x \mapsto x^2$ is a convex function).

However σ Tame $\not\Longrightarrow$ Always Tame (e.g., a small probability where $H \ll E(H)$), and Always Tame $\not\Longrightarrow$ σ Tame (e.g., $H \geq E(H) - \varepsilon$ and with probability $\frac{\varepsilon}{2A}$, $H \geq E(H) + A$ with $A \geq \frac{1}{\varepsilon}$). □

Theorem 15.2 *Let T' be a regular system. Let T be a sub-system of T' where some blocks of T' have been removed. Then:*

- *If T' is Tame, T is Tame.*
- *If T' is Always Tame, T is Always Tame.*
- $\mathrm{W}(T) \leq \mathrm{W}(T')$.

Proof Let $X'_1, X'_2, \ldots, X'_v$ be the variables in T' and not in T. Let $X_1, X_2, \ldots, X_q$ be the variables of T. Let $c'_1, c'_2, \ldots, c'_\mu$ be the constants in T' and not in T. Let $c_1, c_2, \ldots, c_\alpha$ be the constants of T.

Since T' is regular:

$$H(c_1, \ldots, c_\alpha) = \frac{\sum\limits_{c'_1, \ldots, c'_\mu} H(c_1, \ldots, c_\alpha, c'_1, \ldots, c'_\mu)}{[\text{Number of } X'_1, \ldots, X'_v \text{ that satisfy the } \neq \text{ when } X_1, \ldots, X_q \text{ is fixed}]}. \tag{15.1}$$

Therefore $H(c_1, \ldots, c_\alpha)$ of T is proportional (with a fixed constant) to the average value (on $c'_1, \ldots, c'_\mu$) of $H(c_1, \ldots, c_\alpha, c'_1, \ldots, c'_\mu)$ of T'.

By definition,

$$\mathrm{W}(T) = \frac{E(|H - E(H)|)}{E(H)}, \tag{15.2}$$

$$\text{and } \mathrm{W}(T) = \frac{E(|H' - E(H')|)}{E(H')}, \tag{15.3}$$

where H' denotes $H(c_1, \ldots, c_\alpha, c'_1, \ldots, c'_\mu)$. From Theorems 15.1 and 15.2 we get $\mathrm{W}(T) \leq \mathrm{W}(T')$ as claimed (we regroup some terms of H' in H, and according to the absolute value $\mathrm{W}(T) \leq \mathrm{W}(T')$). □

Definition 15.11 *((P1), (P2), and (P3) properties)* We will denote by (P1), (P2), and (P3) these properties (they can be satisfied or not):

> (P1): For all $(c_1, \ldots, c_m)$ compatible by linearity with T we have $H(c_1, \ldots, c_m) \neq 0$.

P1 means that if $(c_1, \ldots, c_m)$ are compatible by linearity, then $(c_1, \ldots, c_m)$ are really compatible.

> (P2): For all $(c_1, \ldots, c_m)$ compatible by linearity with T we have $H(c_1, \ldots, c_m) \geq \tilde{M}$.

Very often we will see that Tame systems T have property (P1) and (P2).

> (P3): When a set A of systems T satisfies: "for all $T \in A$, if T satisfies (P1) then T is Tame" we will say that A satisfies property (P3).

Typical Theorem in Mirror Theory

A typical Theorem in Mirror Theory will be for example: if in all the blocks $\xi \ll$ a value A (or if the average value of ξ is $\leq A$) and the number v of variables is $v \ll |G|$, then the system is Tame. Notice that here for the number of variables we have $|G|$ and not $\sqrt{|G|}$ for example. Moreover, we will generally want precise evaluations for how "Tame" the system is, and for $\ll$, and for the value A. This is what is done for some T systems in Patarin (2009), Patarin (2010a), or Patarin (2013) for example.

15.3 Examples

Example 15.6 (*same example as in Sect. 1)* "$\xi = 3$ *and* $\xi = 2$ *system*") Let T be this system of four equations on $(\mathbb{Z}/2\mathbb{Z})^n$, with pairwise distinct variables $P_1, P_2, P_3, P_4, P_5, P_6, P_7$:

$$P_2 = P_1 \oplus c_1$$

$$P_4 = P_3 \oplus c_2$$

$$\begin{cases} P_6 = P_5 \oplus c_3 \\ P_7 = P_5 \oplus c_4 \end{cases}$$

c_1, c_2, c_3, c_4 are the constants.

We have here three blocks of equations, two blocks with $\xi = 2$, and one block with $\xi = 3$ so $\xi_{\max} = 3$. The block conditions are $c_1 \neq 0$, $c_2 \neq 0$, $c_3 \neq 0$, $c_4 \neq 0$, and $c_3 \neq c_4$.

In $(\mathbb{Z}/2\mathbb{Z})^3$, i.e., on 3 bits

On 3 bits, we have

$$\tilde{M} = \frac{\text{Weight}(T)}{8^4} = \frac{8 \cdot 7 \cdot 6 \cdot 5 \cdot 4 \cdot 3 \cdot 2}{8^4} = \frac{40320}{4096} = 9.84.$$

$\text{Space}(T) = 7^3 \cdot 6 = 2058$ constants (c_1, c_2, c_3, c_4) satisfy the block conditions.

$$\tilde{H} = \frac{\text{Weight}(T)}{\text{Space}(T)} = 19.59.$$

Computer simulations show that here we have

- 924 values (c_1, c_2, c_3, c_4) satisfy the block conditions but have $H = 0$
- 1008 values have $H = 32$;
- 126 values have $H = 64$.

We can check that: $924 + 1008 + 126 = 2058$ (all the constants that satisfy the block conditions) and that: $1008 \cdot 32 + 126 \cdot 64 = 40320$ (all the $(P_1, P_2, \ldots, P_7)$).

We see that here the system T (on 3 bits) is "always wild," i.e., for every constants c_1, c_2, c_3, c_4, H is never $\simeq 9.84$.

In $(\mathbb{Z}/2\mathbb{Z})^4$, i.e., on 4 bits

On 4 bits we have $\tilde{M} = \frac{\text{Weight}(T)}{16^4} = 879.78$.

We have $15^3 \cdot 14 = 47250$ constants that satisfy the block conditions, and $\tilde{H} = \frac{\text{Weight}(T)}{47250} = 1220.26$.

Computer simulations show that here we have

- 0 values $(c_1, \ldots, c_4)$ that satisfy the block conditions with $H = 0$.
- 7560 values have $H = 1024$.
- 20160 values have $H = 1152$.
- 2520 values have $H = 1280$.
- 15120 values have $H = 1344$.
- 1260 values have $H = 1536$.
- 630 values have $H = 1920$.

In particular, the condition $H \geq \tilde{M}$ is always verified. Let $\sigma'(H) = E(|H - \tilde{H}|)$. Here we have $\sigma'(H) \simeq 120 \ll 1220$ and therefore here T is "Tame on average." The standard deviation is $\sigma(H) \simeq 152 \ll 1220$ and therefore here T is also "σ Tame." Moreover here the system T (on 4 bits) is "always Tame," i.e., for every constants c_i compatible with the block conditions we have $H \gtrsim \tilde{H}$ since $1024 \simeq 1220$. Here the system is always Tame but not Homogeneous since 1920 is not $\simeq$ 1220 but this is classical: very often tame systems have very large H on a very small number

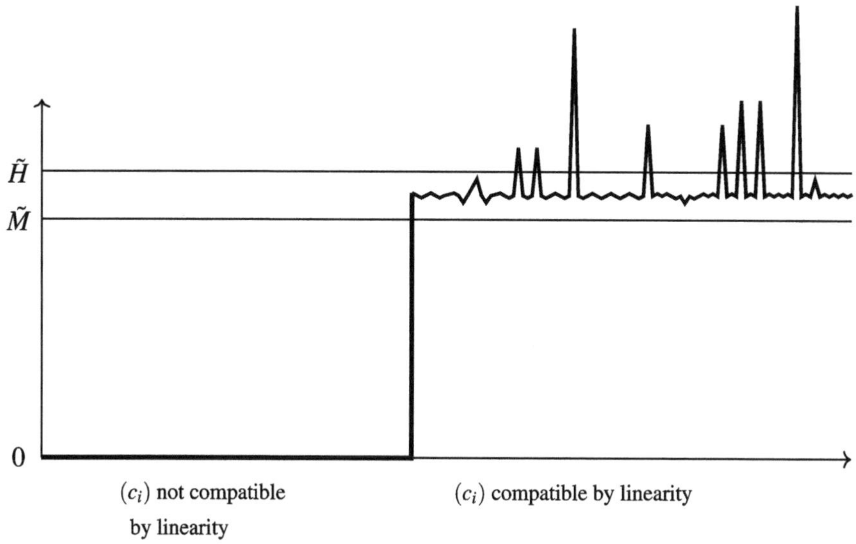

Fig. 15.1 Typical solution H for always Tame systems but not Homogeneous systems, with property (P2)

of variables. We also have here properties (P1) and (P2). Figure reffig:typicalsol illustrates such systems (Fig. 15.1).

Example 15.7 (*"$P_i \oplus Q_j$ with $\xi_{\max} = 2$," or "XOR of Two bijections in H standard"*) Let T be this system of seven equations with pairwise distinct variables P_i, and pairwise distinct variables Q_i:

$$\begin{aligned} P_1 \oplus Q_1 &= c_1 \\ P_2 \oplus Q_2 &= c_2 \\ P_3 \oplus Q_3 &= c_3 \\ P_4 \oplus Q_4 &= c_4 \\ P_5 \oplus Q_5 &= c_5 \\ P_6 \oplus Q_6 &= c_6 \\ P_7 \oplus Q_7 &= c_7. \end{aligned}$$

We have here seven blocks of equations, $\xi = 2$ on each block. Here we have no block conditions on the constants c_i.

Let us consider $G = (\mathbb{Z}/2\mathbb{Z})^3$, i.e., computations are on 3 bits. Here $\text{Weight}(T) = (8!)^2$, $\text{Space}(T) = 8^7$,

$$\tilde{H} = \tilde{M} = \frac{\text{Weight}(T)}{8^7} = 775.19.$$

Computer simulations have been made and show that:

- 0 values $(c_1, \ldots, c_7)$ with $H = 0$.
- 40320 values have $H = 384$.
- 987840 values have $H = 640$.
- 752640 values have $H = 768$.
- 258720 values have $H = 1152$.
- 35280 values have $H = 1408$.
- 18816 values have $H = 1920$.
- 1960 values have $H = 3456$.
- 1568 values have $H = 5760$.
- 8 values have $H = 40320$.

Here the system is not always Tame since $384 \leq \frac{\tilde{H}}{2}$, but it is Tame on average.

Here we have property (P1) (i.e., 0 values block compatible with $H = 0$) and not property (P2). In fact property (P2) was impossible here since $\tilde{H} = \tilde{M}$ and H is not a constant.

Moreover it is interesting to notice that it was possible to see that property (P1) was true without doing any computation. We just have to use a Theorem of 1952 of Marshall Hall Jr: see Hall (1952). We will give more details about this in Sect. 15.5.

Remark 15.6 Let T' be T plus one more equation: $P_8 \oplus Q_8 = c_8$. In $(\mathbb{Z}/2\mathbb{Z})^3$ we will have $\bigoplus_{i=1}^{8} P_i = 0$ and $\bigoplus_{i=1}^{8} Q_i = 0$. Therefore if $\bigoplus_{i=1}^{8} c_i \neq 0$, we will have no solution. And if $\bigoplus_{i=1}^{8} c_i = 0$ then $P_8 \oplus Q_8 = c_8$ can be removed since it is just a consequence of T, and T and T' have the same solutions.

Example 15.8 ("*$P_i \oplus P_j$ with $\xi_{\max} = 2$*").

Seven equations

Let T be this system of seven equations with pairwise distinct variables P_i:

$$\begin{aligned}
P_1 \oplus P_2 &= c_1 \\
P_3 \oplus P_4 &= c_2 \\
P_5 \oplus P_6 &= c_3 \\
P_7 \oplus P_8 &= c_4 \\
P_9 \oplus P_{10} &= c_5 \\
P_{11} \oplus P_{12} &= c_6 \\
P_{13} \oplus P_{14} &= c_7
\end{aligned}$$

We have here seven blocks of equations, $\xi = 2$ on each block. Here the block conditions are $\forall i$, $1 \leq i \leq 7$, $c_i \neq 0$.

Let us consider $G = (\mathbb{Z}/2\mathbb{Z})^4$, i.e., computations are on 4 bits. Here

$$\text{Weight}(T) = \tfrac{16!}{2}, \qquad \text{Space}(T) = 15^7,$$
$$\tilde{H} = \tfrac{\text{Weight}(T)}{15^7} = 61228.10, \qquad \tilde{M} = \tfrac{16!}{2 \cdot 16^7} = 38971.73.$$

Computer simulations have been made and show that here we have (13 cases here):

- 10678710 values $(c_1, \ldots, c_7)$ block compatible give $H = 0$ solutions.
- 40294800 values have $H = 49152$.
- 50803200 values have $H = 57344$.
- 25401600 values have $H = 65536$.
- 11289600 values have $H = 73728$.
- 17992800 values have $H = 81920$.
- 11289600 values have $H = 98304$.
- 2690100 values have $H = 147456$.
- 264600 values have $H = 180224$.
- 141120 values have $H = 245760$.
- 7350 values have $H = 442368$.
- 5880 values have $H = 737280$.
- 15 values have $H = 5160960$.

Here system T is Tame on average but not always Tame. The fact that it is Tame is a very good stability property since here the number of variables (14) is almost the number of elements in G (since here $|G| = 16$) and we have a lot of equations (7). (This result is also compatible with the general analysis of such systems done in Patarin (2005, 2010a), i.e., "Theorem $P_i \oplus P_j$.")

Eight equations

Let T' be T plus one more equation: $P_{15} \oplus P_{16} = c_8$. In $(\mathbb{Z}/2\mathbb{Z})^4$ we will have $\bigoplus_{i=1}^{16} P_i = 0$. Therefore if $\bigoplus_{i=1}^{8} c_i \neq 0$, we will have no solution. And if $\bigoplus_{i=1}^{8} c_i = 0$ then $P_{15} \oplus P_{16} = c_8$ can be removed since it is just a consequence of T, and T' has exactly two times the number of solutions of T (since (P_{15}, P_{16}) and (P_{16}, P_{15}) give the same solution).

Six equations

Let T'' be T without the equation $P_{13} \oplus P_{14} = c_7$. On $(\mathbb{Z}/2\mathbb{Z})^4$ computer simulations show that here we have 0 values $(c_1, \ldots, c_6)$ block compatible with $H = 0$ (i.e., we have property (P1)). We have 19 different values H when $(c_1, \ldots, c_6)$ are block compatible, with $H_{\min} = 57344$ and $H_{\max} = 1290240$, $\tilde{M} = 51962$, and $\tilde{H} = 76535$. In particular, we note that $H_{\min} \geq \tilde{M}$. If we assume $57344 \simeq 76535$ we can say that the system is always Tame (but not homogeneous, as usual).

Four equations

$$T : \begin{cases} P_2 = P_1 \oplus c_1 \\ P_4 = P_3 \oplus c_2 \\ P_6 = P_5 \oplus c_3 \\ P_8 = P_7 \oplus c_4 \end{cases}$$

We want solutions with pairwise distinct P_i, $1 \leq i \leq 8$. On $(\mathbb{Z}/2\mathbb{Z})^4$, computer simulations show that we have 0 values $(c_1, \dots, c_4)$ block compatible with $H = 0$. We have seven different values H when $(c_1, \dots, c_4)$ are block compatible, with $H_{\text{min}} = 9216$, $H_{\text{max}} = 26880$, $\tilde{M} = 7918$, and $\tilde{H} = 10250$. We note that $H_{\text{min}} \geq \tilde{M}$.

Here the system is always Tame ($9216 \simeq 10250$), but not homogeneous (as usual).

We see in these examples with 5, 6, 7, and 8 equations that when we have less blocks the systems are Tamer (as for any system T, see Theorem 15.2).

Example 15.9 ("*σ for the XOR of two bijections*")
Let H be the number of (f_i, g_i, h_i), $1 \leq i \leq m$, $f_i, g_i, h_i \in (\mathbb{Z}/2\mathbb{Z})^n$ such that:

1. All the f_i are pairwise distinct.
2. All the g_i are pairwise distinct.
3. All the h_i are pairwise distinct.
4. All the $f_i \oplus g_i \oplus h_i$ are pairwise distinct.

This system T is associated with the standard deviation of a value in relation with the XOR of two bijections (see Patarin (2008a, b)).

In "standard form" H is also the number of (f_i, g_i, h_i, t_i), $1 \leq i \leq m$, such that:

1. All the f_i are pairwise distinct.
2. All the g_i are pairwise distinct.
3. All the h_i are pairwise distinct.
4. All the t_i are pairwise distinct.
5. $\forall i, 1 \leq i \leq m, f_i \oplus g_i \oplus h_i \oplus t_i = 0$.

Here $\text{Weight}(T) = [2^n(2^n - 1) \cdots (2^n - m + 1)]^4$ and $\text{Space}(T) = 2^{nm}$. Here the minimal equations have four variables and depth 3.

Example 15.10 ("*large ξ*") Let T be this system of six equations with pairwise distinct variables P_i:

$$\begin{cases} P_2 = P_1 \oplus c_1 \\ P_3 = P_1 \oplus c_2 \\ P_4 = P_1 \oplus c_3 \end{cases}$$

$$\text{and} \begin{cases} P_6 = P_5 \oplus c_4 \\ P_7 = P_5 \oplus c_5 \\ P_8 = P_5 \oplus c_6. \end{cases}$$

We have here two blocks of equations, $\xi = 4$ on each block.

Here the block conditions are

$$c_1 \neq 0, c_2 \neq 0, c_3 \neq 0, c_1 \neq c_2, c_1 \neq c_3, c_2 \neq c_3,$$
$$c_4 \neq 0, c_5 \neq 0, c_6 \neq 0, c_4 \neq c_5, c_4 \neq c_6, c_5 \neq c_6.$$

On $G = (\mathbb{Z}/2\mathbb{Z})^4$, we have

$$\begin{aligned}
\text{Weight}(T) &= 16 \cdot 15 \cdot 14 \cdot 13 \cdot 12 \cdot 11 \cdot 10 \cdot 9, \\
\text{Space}(T) &= (15 \cdot 14 \cdot 13)^2 \\
\tilde{H} &= \frac{\text{Weight}(T)}{\text{Space}(T)} = 69.62 \\
\tilde{M} &= \frac{\text{Weight}(T)}{16^6} = 30.93.
\end{aligned}$$

Computer simulations show that:

- 141120 values $(c_1, \ldots, c_6)$ block compatible give $H = 0$ solutions.
- 725760 values have $H = 32$.
- 3991680 values have $H = 64$.
- 1935360 values have $H = 80$.
- 597240 values have $H = 128$.
- 60480 values have $H = 144$.
- 1260 values have $H = 192$.

Here 11% of the values $(c_1, \ldots, c_6)$ block compatible have $H < \frac{\tilde{H}}{2}$, and moreover 1.89% have $H = 0$ (however, as often, when $H \neq 0$, then $H \geq \tilde{M}$). The system is not Tame.

More generally, let $G = (\mathbb{Z}/2\mathbb{Z})^n$ and let us consider number h' of $(P_1, \ldots, P_\alpha)$ such that the P_i are pairwise distinct and:

$$\begin{cases} P_2 = P_1 \oplus c_1 \\ P_3 = P_1 \oplus c_2 \\ \vdots \\ P_{\alpha/2} = P_1 \oplus c_{\alpha/2-1} \end{cases} \quad \textit{and} \quad \begin{cases} P_{\alpha/2+2} = P_{\alpha/2+1} \oplus c_{\alpha/2} \\ \vdots \\ P_\alpha = P_{\alpha/2+1} \oplus c_{\alpha-2}. \end{cases}$$

We have here two blocks of equations and $\xi = \frac{\alpha}{2}$ on these blocks.

For $(P_1, \ldots, P_{\alpha/2})$ we have 2^n possibilities: just fix P_1 to any value.

For $P_{\alpha/2+1}$ we want this value to be different from all the following values (by convention $c_0 = 0$ and $c_{\alpha-1} = 0$):

$$P_1 \oplus c_i \oplus c_j \quad \text{for all } 0 \leq i \leq \frac{\alpha}{2} - 1, \frac{\alpha}{2} \leq j \leq \alpha - 1 \quad \text{(because we want } P_i \neq P_j\text{)}.$$

Now when $\alpha^2 \gg 2^n$ with $\alpha \ll 2^n$, it can occur that $c_i \oplus c_j$ cover all the values of $(\mathbb{Z}/2\mathbb{Z})^n$. Then we will have here $H = 0$ despite the fact that the constants c_i are block compatible.

15.4 About Computer Simulations

When G is very small, it is often possible to effectuate exhaustive search of the solutions on a computer in reasonable time.

However, there are many ways to accelerate the computations (or do them on larger G, or with more variables or equations). Here are some of these ideas (many more exist).

1. In the examples, it is possible to assume $P_1 = 0$.

   Proof If we change all the X_i variables by $X_i \oplus c$, where c is a constant, and since in our example we always have an even number of variables in our equations, then if $(X_1, \ldots, X_m)$ is a solution, $(X_1 \oplus c, \ldots, X_m \oplus c)$ is also a solution. Therefore $H = (H \text{with} P_1 = 0) \cdot 2^n$ when $G = (\mathbb{Z}/2\mathbb{Z})^n$ in our examples. □
2. In Examples 15.7, 15.9, and 15.10, we can assume $c_1 = 1$.

   Proof $c_1 \neq 0$ (from the block conditions). Now in GF(2^n) the value c_1 has an inverse $\beta = \frac{1}{c_1}$.
   If $(X_1, \ldots, X_m)$ is a solution for $(c_1, \ldots, c_m)$, then $(\beta X_1, \ldots, \beta X_m)$ is a solution for $(1, \beta c_2, \ldots, \beta c_m)$. Therefore $(c_1, \ldots, c_m)$ and $(1, \beta c_2, \ldots, \beta c_m)$ have the same number H of solution. So we can assume that $c_1 = 1$ (in Examples 15.7, 15.9, and 15.10) and multiply by $2^n - 1$ the number of $(c_1, \ldots, c_m)$ that give H solutions. □
   Similarly, in Example 15.8 we can compute only for $c_1 = 0$ and for $c_1 = 1$ (all $c_1 \neq 0$ will have the same property as $c_1 = 1$).
3. In Example 15.9, we can assume $P_3 < P_4$, $P_5 < P_6, \ldots, P_{13} < P_{14}$.

   Proof This is obvious by symmetry of the hypothesis. Then we will just multiply H by 2^6 (since P_1 was fixed, from idea 1 above, we did not use $P_1 < P_2$ here). □
   Similarly in all the other examples, there are also many symmetries.
4. We can use symmetries on c_i
   Here we will separate the case where all the c_i are pairwise distinct, then the case where we have exactly one equality, then two cases where we have two equalities (like $c_1 = c_2$ and $c_3 = c_4$ or like $c_1 = c_2 = c_3$), etc.
5. In Patarin (2009, 2008a) some "Orange," "Purple," and "Red" induction equations have been introduced for a theoretical analysis. We will not present these equations here but it is also possible to use them to accelerate the computations on many systems.

15.5 Marshall Hall Jr. Theorem and Conjectures of 2008

In 1952, Marshall Hall Jr. proved (see Hall 1952) that:

$$\forall f \in \mathscr{F}_n, \text{ if } \bigoplus_{x\in\{0,1\}^n} f(x) = 0, \text{ then } \exists(g,h) \in (\mathscr{P}_n)^2 \text{ such that } f = g \oplus h.$$

This theorem was proved again in 1979 in Hall (1979).

Moreover this result was proved on any abelian group, not only for $((\mathbb{Z}/2\mathbb{Z})^n, \oplus)$.

However in Hall (1952, 1979) we just have that there exist $(g,h) \in (\mathscr{P}_n)^2$, but we have no information about the number H of such (g,h) (except that $H \neq 0$).

Remark 15.7 Example 15.7 of Sect. 15.3 (or, more precisely, the system T' given after Example 15.7 of Sect. 15.3) is just a special case on $((\mathbb{Z}/2\mathbb{Z})^3, \oplus)$. Marshall Hall theorem says $H > 0$, and the simulations presented above show that $H \geq 384$.

15.5.1 2008 Conjectures

The following conjectures were made by J. Patarin in 2008. These conjectures are made on any abelian group, not only on $\left((\mathbb{Z}/2\mathbb{Z})^n, \oplus\right)$.

Conjecture 15.1 $\forall f \in F_n$, if $\bigoplus_{x\in\{0,1\}^n} f(x) = 0$, then the number H of $(g,h) \in B_n^2$ such that $f = g \oplus h$ satisfies

$$H \geq \frac{|B_n|^2}{2^{n2^n}}.$$

Conjecture 15.2 The minimum value for H is obtained when f is a bijection.

As far as we know these two conjectures are still open problems.

Remark 15.8 It is however easy to see that the maximum value for H is obtained when f is a constant function. Then $H = |B_n|$ since then for all $g \in B_n$, $f \oplus h$ is a bijection.

For constant functions f the value H is much larger than the average value for H.

15.5.2 Computer Simulations

Example 15.7 of Sect. 15.3 shows that Conjecture 15.1 is true on $(\mathbb{Z}/2\mathbb{Z})^3$ since $384 \geq \frac{775.19}{8} = 96.89$.

Many more computer simulations have been done on various groups in order to test Conjecture 15.1 on various groups. Let $H^* = \frac{|B_n|^2}{2^{n2^n}}$. Here are the results: (where the conjecture means $H_{\min} \geq H^*$):

- $(\mathbb{Z}/2\mathbb{Z})^2 : H_{\min} = 8,\ H^* = 2.$
- $\mathbb{Z}/4\mathbb{Z} : H_{\min} = 8,\ H^* = 2.$
- $\mathbb{Z}/6\mathbb{Z} : H_{\min} = 48,\ H^* = 11.$
- $(\mathbb{Z}/2\mathbb{Z})^3 : H_{\min} = 384,\ H^* = 96$ (same result as Example 15.7).
- $\mathbb{Z}/2\mathbb{Z} \times \mathbb{Z}/4\mathbb{Z} : H_{\min} = 384,\ H^* = 96.$
- $\mathbb{Z}/8\mathbb{Z} : H_{\min} = 512,\ H^* = 96.$
- $\mathbb{Z}/9\mathbb{Z} : H_{\min} = 2025,\ H^* = 340.$
- $\mathbb{Z}/10\mathbb{Z} : H_{\min} = 9280,\ H^* = 1320.$
- $\mathbb{Z}/12\mathbb{Z} : H_{\min} = 210432,\ H^* = 25700.$
- $(\mathbb{Z}/2\mathbb{Z})^4 : H_{\min} = 244744192,\ H^* = 23700000.$

In each case, Conjecture 15.1 is true.

15.6 Examples of Connections Between Mirror Systems and Cryptographic Security of Generic Schemes

15.6.1 XOR of Two Bijections, H Standard Technique

Let f and g be two random bijections from $\{0,1\}^n \to \{0,1\}^n$. We want to distinguish $f \oplus g$ from a random application from $\{0,1\}^n \to \{0,1\}^n$. For this problem, the securities in KPA and CPA are equivalent (see Patarin 2008b, p. 5).

With q queries we have an exact value for the Advantage (see Patarin 2010b, p. 4):

$$\mathscr{A}_q = \frac{1}{2 \cdot 2^{nq}} \sum_{b_1,\dots,b_q \in \{0,1\}^n} \left| \frac{h_q}{\tilde{h}_q} - 1 \right| = \frac{1}{2^{nq}} \sum_{b_1,\dots,b_q \in F} \left(\frac{h_q}{\tilde{h}_q} - 1 \right),$$

where

- h_q is the number of $(P_1, \dots, P_q, Q_1, \dots, Q_q) \in (\{0,1\}^n)^{2q}$ such that
  1. The P_i are pairwise distinct.
  2. The Q_i are pairwise distinct.
  3. $\forall i, 1 \le i \le q, P_i \oplus Q_i = b_i$.
- $\tilde{h}_q$ is the average value of h_q when $(b_1, \dots, b_q) \in_R (\{0,1\}^n)^q$. We have

$$E(h_q) = \tilde{h}_q = \frac{(2^n(2^n-1)\dots(2^n-q+1))^2}{2^{nq}}.$$

- $F = \{(b_1, \dots, b_q) \in (\{0,1\}^n)^q$ such that $h(b_1, \dots, b_q) \ge \tilde{h}_q\}$.

Therefore, we see that the security for this problem is exactly the fact that the system $T: P_i \oplus Q_i = b_i$ for pairwise distinct P_i, and pairwise distinct Q_i is Tame on average, and $\mathrm{Adv}_q = \frac{\mathrm{W}(T)}{2}$, where $\mathrm{W}(T)$ is the "wild coefficient" of the system T.

From this in Patarin (2008b) security for this problem is proved when $q \ll 2^n$.

15.6.2 XOR of Two Bijections, H_σ Technique

With the same notations as above, we have

$$\mathrm{Adv}_q \leq 2\left(\frac{\sigma(h_q)}{E(h_q)}\right)^{2/3} \qquad \text{(see [6]).}$$

Let λ_q be the number of $(f_i, g_i, h_i) \in (\{0, 1\}^n)^{3q}$ such that:

1. The f_i are pairwise distinct.
2. The g_i are pairwise distinct.
3. The h_i are pairwise distinct.
4. The $f_i \oplus g_i \oplus h_i$ are pairwise distinct.

Let T' be these sets of equalities and non-equalities. Let

$$U_m = \frac{(2^n(2^n-1)\cdots(2^n-q+1))^4}{2^{nq}} = \tilde{H}(T').$$

Then

$$\mathrm{Adv}_q \leq 2\left(\frac{\lambda_q}{U_q} - 1\right)^{1/3} \qquad [6].$$

Here we have no more c_i values (only the constant 0), and we introduce $z_i = f_i \oplus g_i \oplus h_i$, we have equations involving four variables. The security is directly related to the mirror system T': we have security if $\lambda_q \simeq U_q$, i.e., $H(T') \simeq \tilde{H}(T')$, i.e., if T' is Tame (for the constants 0).

From this security for this problem is proved in Patarin (2008b) when $q \ll 2^n$ (as with classical H technique).

15.6.3 Security of Balanced Feistel Schemes

As shown in Patarin (2005), the security (for 4 rounds in KPA, 5 or 6 rounds in CPA) is related to this system T of Mirror Theory (called "problem $P_i \oplus P_j$"):

T: The P_i variables are pairwise distinct variables of $\{0, 1\}^n$, and we have some equalities $P_i \oplus P_j = c_{ij}$.

The number of variables P_i is smaller than the number of queries q (it is about $\frac{q^2}{2^n}$), the average value of the number $\xi(A)$ for a block A is about 2.

The security in KPA is related to the fact that T is tame on average, and a sufficient condition for CPA security is for T to be always Tame.

From this security when $q \ll 2^n$ is given in Patarin (2010b).

15.6.4 Security of $f(x||0) \oplus f(x||1)$ When f is a Bijection

This problem can be seen as a variant of the XOR of two bijections, but here the two bijections are not independent: we use only one bijection f, and the last bit of the input is fixed to be 0 in the first term, and 1 in the second term. Again, the problem is to distinguish the XOR of these two bijections from a random function from $\{0, 1\}^n \to \{0, 1\}^n$. In fact, this problem is exactly equivalent to the "problem $P_i \oplus P_j$" (seen above) when $\xi_{\max} = 2$, i.e., for this mirror system T:

T: the P_i variables are pairwise distinct variables of $\{0, 1\}^n$, (we write them $P_{j,1}$ or $P_{j,2}$) and we have some equalities: $P_{i,1} \oplus P_{i,2} = c_i$ ($c_i \neq 0$).

The security of $f(x||0) \oplus f(x||1)$ is directly related to the analysis if T is Tame on average.

This problem is slightly more difficult than the XOR of two random permutations (we have $P_i \oplus P_j$ instead of $P_i \oplus Q_j$ but we can proceed similarly) and simpler than the general $P_i \oplus P_j$ problem related with the security of classical Feistel schemes (since here ξ is always 2).

15.6.5 Other Schemes

We can also get similar connections for generic Benes schemes, Misty L schemes, unbalanced Feistel schemes, Feistel schemes with internal bijections (instead of internal functions), etc.

15.7 Conclusion

In this chapter, we have defined the mirror systems, and given some of their properties. We also have shown many examples, some computer simulations, and the connections between these systems and some generic cryptographic constructions. It is interesting to notice how complexity and order quickly appear even in very small examples. This area of research is still in progress. The fact that we have often an equivalence between the security of some generic cryptographic schemes and the property "Tame" or not of the related system is a strong motivation to study the property of these systems.

References

Hall, M., Jr. 1952. A combinatorial problem on abelian groups. *Proceedings of the American Mathematical Society* 3 (4): 584–587.

Hall, M., Jr. 1979. A problem in combinatorial group theory. *Ars Combinatoria* 7: 3–5.

Patarin, J. 2004. Security of random feistel schemes with 5 or more rounds. In *Advances in cryptology—CRYPTO '04*, ed. M. Franklin. Lecture notes in computer science, 106–222. Berlin: Springer.

Patarin, J. 2005. On linear systems of equations with distinct variables and small block size. In *Information and communications security—ICISC '05*, vol. 3935, ed. D. Won and S. Kim. Lecture notes in computer science, 299–321. Berlin: Springer.

Patarin, J. 2008a. A proof of security in $O(2^n)$ for the Xor of two random permutations. In *Information and communications security—ICITS 2008*, vol. 5155, ed. R. Safavi-Naini. Lecture notes in computer science, 232–248. Berlin: Springer.

Patarin, J. 2008b. A proof of security in $O(2^n)$ for the Xor of two random permutations—proof with the H_σ technique. In *Cryptology ePrint Archive: Report 2008/010.*

Patarin, J. 2009. The "Coefficients H" technique. In *Selected areas in cryptography—SAC '08*, vol. 5381, ed. R. Avanzi, L. Keliher, and F. Sica. Lecture notes in computer science, 328–345. Berlin: Springer.

Patarin, J. 2010a. Introduction to mirror theory: analysis of systems of linear equalities and linear non equalities for cryptography. In *Cryptology ePrint Archive: Report 2010/287.*

Patarin, J. 2010b. Security of balanced and unbalanced feistel schemes with linear non equalities. In *Cryptology ePrint Archive: Report 2010/293.*

Patarin, J. 2013. Security in $O(2^n)$ for the Xor of two random permutations—proof with the standard H technique. In *Cryptology ePrint Archive: Report 2013/368.*

References

Chapter 16
“$P_i \oplus P_j$ Theorem” for a Wide Range of ξ_{max}

Abstract In this chapter, we will study and prove the so-called “ $P_i \oplus P_j$ Theorem” of Patarin (2005). More precisely, we will study the case for a wide range of ξ_{max} values that is sufficient for most cryptographic applications. Then, in Chap. 17, we will use this result to prove some very strong security bound on generic Feistel ciphers. We also illustrate the usefulness of the result with the case $\xi_{max} = 2$, which has its own interest from a cryptographic point of view. Indeed, as we will see, it is closely related to the problem of distinguishing $f(x\|0) \oplus f(x\|1)$ where f is a random permutation on n bits from a random function. The proof presented in this chapter follows the recent work by Cogliati et al. (2023), in which a complete and compact proof of the result is provided. We extend this paper by providing additional explanations and clarifications.

16.1 Presentation of “$P_i \oplus P_j$ Theorem”

Let a be an integer. We denote by J_a the number of variables $P_1, \ldots, P_p$ pairwise distinct of $\{0, 1\}^n$, i.e., $J_p = 2^n(2^n - 1) \ldots (2^n - p + 1)$. J_p is also the number of $(P_1, \ldots, P_p)$ such that the P_i values are pairwise distinct and $P_i \in \{0, 1\}^n$.

Let us now consider a set (A) of m equations in $(P_1, \ldots, P_p)$ of the form:

$$P_i \oplus P_j = \lambda_k.$$

The aim of this chapter is to prove that, as long as (A) satisfies appropriate conditions, the number of solutions of (A) such that the P_i are also pairwise distinct is greater than $J_p/2^{nm}$. We now introduce a few definitions that will allow us to characterize these conditions.

Definition 16.1 (“*Circles*”) Let (A) be a set of equations $P_i \oplus P_j = \lambda_k$, with $P_i, P_j \in \{0, 1\}^n$. If by linearity from (A), we cannot generate an equation in only the λ_k, we say that (A) has no “circle in P,” or that the equations of (A) are “linearly independent in P.” In particular, it is also impossible to generate the equation $0^n = 0^n$ with a non-trivial linear combination of the equations from (A).

J. Patarin et al., *Guide to Feistel Ciphers*, Information Security and Cryptography,
https://doi.org/10.1007/978-3-031-99850-8_16

Let m be the number of equations in (A), and let p be the number of variables P_i in (A). Therefore, we have parameters $\lambda_1, \lambda_2, \ldots, \lambda_m$ and $m + 1 \leq p \leq 2m$.

Definition 16.2 ("*Blocks*")
We will say that two indices i and j are "in the same block" if by linearity from the equations of (A), it is possible to obtain $P_i \oplus P_j$ as an expression in $\lambda_1, \lambda_2, \ldots, \lambda_m$, or if $i = j$.

Remark 16.1 We remark that being in the same block is an equivalence relation on the set of indices, which means that blocks form a partition of $\{1, \ldots, p\}$. Moreover, it is clear by definition that no equation from (A) can involve variables that are in different blocks, and that it is impossible to generate such an equation by linearity from (A). This also means that the equations from (A) can similarly be partitioned in blocks.

Definition 16.3 The number of blocks in (A) is denoted by α, and ξ_{max} represents the maximal number of indices that are in the same block.

Example 16.1 If $A = \{P_1 \oplus P_2 = \lambda_1,\ P_1 \oplus P_3 = \lambda_2,\ P_4 \oplus P_5 = \lambda_3\}$, we have $\alpha =$ 2 blocks of indices ($\{1, 2, 3\}$ and $\{4, 5\}$, $\xi_{max} = 3$, $p = 5$ (variables) and $m = 3$ (equations).

Definition 16.4 Let (A) be a system of equations, where $\lambda_1, \lambda_2, \ldots, \lambda_m$ are fixed, we will denote by $h_\alpha(A)$ the number of $P_1, P_2, \ldots, P_p$ solutions of (A) such that: $\forall i, j,\ i \neq j \Rightarrow P_i \neq P_j$. We will also denote $H_p(A) = 2^{nm} h_\alpha(A)$. We will generally denote $H_p(A)$ simply by $H(A)$ and $h_\alpha(A)$ simply by $h(A)$. We will also denote by J_p the number of $P_1, P_2, \ldots, P_p$ such that $\forall i, j,\ i \neq j \Rightarrow P_i \neq P_j$. Therefore, $J_p = 2^n(2^n - 1)\ldots(2^n - p + 1)$.

We are now ready to express the main result of this chapter.

Theorem 16.1 *"Theorem $P_i \oplus P_j$" for any ξ_{max}.*
Let (A) be a set of equations $P_i \oplus P_j = \lambda_k$, with p variables such that:

1. *There are no circle in P in the equations (A).*
2. *There are no more than ξ_{max} indices in the same block.*
3. *By linearity from (A), it is not possible to generate an equation $P_i = P_j$ with $i \neq j$.*

Then, if $p \leq \sqrt{N}$ or $\sqrt{N} \geq \xi_{\max}^2 n + \xi_{\max}$, and $1 \leq p \leq 2^n/12\xi_{\max}^2$, we have $H_p(A) \geq J_p$.

Let us consider the case where $\xi_{max} = 2$ in order to illustrate this result. Here $\xi_{max} = 2$ means that when we fix one variable P_i, then at most one other variable P_j is fixed from the equations (A). We can notice that $\frac{J_p}{2^{nm}}$ is the average number of solutions on all 2^{nm} values $\lambda_1, \lambda_2, \ldots, \lambda_m$ (including values $\lambda_i = 0$ where $h_\alpha = 0$). $\frac{J_p}{(2^n-1)^m}$ is the average number of solutions on all the $(2^n - 1)^m$ nonzero values $\lambda_1, \lambda_2, \ldots, \lambda_m$. Therefore, Theorem 16.1 means that when $p \ll 2^n$, and when the λ_i values are compatible by linearity with the P_i pairwise distinct (i.e., $\lambda_i \neq 0$), then the number of solutions is always greater than the average. It is like if, in a classroom all the students have either the grade 0, or a grade larger than the average grade. When $h_\alpha \neq 0$, and $p \ll 2^n$, then h_α is always greater than the average and sometimes much greater than the average.

Remark 16.2 Note that, in most cryptographic applications (where $n \geq 64$), $\xi_{\max}$ is either a small constant, or can be shown to be smaller than n with overwhelming probability. Typically, this is sufficient to prove that the cryptographic scheme is secure as long as the number of adversarial queries is upper bounded by $2^n/12\xi_{\max}^2$, as $n^3 + n \leq \sqrt{2^n}$ for $n \geq 30$.

The goal of this chapter is to provide a complete and easily verifiable proof of Theorem 16.1. In order to do so, we adapt the proof from Cogliati et al. (2023) to our notation, and provide additional explanations and clarifications.

16.2 Reformulation of the Problem

In order to streamline the proof of Theorem 16.1, we will operate two simple changes. First, we prove the following Lemma.

Lemma 16.1 *Let (A) be a set of equations $P_i \oplus P_j = \lambda_k$, with p variables, m equations, and α blocks, such that:*

1. *There are no circle in P in the equations (A).*
2. *There are no more than ξ_{max} indices in the same block.*
3. *By linearity from (A), it is not possible to generate an equation $P_i = P_j$ with $i \neq j$.*

Let also (B) be a block of (A) that involves ξ indices. Then the number of equations in (B) is exactly $\xi - 1$. As a consequence, one has $p = m + \alpha$.

Proof First, note that if (B) contains a single equation, then it necessarily involves exactly two indices, and there is nothing to prove. Let us assume that (B) involves at least two equations. In order to prove the first part of the result, it is sufficient to notice that the equations from (B) can always be ordered as follows:

- the first equation is chosen arbitrarily;
- each subsequent equation involves exactly one variable from the previous equations, and a new one.

Indeed, let us assume that such an ordering cannot be built. It means that there exists $i \geq 1$ such that i equations were ordered as described, but no suitable $i+1$-th equation can be found in the remaining equations. First, note that no remaining equation can solely involve indices from the first i ones, as this would imply that (B) contains a circle in P. Hence, this means that all remaining equations only involve new indices. However, this would contradict the fact that these indices are in the same block, as it would be impossible to generate, by linearity from (A), a bivariate equation with one variable from the first i equations, and another one from the remaining equations.

Finally, summing over the α blocks in (A), we get

$$p = \sum_{i=1}^{\alpha} \xi_i = \sum_{i=1}^{\alpha} (\xi_i - 1) + \alpha = m + \alpha,$$

which ends the proof of Lemma 16.1. □

Let us fix a system (A) of equations $P_i \oplus P_j = \lambda_k$, with a variables, α blocks, and b equations. We assume that (A) satisfies the three properties from Theorem 16.1 and Lemma 16.1. For $i = 1, \ldots, \alpha$, let us denote $P_{i,1}, \ldots, P_{i,\xi_i}$ the variables from the i-th block using an arbitrary ordering of the indices. By linearity from (A), it is possible to generate a unique (up to the ordering of the $P_{i,j}$ variables) system of the form:

$$\begin{cases} P_{i,1} \oplus P_{i,2} = \lambda_{i,2} \\ \qquad\vdots \\ P_{i,1} \oplus P_{i,\xi_i} = \lambda_{i,\xi_i}. \end{cases} \tag{16.1}$$

Indeed, since $P_{i,1}$ and $P_{i,j}$ for $j > 1$ are in the same block, one can generate an equation that only involves these two variables. The corresponding right-hand side of the equation is necessarily unique, or it would imply that (A) contains a circle in P. Moreover, from such a system, it is always possible to come back to the original set of equalities. Overall, this means that all the blocks of (A) can always be equivalently written as in Eq. (16.1), and we will now always assume that (A) is in this form.

Second, note that, in order to have solutions, the system has to satisfy the third condition from Theorem 16.1, which corresponds to two distinct conditions: $\lambda_{i,j} \neq 0^n$ for all $i \in \{2, ..., \xi_i\}$, and for all $j \neq j' > 1, \lambda_{i,j} \neq \lambda_{i,j'}$. While easy to manipulate, both conditions have to be handled in a different way, which complicates the proof. The simplest fix is to introduce, for every block, an additional λ_{i_1} value that can be thought to be 0^n.

We now introduce several notations and definitions that will help us formalize the equivalent formulation of Theorem 16.1. For integers $a \leq b$, the set $\{a, a+1, \ldots, b\}$ is denoted as $[a..b]$ (or simply $[b]$, when $a = 1$). We write X^q to denote a q-tuple $(X_1, \ldots, X_q)$. For $x \in S$, we write $S \setminus x$ to mean $S \setminus \{x\}$. We use $A \sqcup B$ to denote the disjoint union of A and B (which implicitly means that A and B are disjoint). We denote by $N := 2^n$, the number of elements in $\{0, 1\}^n$.

A multiset γ is a collection of elements that can repeat. In other words, multiset is an unordered version of a tuple. For $S \in \gamma$, we write γ_{-S} to denote the multiset formed by removing S from γ. We similarly write γ_{+T} to denote the multiset formed by adding an element T to γ. For $S \in \gamma$, we also write γ_{-S+T} to denote the resulting multiset after deleting S and adding T to γ. We say that γ is a **set-system** if it is a multiset of sets. When we want to emphasize an ordering of the elements of γ, we also write the set-system as $\gamma^{[\alpha]} = (\gamma_1, \ldots, \gamma_\alpha)$, which is an enumeration of the sets in γ. In this chapter, we consider set-systems γ of non-empty subsets of $\{0, 1\}^n$.

More formally, given a set-system $\gamma = \{\gamma_i : i \in [\alpha]\}$, we define the following set:

$$\mathsf{Disj}(\gamma) := \left\{(r_1, \ldots, r_\alpha) \in \left(\{0, 1\}^n\right)^\alpha \ : \ \gamma_1 \oplus r_1, \ldots, \gamma_\alpha \oplus r_\alpha \text{ are disjoint }^1\right\}$$

[1]along with its so-called h coefficient:

$$h(\gamma) = |\mathsf{Disj}(\gamma)|.$$

We write $\|\gamma\| := \sum_{i=1}^{\alpha} |\gamma_i|$ and $\|\gamma\|_{\max} = \max_i |\gamma_i|$. It is easy to see that the h coefficient is invariant under any translation of the sets, i.e., $h(\gamma) = h(\gamma')$ where $\gamma_i' = \gamma_i \oplus a_i$ for $a_1, \ldots, a_\alpha \in \{0, 1\}^n$.

Theorem 16.1 can be rephrased in the following way.

Theorem 16.2 (Equivalent Formulation) *Let γ be a set-system of elements of $\{0, 1\}^n$ such that $\xi_{\max} = \|\gamma\|_{\max}$. If $\|\gamma\| \le \sqrt{N}$ or $\sqrt{N} \ge \xi_{\max}^2 \log_2 N + \xi_{\max}$, and $1 \le \|\gamma\| \le N/12\xi_{\max}^2$, then*

$$h(\gamma) \ge \frac{J_{\|\gamma\|}}{N^{\|\gamma\| - |\gamma|}}.$$

The equivalence between both statements is proven in Sect. 16.2.1. From a high level, the proof of Theorem 16.2 relies on the following three facts:

1. if γ is small ($\|\gamma\| \le \sqrt{N}$), then simple calculations show that Theorem 16.2 holds;
2. if there exists $T \in \gamma$ such that $|T| = 1$, it is clear that by the definition of Disj that
$$h(\gamma) = (N - (\|\gamma\| - 1))\, h(\gamma \setminus T);$$
3. otherwise, we prove that, for a well-chosen $a \in T \in \gamma$, one has
$$h(\gamma) \ge \left(1 - \frac{\|\gamma\| - 1}{N}\right) h(\gamma'),$$
where γ' is a set-system containing exactly the same sets as γ, except that the set T has been replaced with $T \setminus \{a\}$.

Applying points 2 and 3 repeatedly until $\|\gamma\| \le \sqrt{N}$ allows us to conclude the proof of Theorem 16.2. Intuitively, the element that we remove from γ is the one that appears, in the associated system of equations, with maximum multiplicity.

[1] For a set $A \subseteq \{0, 1\}^n$ and a n-bit number $x \in \{0, 1\}^n$, $x \oplus A := \{x \oplus a \mid a \in A\}$.

More formally, given $z \in \{0, 1\}^n \setminus \{0^n\}$, and a set S, we define $\delta_S(z)$ as the number of 2-subsets $\{a, b\}$ of S with $a \oplus b = z$. For a set-system γ, we define

$$\delta_\gamma(z) := \sum_{S \in \gamma} \delta_S(z), \quad \Delta_\gamma := \max_{z \in \{0,1\}^n} \delta_\gamma(z).$$

Clearly, for any set-system γ, $\Delta_\gamma \geq 1$. The underlying statement behind the second point of the proof strategy is the following one.

Proposition 16.1 *Let λ be a set-system with $\sqrt{N} \leq \|\lambda\| \leq N/12\xi_{\max}^2$ where $\xi_{\max} = \|\lambda\|_{\max}$ satisfies the bound given in Theorem 16.2, i.e., $\sqrt{N} \geq \xi_{\max}^2 n + \xi_{\max}$. Suppose the maximum Δ_λ is attained for $a \oplus b$ with $\{a, b\} \subseteq T \in \lambda$. Then,*

$$h(\lambda) \geq \left(1 - \frac{\|\lambda\| - 1}{N}\right) \cdot h(\lambda_{-a|T}),$$

where $\lambda_{-a|T} = \lambda_{-T+T\setminus a}$ (i.e., replacing the element T by $T \setminus a$).

The proof of Proposition 16.1 is given in Sect. 16.3, and we explain how to derive Theorem 16.2 from Proposition 16.1 in Sect. 16.2.2.

16.2.1 Proof of Equivalence

Here we prove why Theorem 16.2 is an equivalent statement of Theorem 16.1. First, we establish a one-to-one relationship between the number of disjoint favorable solutions r^q with the number of solutions of systems of equations.

Let (A) be a system of equations in the form of Eq. (16.1) that also satisfies the conditions from Theorem 16.1. Recall that (A) consists of m equations that involve p variables in α blocks. For $i = 1, \ldots, \alpha$, let L_i be the set of all associated $\lambda_{i,j}$ values. As previously stated, all elements of L_i are distinct (and hence it is a set of size $\xi_i - 1$, where ξ_i is the number of variables in the ith block) nonzero elements. For $j = 2, \ldots, \xi_i$, we have an equation of the form $P_{i,1} \oplus P_{i,j} = \lambda_k$ for some k. Now we consider a set-system γ containing all sets of the form $S_i := L_i \cup \{0\}$. Thus, $\|\gamma\| = \sum_{i=1}^{\alpha} |S_i| = \sum_{i=1}^{\alpha} \xi_i = p$ and $|\gamma| = \alpha$. Now consider a map f, mapping a pairwise distinct solution of the system to r^α, where $r_i = P_{i,1}$ for all $i \in [\alpha]$. It is easy to see that the $S_i \oplus r_i$ are disjoint sets (as these represent all P values). Moreover, f is clearly injective as a solution is uniquely determined by the tuple $(P_{i,1}, \ldots, P_{\alpha,1})$. Conversely, for any r^α with disjoint $S_i \oplus r_i$ sets, we can define P^a consisting of all values from the set $\sqcup_i (S_i \oplus r_i)$ in an appropriate order (with $P_{i,1} = r_i$). Clearly, this map is f^{-1} and so f is a bijective function. Hence, the number of pairwise distinct solutions of (A) is same as the number of solutions of r^α so that $\mathsf{Disj}(\gamma)$

holds. Second, we note that Theorem 16.2 can be simply restated as the number of solutions $r^{|\gamma|}$ so that the sets $(\gamma_i \oplus r_i)$ are disjoint for all $i \in [\alpha]$ is at least

$$\frac{J_{\|\gamma\|}}{N^{\|\gamma\|-|\gamma|}} = \frac{J_p}{N^{p-\alpha}} = \frac{J_p}{N^m},$$

where $p - \alpha = m$ corresponds to the number of equations in the system (A) by Lemma 16.1. Using the fact that $H_p(A) = N^m h_\alpha(A)$ proves the equivalence between Theorem 16.1 and the equivalent formulation.

16.2.2 *Proof of Theorem 16.2*

We first prove the statement when $\|\gamma\| \leq \sqrt{N}$. In this case we remove elements from γ one by one until we end up with a single element. We first note that

$$h(\gamma) = h(\gamma_{-S}) \times (N - (\|\gamma\| - 1)) \qquad \textit{if } |S| = 1 \tag{16.2}$$

$$h(\gamma) \geq h(\gamma_{-S}) \times (N - |S| \times \|\gamma_{-S}\|) \qquad \textit{if } |S| \geq 2 \tag{16.3}$$

where $S \in \gamma$. Indeed, let us assume $S = \gamma_1$. Then, once $r_2, \ldots, r_{|\gamma|}$ are chosen such that the equations from $\mathsf{Disj}(\gamma_{-S})$ are satisfied, $\mathsf{Disj}(\gamma)$ adds the following restrictions on r_1:

$$r_1 \oplus x \neq r_i \oplus y \text{ for all } x \in S,\ i \neq 1,\ y \in \gamma_i.$$

Hence, if $|S| = 1$, r_1 has to be different from exactly $\|\gamma\| - 1$ values, while, if $|S| \neq 1$, it has to avoid at most $|S| \times \|\gamma_{-S}\|$ group elements.

Let us write $W_i := (1 - \frac{i}{N})$, so that $\prod_{i=1}^{k-1} W_i = J_k/N^k$. Now we claim that, for $\|\gamma\| \leq \sqrt{N}$:

$$\left(1 - \frac{|S| \times \|\gamma_{-S}\|}{N}\right) \geq \prod_{i=\|\gamma_{-S}\|}^{\|\gamma\|-1} W_i \tag{16.4}$$

and hence $h(\gamma) \geq N \times h(\gamma_{-S}) \times \prod_{i=\|\gamma_{-S}\|}^{\|\gamma\|-1} W_i$. After repeatedly removing an element one by one, we have $h(\gamma) \geq N^{|\gamma|} \prod_{i=1}^{\|\gamma\|-1} W_i$ which proves the theorem. Now we prove Eq. (16.4). It is sufficient to show that

$$1 - \frac{xr}{N} \geq \left(1 - \frac{x}{N}\right) \cdots \left(1 - \frac{x + r - 1}{N}\right)$$

where $x + r \leq \sqrt{N}$. This can be easily shown by induction on r. For $r = 1$, it is obvious. Now by applying induction hypothesis for r, we obtain

$$\left(1 - \frac{x}{N}\right) \cdots \left(1 - \frac{x + r - 1}{N}\right)\left(1 - \frac{x + r}{N}\right) \leq \left(1 - \frac{xr}{N}\right)\left(1 - \frac{x + r}{N}\right)$$
$$\leq 1 - \frac{xr + x}{N} - \frac{r}{N}\left(1 - \frac{x(x + r)}{N}\right) \leq 1 - \frac{xr + x}{N}.$$

For the last inequality we use the fact that $x + r + 1 \leq \sqrt{N}$.

For the next case, we assume that $\sqrt{N} \leq \|\gamma\| \leq N/12\xi_{\max}^2$, i.e., $\|\gamma\|$ is within the required bounds for which Proposition 16.1 holds. We can create a sequence of nested set-systems $\{\gamma^{(i)}\}_{i=0}^{\sigma}$, with

$$\gamma^{(0)} := \gamma, \quad \|\gamma^{(i+1)}\| = \|\gamma^{(i)}\| - 1, \ \forall i \in [\sigma - 1], \quad \|\gamma^{(\sigma)}\| \leq \sqrt{N},$$

in the following manner: let $\{x_i, y_i\} \subseteq S_i \in \gamma^{(i)}$ such that $x_i \oplus y_i$ attains the highest multiplicity in $\gamma^{(i)}$, $\Delta_{\gamma^{(i)}}$. We choose one arbitrarily if there exists more than one choice. We define $\gamma^{(i+1)} := \gamma^{(i)}_{-x_i|S_i}$. Now for every $i \in [\sigma - 1]$, if $|S_i| = 1$ we apply Eq. (16.2), and if $|S_i| \geq 2$, we apply Proposition 16.1, to obtain

$$h(\gamma) \geq \frac{h(\gamma^{(\sigma)})}{N^{\|\gamma\| - \|\gamma^{(\sigma)}\| - (|\gamma| - |\gamma^{(\sigma)}|)}} \prod_{i=1}^{\sigma} (N - (\|\gamma\| - i)).$$

We already have shown the result for $\gamma^{(\sigma)}$ that

$$h(\gamma^{(\sigma)}) \geq J_{\|\gamma^{(\sigma)}\|}/N^{\|\gamma^{(\sigma)}\| - |\gamma^{(\sigma)}|},$$

which completes the proof.

16.3 Proof of Proposition 16.1

16.3.1 Notations and Conventions

In the Proposition statement, $\{a, b\} \subseteq T \in \lambda$ and $\Delta_\lambda = \sum_{S \in \lambda} \delta_S(a \oplus b)$. Let $\lambda = \{\lambda_i : i \in [q]\}$ and we write $|\lambda_i| = \xi_i$, $\xi_{\max} = \max_i \xi_i$ and $\sigma := \sum_i \xi_i$. We also write Δ to denote Δ_λ. Throughout the section we follow this notation. Moreover, we use the notation γ to denote a set-system such that $\gamma \subseteq \lambda$ (as a multiset).

16.3.2 Initial Condition

Note that, after applying Eq. (16.3) repeatedly (or by applying induction on $|\lambda \setminus \gamma|$) for $\gamma \subseteq \lambda$, we have

$$\frac{h(\lambda)}{h(\gamma)} \geq \left(N - q\xi_{\max}^2\right)^{|\lambda\setminus\gamma|}. \tag{16.5}$$

We call this an initial condition that would be used later to prove Proposition 16.1.

16.3.3 Link-Deletion Equation and Proof Overview

16.3.3.1 Link-Deletion Equation

Let $x \in S \in \gamma \subseteq \lambda$. Let us write

$$\gamma = \{\gamma_1, \ldots, \gamma_\alpha\}$$

using an arbitrary ordering of the multiset γ, and let us assume $S = \gamma_1$ and $x = \gamma_{1,1}$. Then, the set $\mathsf{Disj}(\gamma)$ corresponds to the tuples $(r_1, \ldots, r_{|\gamma|})$ that all the $r_i \oplus \gamma_{i,j}$ values are pairwise distinct, and the set $\mathsf{Disj}(\gamma_{-x|S})$ corresponds to the same set, where the conditions involving $r_1 \oplus \gamma_{1,1}$ are ignored. Hence, one has $\mathsf{Disj}(\gamma) \subset \mathsf{Disj}(\gamma_{-x|S})$. Our goal here is to prove a recursive relation between $\mathsf{Disj}(\gamma)$ and $\mathsf{Disj}(\gamma_{-x|S})$.

Let us fix $(r_1, \ldots, r_{|\gamma|}) \in \mathsf{Disj}(\gamma_{-x|S}) \cap \overline{\mathsf{Disj}(\gamma)}$. Then, there must exist $y \in S' \in \gamma_{-\gamma_1}$ such that $S' = \gamma_i$ for some integer $i \neq 1$, and $y \oplus r_i = x \oplus r_1$. Since $(r_1, \ldots, r_{|\gamma|}) \in \mathsf{Disj}(\gamma_{-x|S})$, $(S \setminus x) \oplus r_1$ is disjoint from $S' \oplus r_i = S' \oplus (x \oplus y \oplus r_1)$. Thus, in particular, $S \setminus x$ should be disjoint from $S' \oplus x \oplus y$. Moreover, we argue that the pair $(x \oplus y, S')$ is necessarily unique. Indeed, if one simultaneously has $r_1 \oplus x = r_i \oplus y = r_j \oplus y'$ for some $y' \in \gamma_j \in \gamma_{-S}$, the disjointness of $\gamma_{-x|S}$ cannot hold. This allows us to write $\mathsf{Disj}(\gamma_{-x|S}) \setminus \mathsf{Disj}(\gamma)$ as a disjoint union.

Let

$$I := \{(x \oplus y, S') : y \in S' \in \gamma_{-S},\ \ S' \oplus (x \oplus y) \text{ is disjoint with } S \setminus x\}.$$

For each $(r_1, \ldots, r_{|\gamma|}) \in \mathsf{Disj}(\gamma_{-x|S})$, there are two distinct possibilities:

- either adding x to the set S does not create any collision (there are exactly $h(\gamma)$ such tuples), or
- a collision is created; all those collision events are disjoint, and correspond to a unique element from the set I. Each pair $(\delta, \gamma_i) \in I$ corresponds to the tuples that satisfy $\delta = r_1 \oplus r_i$ and such that all disjointness conditions from $\mathsf{Disj}(\gamma_{-x|S})$ hold. This can be equivalently described as the tuples $(r_1, \ldots, r_{|\gamma|})$ such that $r_i = r_1 \oplus \delta$ and $(r_1, \ldots, r_{i-1}, r_{i+1}, \ldots, r_\gamma) \in \mathsf{Disj}(\gamma_{\delta,S'})$, where $\gamma_{\delta,S'} = \gamma_{-S-S'+S_1}$ and

$S_1 = (\delta \oplus S') \sqcup (S \setminus x)$. This corresponds to exactly $h(\gamma_{\delta,S'})$ tuples, as r_i is fixed once the other variables are fixed.

This yields a useful recursive relation on h coefficients, called **link-deletion equation** or **half-orange equation** in relation to previous proofs:

$$h(\gamma) = h(\gamma_{-x|S}) - \sum_{(\delta,S')\in I} h(\gamma_{\delta,S'}). \tag{16.6}$$

16.3.3.2 Proof Strategy

In order to prove Proposition 16.1, we will prove that $|h(\gamma_{\delta,S'}) - h(\gamma_{-x|S})/N|$ is small enough in front of $h(\gamma_{-x|S})$, for all $(\delta, S') \in I$. This will be done in the following steps:

1. Upper bound the size of the set I (in Sect. 16.3.4).
2. Establish a recursive inequality between the maximum difference between terms of the form $h(\gamma'_{-x|S})/N$, and terms of the form $h(\gamma'_{\delta,S'})$, with $\gamma'_{-S} \subset \lambda$, and S an arbitrary set of some fixed size (in Sect. 16.3.7). This will be done by applying the link-deletion equation to the two probabilities that maximize the difference term, thus introducing new difference terms and an error term.
3. After applying this inequality a logarithmic number of times along with simple bounds on the probability ratios, prove that remaining terms become sufficiently small thanks to the geometric reduction offered by the recursive inequality (Sects. 16.3.6 and 16.3.7).

16.3.3.3 Comparison with Previous Proofs

The main difference with previous proof strategies is centered around the link-deletion equation. Indeed, previous works started with the introduction of the so-called orange equation, which can be seen as two consecutive applications of the link-deletion equations. Hence, instead of always merging a single set $S' \in \gamma$ with the final set S, this could be seen as merging two distinct sets S', $S'' \in \gamma$, which leads to a more complicated analysis.

16.3.4 Size Lemma

We also write the above set I as $I_{x|S}$ to emphasize that I depends on x, S. Clearly, for all $x \in S \in \gamma$, $|I| \leq \|\gamma\|$. However, we establish an improved upper bound for the size of $I_{a|T}$ where a and T are described in the statement of the Proposition.

Lemma 16.2 (Size Lemma) *For a given $a \in T \in \lambda$ as described in Proposition 16.1 statement, we have*

$$|I_{a|T}| \leq \|\lambda\| - \Delta_\lambda - \frac{|T|}{2}.$$

Proof Take any $S \in \lambda_{-T}$. By definition of δ_S, there are exactly $\delta_S(a \oplus b)$ many 2-sets $\{w_1, w_2\} \subseteq S$ such that $w_1 \oplus w_2 = a \oplus b$. Note that for such a 2-set, one has $b = w_2 \oplus (a \oplus w_1) \in S \oplus (a \oplus w_1)$. This implies that $b \in (S \setminus a) \bigcap (S \oplus (a \oplus w_1))$, and $(a \oplus w_1, S) \notin I_{a|T}$ by definition of $I_{a|T}$. So,

$$|I_{a|T}| \leq \sum_{S \in \lambda_{-T}} (|S| - \delta_S(a \oplus b)) = \|\lambda\| - |T| - \sum_{S \in \lambda_{-T}} \delta_S(a \oplus b). \tag{16.7}$$

By assumption, a, b was chosen such that $a \oplus b$ realizes the maximum Δ_λ. This implies that

$$\Delta_\lambda = \sum_{S \in \lambda} \delta_S(a \oplus b) = \sum_{S \in \lambda_{-T}} \delta_S(a \oplus b) + \delta_T(a \oplus b).$$

Finally, for every element, $x \in T$, there exists at most one y in T such that $x \oplus y = a \oplus b$. In the case where it exists, then neither x nor y can be part of a different 2-set. This implies that $\delta_T(a \oplus b) \leq |T|/2$, and thus

$$\Delta_\lambda \leq \sum_{S \in \lambda_{-T}} \delta_S(a \oplus b) + \frac{|T|}{2}.$$

Combining the previous upper bound with Eq. (16.7) ends the proof of Lemma 16.2. □

16.3.5 Recursive Inequality of D-Terms

In this section, we introduce D-terms, which correspond to the maximum difference between the two types of terms that can appear in the link-deletion equation. Let $\tau = \gamma_{+U}$ with $\gamma \subseteq \lambda$ where $|\gamma| = \alpha$ and $|U| = \ell + 2$. Recall that, from the link-deletion equation Eq. (16.6) applied to $x \in U \in \tau$, one has

$$h(\tau) = h(\tau_{-x|U}) - \sum_{(\delta, S) \in I} h(\tau_{\delta,S}),$$

where

$$I := \{(x \oplus y, S) : y \in S \in \gamma_{-U},\ \ S \oplus (x \oplus y) \text{ is disjoint with } U \setminus x\}.$$

The goal here is to upper bound the maximum difference between terms of the form $h(\tau_{-x|U})/N$ and $h(\tau_{\delta,S})$ for any $(\delta, S) \in I$. Let us fix any $(\delta, S) \in I$, and let us denote $U' = U_{-x}$, $U'' = \delta + U'$, and $S' = \delta + S$. By definition of I, U' is disjoint from S', which also means that U'' is disjoint from S. Moreover, one has the following equalities:

$$\begin{aligned}\tau_{-x|U} &= \gamma_{+U_{-x}} = \gamma_{+U'},\\ \tau_{\delta,S} &= \gamma_{-S+((U_{-x})\sqcup(\delta+S))} = \gamma_{-S+(U'\sqcup S')}.\end{aligned}$$

Given the fact that the h coefficient is invariant under translation of the sets, one clearly has

$$\begin{aligned}h(\tau_{\delta,S}) &= h(\gamma_{-S+(U'\sqcup S')}) = h(\gamma_{-S+(\delta+(U'\sqcup S'))}), \text{ and}\\ h(\tau_{-x|U}) &= h(\gamma_{+(\delta+U')}) = h(\gamma_{+U''}).\end{aligned}$$

Since $\delta + (U' \sqcup S') = U'' \sqcup S$, one has

$$h(\tau_{\delta,S}) = h(\gamma_{-S+(U''\sqcup S)}),$$

and thus

$$\left|\frac{h(\tau_{-x|U})}{N} - h(\tau_{\delta,S})\right| = \left|\frac{h(\gamma_{+U''})}{N} - h(\gamma_{-S+(U''\sqcup S)})\right|, \tag{16.8}$$

where $|U''| = \ell + 1$, and $S \in \gamma$ is disjoint with U''.

This motivates the following definition.

Definition 16.5 Let $\tau = \gamma_{+U}$ with $\gamma \subseteq \lambda$ where $|\gamma| = \alpha$ and $|U| = \ell + 1$. For any $S \in \gamma$ disjoint with U, let $\tau' := \gamma_{-S+(S\sqcup U)}$ (same as $\tau_{-S-U+S\sqcup U}$, i.e., we merge two disjoint elements of τ). We define

$$D(\alpha, \ell) = \max_{\gamma,U,S} \left|h(\tau)/N - h(\tau')\right|, \tag{16.9}$$

where the maximum is taken over all choices of $\gamma \subseteq \lambda$ of size α, $S \in \gamma$ and a set U of size $\ell + 1$ disjoint with S. For all $\ell < 0$, we define $D(\alpha, \ell) = 0$.

Lemma 16.3 *Assuming $\tau = \gamma_{+U}$ with $\gamma \subseteq \lambda$ where $|\gamma| = \alpha$ and $|U| = \ell + 2$, we have*

$$\left|\frac{h(\tau_{-x|U})}{N} - h(\tau_{\delta,S})\right| \leq D(|\gamma|, |U| - 2).$$

Proof The lemma directly follows from the previous definition and Eq. (16.8). □

Now we state and prove the recursive inequality for D-terms:

Lemma 16.4 (Recursive Inequality of D-Terms) *Let $\alpha \le q \le \frac{N}{12\xi_{\max}^2}$, $\ell \ge 0$. We write $\beta := \xi_{\max}/N$. Then,*

$$D(\alpha, \ell) \le D(\alpha, \ell - 1) + \xi_{\max} \sum_{i=1}^{q} D(\alpha - 1, \ell + \xi_i - 1) + \frac{2\Delta\xi_{\max} \cdot h(\lambda)}{N \cdot \left(N - q\xi_{\max}^2\right)^{q-\alpha}}. \tag{16.10}$$

Denoting $\beta := \xi_{\max}/N$, and $a_{d,\ell} = \frac{(N\beta)^d}{2h(\lambda)} D(q - d, \ell)$, we have

$$a_{d,\ell} \le \frac{1}{(4e\xi_{\max} q)^d}, \tag{16.11}$$

and, writing $\ell_i = \xi_i - 1$,

$$a_{d,\ell} \le a_{d,\ell-1} + \sum_{i=1}^{q} a_{d+1,\ell+\ell_i} + \beta\Delta \left(4e\xi_{\max} q\right)^{-d}. \tag{16.12}$$

Remark 16.3 Note that the r.h.s. of the inequality contains three types of terms:

- $D(\alpha, \ell - 1)$ which will disappear after $\ell - 1$ applications of the recursive inequality;
- terms of the form $D(\alpha - 1, \ell + \xi_i - 1)$ which involve a smaller set-system, but a larger U set, which will ensure their geometric reduction;
- a parasite term that, as we will see, is small enough not to cause an issue after a logarithmic number of iterations.

Proof (of Lemma 16.4) We fix $S \in \gamma \subseteq \lambda$ where $|\gamma| = \alpha$ and a set U with $|U| = \ell + 1$ disjoint with S. Let $\tau := \gamma_{+U}$ and $\tau' := \gamma_{-S+(S \sqcup U)}$. In words, γ is a set-system that is included in λ, U is any subset of $\{0, 1\}^n$ of size $\ell + 1$, and S is an element of γ. Then, τ corresponds to the $\gamma \cup \{U\}$, while τ' corresponds to τ after S and U have been merged. Looking back at Fig. 16.1, τ and τ' would correspond, respectively, to the second and third graphs. We assume that γ, U, S are chosen in such a manner that $|h(\tau)/N - h(\tau')| = D(\alpha, \ell)$. Now we prove the inequality in two cases.

Case $|U| = 1$. In this case, let $U = \{x\}$. Then $h(\tau) = h(\gamma) \cdot (N - \|\gamma\|)$ from Eq. (16.2). Also $\tau'_{-x|S \sqcup U} = \gamma$. Hence from link-deletion equation, Eq. (16.6),

$$h(\tau') = h(\gamma) - \sum_{(\delta, S') \in I} h(\tau'_{\delta, S'})$$

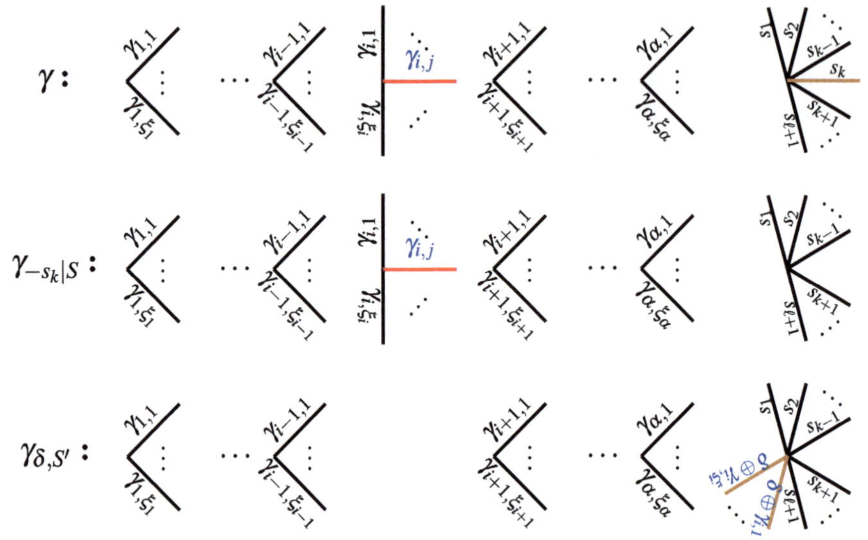

Fig. 16.1 Graphical depiction of the link-deletion operation from Cogliati et al. (2023). Here, graphs corresponding to the three types of terms appearing in the link-deletion equation are represented, with $x = s_k$, $y = \gamma_{i,j}$, $\delta = s_k \oplus \gamma_{i,j}$, $S = \{s_1, \ldots, s_{\ell+1}\}$, and $S' = \gamma_i$. Central vertices correspond to the $r_1, \ldots, r_\alpha, r$ variables

where $I := I_{x,S} = \{(\delta, S') : x \oplus \delta \in S' \in \gamma_{-S},\ S' \oplus \delta \text{ is disjoint with } S\}$. For $z' \in S' \in \gamma_{-S}$, by definition of I, it is clear that $(x \oplus z, S') \notin I$ if and only if there exists $y \in S$ and $w \in S'$ such that $x \oplus y = z \oplus w$. Thus

$$
\begin{aligned}
|I| &\geq \sum_{S' \in \gamma_{-S}} \left(|S'| - \sum_{y \in S} 2\delta_{S'}(x \oplus y) \right) \\
&= \|\gamma\| - |S| - \sum_{y \in S} 2\delta_{\gamma_{-S}}(x \oplus y) \\
&\geq \|\gamma\| - \|\gamma\|_{\max} \cdot 2\Delta_\gamma .
\end{aligned}
$$

Hence

$$
\begin{aligned}
D(\alpha, 0) = |h(\tau)/N - h(\tau')| &= \left| \|\gamma\| h(\gamma)/N - \sum_{(\delta, S') \in I} h(\tau'_{\delta, S'}) \right| \\
&\leq \sum_{(\delta, S') \in I} \left| h(\gamma)/N - h(\tau'_{\delta, S'}) \right| + (\|\gamma\| - |I|) \cdot h(\gamma)/N \\
&\leq \sum_{(\delta, S') \in I} \left| h(\gamma)/N - h(\tau'_{\delta, S'}) \right| + 2\Delta_\gamma \|\gamma\|_{\max} \cdot h(\gamma)/N .
\end{aligned}
$$

Using the initial condition Eq. (16.5) gives

$$\begin{aligned}
D(\alpha, 0) &\le \sum_{(\delta,S')\in I} |h(\gamma)/N - h(\tau'_{\delta,S'})| + \frac{2\Delta_\gamma \|\gamma\|_{\max} \cdot h(\lambda)}{N \cdot \left(N - q\xi_{\max}^2\right)^{|\lambda\setminus\gamma|}} \\
&\le \|\gamma_{-S}\|_{\max} \sum_{S'\in\gamma\setminus S} D(\alpha - 1, |S'| - 1) + \frac{2\Delta_\gamma \|\gamma\|_{\max} \cdot h(\lambda)}{N \cdot \left(N - q\xi_{\max}^2\right)^{|\lambda\setminus\gamma|}},
\end{aligned}$$

where the last term in ($\star$) is obtained from the initial condition Eq. (16.5).

Case $|U| \ge 2$. Fix $x \in U$. By link-deletion equation, we have

$$\begin{aligned}
h(\tau) &= h(\tau_{-x|U}) - \sum_{(\delta,S')\in I} h(\tau_{\delta,S'}) \\
h(\tau') &= h(\tau'_{-x|S\sqcup U}) - \sum_{(\delta,S')\in I'} h(\tau'_{\delta,S'}),
\end{aligned}$$

where

$$\begin{aligned}
I &:= I_{x|U} = \{(\delta, S') : x \oplus \delta \in S' \in \gamma,\ \ S' \oplus \delta \text{ is disjoint with } U \setminus x\}, \\
I' &:= I_{x|S\sqcup U} = \{(\delta, S') : x \oplus \delta \in S' \in \gamma_{-S},\ \ S' \oplus \delta \text{ is disjoint with } S \sqcup U \setminus x\}.
\end{aligned}$$

It is easy to see that $I' \subseteq I$. If $(\delta, S') \in I \setminus I'$, then,

- either $S' = S$ and $\delta = x \oplus y$ for some $y \in S$, such that $S \oplus (x \oplus y)$ is disjoint with $U \setminus x$ or
- $S' \in \gamma \setminus S$ and $\delta = x \oplus z$ for some $z \in S'$, such that $S' \oplus (x \oplus z)$ is disjoint with $U \setminus x$ but not disjoint with $S \sqcup (U \setminus x)$.

The first case can contribute at most $|S|$. The second case will happen if for some $z, w \in S'$, and $y \in S$, $z \oplus w = x \oplus y$. Thus

$$|I \setminus I'| \le |S| + \sum_{y\in S} \delta_{\gamma_{-S}}(x \oplus y) \le \|\gamma\|_{\max} \cdot 2\Delta_\gamma.$$

Hence, we have the following:

$$\begin{aligned}
&D(\alpha, \ell) = |h(\tau)/N - h(\tau')| \\
&\le \left|h(\tau_{-x})/N - h(\tau'_{-x})\right| + \sum_{(\delta,S')\in I'} \left|h(\tau_{\delta,S'})/N - h(\tau'_{\delta,S'})\right| + \sum_{(\delta,S')\in I\setminus I'} h(\tau_{\delta,S'})/N \\
&\le D(\alpha, \ell - 1) + \|\gamma_{-S}\|_{\max} \sum_{S'\in\gamma_{-S}} D(\alpha - 1, \ell + |S'| - 1) + \frac{2\Delta_\gamma \|\gamma\|_{\max} \cdot h(\lambda)}{N \cdot \left(N - q\xi_{\max}^2\right)^{|\lambda\setminus\gamma|}}.
\end{aligned} \tag{16.13}$$

The last inequality follows from the observation that $\tau_{\delta,S'}$ and $\tau'_{\delta,S'}$ are considered when we take maximum to compute $D(\alpha - 1, \ell + |S'| - 1)$. Moreover, from the initial condition Eq. (16.5):

$$h(\tau_{\delta,S'}) \leq h(\gamma) \leq h(\lambda) / \left(N - q\xi_{\max}^2\right)^{|\lambda \setminus \gamma|}.$$

Now, taking upper bounds of the total size terms, and adding some positive terms in the middle sum, and noting that $\Delta_\gamma \leq \Delta_\lambda$,[2] the inequality, Eq. (16.13), can be easily modified to the theorem statement, i.e.:

$$D(\alpha, \ell) \leq D(\alpha, \ell - 1) + \xi_{\max} \sum_{i=1}^{q} D(\alpha - 1, \ell + \xi_i - 1) + \frac{2\Delta\xi_{\max} \cdot h(\lambda)}{N \cdot \left(N - q\xi_{\max}^2\right)^{q-\alpha}}.$$

Let $d = q - \alpha$. Then, one has

$$\begin{aligned} D(q-d, \ell) \leq& D(q-d, \ell-1) + \xi_{\max} \sum_{i=1}^{q} D(q-d-1, \ell+\xi_i-1) \\ &+ \frac{2\Delta\xi_{\max} \cdot h(\lambda)}{N \cdot \left(N - q\xi_{\max}^2\right)^d}. \end{aligned}$$

Multiplying the equation by $\frac{\xi_{\max}^d}{2h(\lambda)}$ yields

$$\begin{aligned} \frac{\xi_{\max}^d}{2h(\lambda)} D(q-d, \ell) \leq& \frac{\xi_{\max}^d}{2h(\lambda)} D(q-d, \ell-1) + \frac{\xi_{\max}^{d+1}}{2h(\lambda)} \sum_{i=1}^{q} D(q-d-1, \ell+\xi_i-1) \\ &+ \frac{\Delta\xi_{\max}^{d+1}}{N \cdot \left(N - q\xi_{\max}^2\right)^d}. \end{aligned}$$

Note, for $q \leq \|\lambda\| \leq N/12\xi_{\max}^2$, $\frac{\xi_{\max}}{(N - q\xi_{\max}^2)} \leq (4\xi_{\max} eq)^{-1}$, which gives Eq. (16.12). Besides, in addition to the above recursive inequality, we also have the following bound, which follows from Eq. (16.5):

$$\begin{aligned} D(\alpha, \ell) &= |h(\tau)/N - h(\tau')| \\ &\leq \frac{h(\gamma_{+U})}{N} + h(\gamma_{-S+(S \sqcup U)}) \\ &\leq 2h(\gamma) \leq \frac{2h(\lambda)}{(N - q\xi_{\max}^2)^{q-\alpha}}, \end{aligned}$$

[2] Since $\gamma \subseteq \lambda$, we have $\sum_{S \in \gamma} \delta_S(z) \leq \sum_{S \in \lambda} \delta_{S'}(z)$ for every $z \in \{0, 1\}^n$, since every $S \in \gamma$ is subset of some $S \in \lambda$. So taking maximum over all $z \in \{0, 1\}^n$, on both sides would give us $\Delta_\gamma \leq \Delta_\lambda$.

and so

$$a_{d,\ell} = \frac{(N\beta)^d}{2h(\lambda)} D(q-d,\ell) \le \left(\frac{\xi_{\max}}{N - q\xi_{\max}^2}\right)^d \le 1/(4e\xi_{\max}q)^d,$$

which ends the proof of Lemma 16.4. □

16.3.6 Final Wrap Up of the Proof

We can conclude the proof of Proposition 16.1 using Lemmas 16.2 and 16.4, along with the following result that will be proven in Sect. 16.3.7.

Lemma 16.5 (Recursive Inequality Lemma) *Suppose $a_{d,\ell} \ge 0$ such that: (i) $a_{d,k} := 0$ for all $k < 0$ and (ii) for all $0 \le d \le \xi n$ and $0 \le \ell_i \le \xi - 1$ for $i \in [q]$, we have*

$$a_{d,\ell} \le (4\xi eq)^{-d} \qquad \text{(initial bound)} \qquad (16.14)$$

$$a_{d,\ell} \le a_{d,\ell-1} + \sum_{i=1}^{q} a_{d+1,\ell+\ell_i} + C \cdot (4\xi eq)^{-d} \qquad \text{(recursive inequality)} \qquad (16.15)$$

for some $C > 0$. Then, for every $\ell \in [\xi - 2]$,

$$a_{0,\ell} \le \frac{4}{N} + 4C\xi.$$

Let a, b, T, λ be as in the statement of Proposition 16.1, and let $\lambda_0 = \lambda_{-T}$. Note that one has $\xi_{\max}^2 n \le \sqrt{N} - \xi_{\max} \le \|\lambda_0\| \le N/12\xi_{\max}^2$. Moreover, let $q = |\lambda_0|$. Similarly, one has $\xi_{\max} q \ge \|\lambda_0\| \ge \xi_{\max}^2 n$, which means that $q \ge \xi_{\max} n$. We are going to apply Lemma 16.5 to λ_0 as follows.

Let us take, $\xi = \xi_{\max}$, $C = \beta\Delta = \Delta_\lambda \xi_{\max}/N$ in the statement of Lemma 16.5. From the definition of $a_{d,\ell} = \frac{(N\beta)^d}{2h(\lambda_0)} D(q-d,\ell)$, we must ensure that $q \ge d$ in order to apply Lemma 16.5. This can easily be seen to be true as $q \ge \xi n$ and $d \le \xi n$. Then, for $(\delta, S) \in I_{a|T}$, Lemma 16.3 yields

$$|h(\lambda_{\delta,S}) - h(\lambda_{-a|T})/N| \le D(q, |T|-2) \le 2h(\lambda_0) a_{0,|T|-2} \le \frac{8h(\lambda_0)}{N}(\Delta\xi_{\max}^2 + 1).$$

Note that one has

$$h(\lambda_{-a|T}) \ge h(\lambda_0) N \left(1 - \frac{\|\lambda_0\|\xi_{\max}}{N}\right) \ge h(\lambda_0) N \left(1 - \frac{1}{12\xi_{\max}}\right) \ge h(\lambda_0)\frac{23N}{24}.$$

Thus, one has

$$h(\lambda_{\delta,S}) \leq \frac{8h(\lambda_0)}{N}(\Delta\xi_{\max}^2+1)+\frac{h(\lambda_{-a|T})}{N} \leq \left(\frac{8h(\lambda_0)(\Delta\xi_{\max}^2+1)}{h(\lambda_{-a|T})}+1\right)\frac{h(\lambda_{-a|T})}{N}$$
$$\leq \left(\frac{24\cdot 8}{23\cdot N}(\Delta\xi_{\max}^2+1)+1\right)\frac{h(\lambda_{-a|T})}{N} \leq \left(\frac{C'\Delta}{N}+1\right)\frac{h(\lambda_{-a|T})}{N},$$

where $C' = 9(\xi_{\max}^2+1)$, as $\Delta \geq 1$. Using this bound in the appropriate link-deletion equation we have

$$\begin{aligned} h(\lambda) &= h(\lambda_{-a|T}) - \sum_{(\delta,S)\in I_{a|T}} h(\lambda_{\delta,S}) \quad \text{(From Eq. (16.6))} \\ &\geq h(\lambda_{-a|T}) - \sum_{(\delta,S)\in I_{a|T}} \frac{h(\lambda_{-a|T})}{N}(1+C'\Delta/N) \\ &\geq h(\lambda_{-a|T})\left(1-\frac{\|\lambda\|-\Delta-|T|/2}{N}\left(1+\frac{C'\Delta}{N}\right)\right) \quad \text{(From Lemma 16.2)} \\ &\geq h(\lambda_{-a|T})\left(1-\frac{\|\lambda\|-1}{N}+\frac{\Delta}{N}\left(1-\frac{C'(\|\lambda\|-\Delta-1)}{N}\right)\right) \\ &\geq h(\lambda_{-a|T})\left(1-\frac{\|\lambda\|-1}{N}\right). \end{aligned}$$

The last inequality follows as $C'\|\lambda\| \leq N$, for $\|\lambda\| \leq N/12\xi_{\max}^2$, which concludes the proof of Proposition 16.1. □

Remark 16.4 Note that the initial bound ensures only that $a_{0,\ell} \leq 1$. However, the presence of recursive inequality forces the value of $a_{0,\ell}$ to be very small.

16.3.7 Proof of Recursive Inequality Lemma

Let us denote by an ordered tuple of integers from $[q]$, as $i^k := (i_1, \ldots, i_k) \in [q]^k$. Note that, for all positive integer j, $e^j \geq \frac{j^j}{j!}$ and so $1/j! \leq (e/j)^j$, and we have

$$\binom{m}{j} \leq \frac{m^j}{j!} \leq (em/j)^j. \tag{16.16}$$

This inequality will be frequently used for the proof of this lemma. We also use the following fact extensively: for $r < 1$, $\sum_{j\geq i} r^j \leq \frac{r^i}{1-r}$.

We state the following claim, which follows from iterated applications of the recursive inequality.

Claim For any $0 \le d \le \xi n$, and $0 \le \ell < \xi - 1$ we have

$$a_{0,\ell} \le \sum_{k=\lceil \frac{d-\ell}{\xi} \rceil}^{d} \binom{d}{k} \sum_{i^k \in [q]^k} a_{k,\ell+k+\sum_{j=1}^{k} \ell_{i_j}-d} + C \sum_{i=0}^{d-1} \sum_{j=\lceil \frac{i-\ell}{\xi} \rceil}^{i} \binom{i}{j} (4\xi e)^{-j}. \quad (16.17)$$

Proof (of Claim 16.3.6) We prove the claim by induction on d. The result holds trivially for $d = 1$ (by applying $d = 0$ in Eq. (16.15)). Now we prove the statement for $d_0 + 1$, assuming it true for d_0. Applying the recursive inequality to the induction hypothesis yields

$$\begin{aligned}
a_{0,\ell} &\le \sum_{k=\lceil \frac{d_0-\ell}{\xi} \rceil}^{d_0} \binom{d_0}{k} \sum_{i^k \in [q]^k} a_{k,\ell+k+\sum_{j=1}^{k} \ell_{i_j}-d_0} + C \sum_{i=0}^{d_0-1} \sum_{j=\lceil \frac{i-\ell}{\xi} \rceil}^{i} \binom{i}{j} (4\xi e)^{-j} \\
&\le \underbrace{\sum_{k=\lceil \frac{d_0-\ell}{\xi} \rceil}^{d_0} \binom{d_0}{k} \sum_{i^k \in [q]^k} \sum_{i_{k+1} \in [q]} a_{k+1,\ell+k+1+\sum_{j=1}^{k+1} \ell_{i_j}-(d_0+1)}}_{A_1} \\
&\quad + \underbrace{\sum_{k=\lceil \frac{d_0-\ell}{\xi} \rceil}^{d_0} \binom{d_0}{k} \sum_{i^k \in [q]^k} a_{k,\ell+k+\sum_{j=1}^{k} \ell_{i_j}-(d_0+1)}}_{A_2} \\
&\quad + \underbrace{C \sum_{k=\lceil \frac{d_0-\ell}{\xi} \rceil}^{d_0} \binom{d_0}{k} \sum_{i^k \in [q]^k} (4\xi e q)^{-k}}_{A_3} + \underbrace{C \sum_{i=0}^{d_0-1} \sum_{j=\lceil \frac{i-\ell}{\xi} \rceil}^{i} \binom{i}{j} (4\xi e)^{-j}}_{A_4}. \quad (16.18)
\end{aligned}$$

First, A_1 can be rewritten as follows:

$$\begin{aligned}
A_1 &= \sum_{k=\lceil \frac{d_0-\ell}{\xi} \rceil}^{d_0} \binom{d_0}{k} \sum_{i^k \in [q]^k} \sum_{i_{k+1} \in [q]} a_{k+1,\ell+k+1+\sum_{j=1}^{k+1} \ell_{i_j}-(d_0+1)} \\
&= \sum_{k=\lceil \frac{d_0-\ell}{\xi} \rceil}^{d_0} \binom{d_0}{k} \sum_{i^{k+1} \in [q]^{k+1}} a_{k+1,\ell+k+1+\sum_{j=1}^{k+1} \ell_{i_j}-(d_0+1)} \\
&= \sum_{k=\lceil \frac{d_0-\ell}{\xi} \rceil+1}^{d_0+1} \binom{d_0}{k-1} \sum_{i^k \in [q]^k} a_{k,\ell+k+\sum_{j=1}^{k} \ell_{i_j}-(d_0+1)}.
\end{aligned}$$

Remark that $\lceil (d_0 + 1 - \ell)/\xi \rceil \leq \lceil (d_0 - \ell)/\xi \rceil + 1$, which yields the following upper bound:

$$A_1 \leq \sum_{k=\lceil (d_0+1-\ell)/\xi \rceil}^{d_0+1} \binom{d_0}{k-1} \sum_{i^k \in [q]^k} a_{k,\ell+k+\sum_{j=1}^k \ell_{i_j} - (d_0+1)}. \tag{16.19}$$

As for A_2, the summation starts with $k = \left\lceil \frac{d_0-\ell}{\xi} \right\rceil$. However, we argue that, when $k < (d_0 + 1 - \ell)/\xi$, one has $a_{k,\ell+k+\sum_{j=1}^k \ell_{i_j} - (d_0+1)} = 0$. Indeed, in that case, one has

$$\begin{aligned} \ell + k + \sum_{j=1}^{k} \ell_{i_j} - (d_0 + 1) &< \ell + \frac{d_0 + 1 - \ell}{\xi} + \sum_{j=1}^{k} (\xi_{i_j} - 1) - (d_0 + 1) \\ &< \ell + \frac{d_0 + 1 - \ell}{\xi} + \frac{d_0 + 1 - \ell}{\xi} (\xi - 1) - (d_0 + 1) = 0. \end{aligned}$$

Hence, the summation in A_2 can actually start from $\lceil (d_0 + 1 - \ell)/\xi \rceil$, which yields

$$\begin{aligned} A_2 &= \sum_{k=\left\lceil \frac{d_0-\ell}{\xi} \right\rceil}^{d_0} \binom{d_0}{k} \sum_{i^k \in [q]^k} a_{k,\ell+k+\sum_{j=1}^k \ell_{i_j} - (d_0+1)} \\ &\leq \sum_{k=\lceil (d_0+1-\ell)/\xi \rceil}^{d_0+1} \binom{d_0}{k} \sum_{i^k \in [q]^k} a_{k,\ell+k+\sum_{j=1}^k \ell_{i_j} - (d_0+1)}. \end{aligned} \tag{16.20}$$

Finally, combining Eqs. (16.19) and (16.20), we can observe that the coefficient of $\sum_{i^k \in [q]^k} a_{k,k+\sum_{j=1}^k - (d_0+1)}$ in the upper bound of $A_1 + A_2$ is equal to $\binom{d_0}{k-1} + \binom{d_0}{k} = \binom{d_0+1}{k}$, which yields

$$A_1 + A_2 \leq \sum_{k=\lceil (d_0+1-\ell)/\xi \rceil}^{d_0+1} \binom{d_0+1}{k} \sum_{i^k \in [q]^k} a_{k,\ell+k+\sum_{j=1}^k \ell_{i_j} - (d_0+1)}. \tag{16.21}$$

Then, note that A_3 satisfies

$$A_3 = C \sum_{k=\left\lceil \frac{d_0-\ell}{\xi} \right\rceil}^{d_0} \binom{d_0}{k} \sum_{i^k \in [q]^k} (4\xi e q)^{-k} \leq C \sum_{k=\left\lceil \frac{d_0-\ell}{\xi} \right\rceil}^{d_0} \binom{d_0}{k} (4\xi e)^{-k},$$

which means that one has

$$A_3 + A_4 \le C \sum_{i=0}^{d_0} \sum_{j=\left\lceil \frac{i-\ell}{\xi} \right\rceil}^{i} \binom{i}{j} (4\xi e)^{-j}. \tag{16.22}$$

Combining Eqs. (16.21) and (16.22) yields

$$\begin{aligned} a_{0,\ell} \le & \sum_{k=\lceil (d_0+1-\ell)/\xi \rceil}^{d_0+1} \binom{d_0+1}{k} \sum_{i^k \in [q]^k} a_{k,\ell+k+\sum_{j=1}^{k} \ell_{i_j} - (d_0+1)} \\ & + C \sum_{i=0}^{d_0} \sum_{j=\left\lceil \frac{i-\ell}{\xi} \right\rceil}^{i} \binom{i}{j} (4\xi e)^{-j}, \end{aligned}$$

which concludes the proof of Claim 1. □

16.3.7.1 Concluding the Proof of Lemma 16.5

Let us take $d = \xi n$. In that case, Claim 1 becomes

$$a_{0,\ell} \le \sum_{k=\left\lceil \frac{\xi n-\ell}{\xi} \right\rceil}^{\xi n} \binom{\xi n}{k} \sum_{i^k \in [q]^k} a_{k,\ell+k+\sum_{j=1}^{k} \ell_{i_j} - \xi n} + C \sum_{i=0}^{\xi n-1} \sum_{j=\left\lceil \frac{i-\ell}{\xi} \right\rceil}^{i} \binom{i}{j} N^j (4\xi e)^{-j}.$$

We are going to upper bound both terms of the sum in subsequent turns. For the first term, note that one has $k \ge n - \frac{\ell}{\xi} > n - 1$ since $\ell < \xi - 1$ by definition. This implies that

$$\binom{\xi n}{k} \le \left(\frac{e\xi n}{k}\right)^k \le \left(\frac{e\xi n}{n-1}\right)^k \le (2e\xi)^k.$$

Hence, using the initial bound, one has

$$\sum_{k=\left\lceil \frac{\xi n-\ell}{\xi} \right\rceil}^{\xi n} \binom{\xi n}{k} \sum_{i^k \in [q]^k} a_{k,\ell+k+\sum_{j=1}^{k} \ell_{i_j} - \xi n} \le \sum_{k=\left\lceil \frac{\xi n-\ell}{\xi} \right\rceil}^{\xi n} (2e\xi)^k q^k (4\xi e q)^{-k} \le \frac{4}{2^n} \le \frac{4}{N}.$$

As for the second term, we make the following observation: For $\xi k < i \le \xi(k+1)$, $k \in (n-1]$, $j \ge \lceil \frac{i-\ell}{\xi} \rceil \ge k$, and hence

$$\binom{i}{j} \le \left(\frac{ei}{j}\right)^j \le \left(\frac{e\xi(k+1)}{k}\right)^j \le (2e\xi)^j.$$

For $0 \leq i \leq \xi$ and $j \geq 1$, $\binom{i}{j} \leq \left(\frac{ei}{j}\right)^j \leq (e\xi)^j$. Thus, we are going to break the sum into two parts:

$$\begin{aligned}
\sum_{i=0}^{\xi n-1} \sum_{j=\left\lceil \frac{i-\ell}{\xi} \right\rceil}^{i} \binom{i}{j} (4\xi e)^{-j} &= \sum_{i=0}^{\xi} \sum_{j=\left\lceil \frac{i-\ell}{\xi} \right\rceil}^{i} \binom{i}{j} (4\xi e)^{-j} + \sum_{i=\xi+1}^{\xi n-1} \sum_{j=\left\lceil \frac{i-\ell}{\xi} \right\rceil}^{i} \binom{i}{j} (4\xi e)^{-j} \\
&\leq \xi + 1 + \sum_{i=0}^{\xi} \sum_{j=1}^{i} (e\xi)^j (4e\xi)^{-j} \\
&\quad + \sum_{i=\xi+1}^{\xi n-1} \sum_{j=\lceil i/\xi \rceil -1}^{i} (2e\xi)^j (4e\xi)^{-j} \\
&\leq \xi + 1 + \frac{\xi+1}{3} + 4 \sum_{i=\xi+1}^{\xi n-1} \frac{1}{2^{\lceil i/\xi \rceil}} \\
&\leq \frac{4}{3}(\xi + 1) + 2\xi \leq 4\xi,
\end{aligned}$$

where the last inequality follows from the fact that $\xi \geq 2$.

Problems

16.1 Let $\|$ be the concatenation function. When f is a permutation of $\mathcal{P}_n$, we denote by $f(x\|0) \oplus f(x\|1)$ the function F from $\{0, 1\}^{n-1}$ to $\{0, 1\}^{n-1}$ such that for all $x \in \{0, 1\}^{n-1}$, $F(x) = f(x\|0) \oplus f(x\|1)$.

In this exercise, we ask to prove the following theorem (from the results of this chapter and the "Coefficient H theorem" in CPA, i.e., Theorem 3.4).

Theorem 16.3 *For all CPA distinguisher* D *on a function G of* $\mathcal{F}_n$ *with q chosen plaintexts, we have* $\mathcal{A}_{\mathsf{D}}^{PRF} \leq O(\frac{q}{2^n})$ *where* $\mathcal{A}_{\mathsf{D}}^{PRF}$ *denotes the advantage to distinguish* $f(x\|0) \oplus f(x\|1)$ *with* $f \in \mathcal{P}_n$ *from a random function* $g : \{0, 1\}^{n-1} \rightarrow \{0, 1\}^n$. *A precise bound for* $O(\frac{q}{2^n})$ *can be given.*

16.2 Show that there is an attack in $q = 0(2^n)$ to distinguish $f(x\|0) \oplus f(x\|1)$ with $f \in \mathcal{P}_n$ from a random function $g : \{0, 1\}^{n-1} \rightarrow \{0, 1\}^n$. Therefore, the bound in $O(\frac{q}{2^n})$ in Theorem 16.3 is optimal.

Theorem 16.3 can be seen as "Luby–Rackoff backward" result. From Luby–Rackoff theorem, we are able to build pseudo-random permutations from pseudo-random functions. From Theorem 16.3, we are able to build pseudo-random functions from pseudo-random permutations, with a security "above the birthday bound

$q \ll \sqrt{2^n}$" since here we have security when $q \ll O(2^n)$. Any pseudo-random permutation can be seen as a pseudo-random function when $q \ll \sqrt{2^n}$, i.e., below and not above the birthday bound.

Remark 16.5 The problem of distinguishing $f(x\|0) \oplus f(x\|1)$ when f is a permutation on $n-1$ bits from a random function on n bits was first studied to be Bellare and Impagliazzo (1999). They obtained a similar security result with a completely different proof technique.

References

Bellare, M., and R. Impagliazzo. 1999. A tool for obtaining tighter security analyses of pseudorandom function based constructions, with applications to PRP to PRF conversion. In *Cryptology ePrint Archive: Report 1999/024*.

Cogliati, B., A. Dutta, M. Nandi, J. Patarin, and A. Saha. 2023. Proof of mirror theory for a wide range of $\xi_{\max}$. In *Advances in cryptology—EUROCRYPT 2023. EUROCRYPT 2023*, vol. 14007, ed. C. Hazay and M. Stam. Lecture notes in computer science. Cham: Springer.

Patarin, J. 2005. On linear systems of equations with distinct variables and small block size. In *Information and communications security—ICISC '05*, vol. 3935, ed. D. Won and S. Kim. Lecture notes in computer science, 299–321. Berlin: Springer.

Chapter 17
Proofs Beyond the Birthday Bound on Ψ^k with the H-Coefficient Method

Abstract In this chapter, we will use the result obtained in Chap. 16 in order to prove some security results on Generic balanced Feistel ciphers Ψ^k. The main results will be the proof of security for Ψ^4 in KPA, for Ψ^5 in CPA and CCA, and for Ψ^6 in CCA, when $q \ll 2^n$. We will also see what kind of bound we obtain from the results on Ψ^6 for Ψ^k, $k \geq 6$, by using some compositions theorem. Finally, at the end of this chapter, we will compare the results obtained with proofs from Mirror theory and H-coefficient technique from the results obtained in Chap. 14 with the coupling technique.

17.1 Exact Fomulas for H and Ψ^k with "Frameworks"

Definition of Ψ^k

We recall the definition of the balanced Feistel Schemes, i.e., the classical Feistel schemes. Let L, R, S, and T be four n-bit strings in $\{0, 1\}^n$. Let $\Psi(f_1)$ denotes the permutation of $\mathscr{P}_{2n}$ such that:

$$\Psi(f_1)[L, R] = [S, T] \overset{\text{def}}{\Leftrightarrow} \begin{cases} S = R \\ T = L \oplus f_1(R) \end{cases}$$

More generally if $f_1, f_2, \ldots, f_k$ are k functions of $\mathscr{F}_n$, let $\Psi^k(f_1, \ldots, f_k)$ denotes the permutation of $\mathscr{P}_{2n}$ such that:

$$\Psi^k(f_1, \ldots, f_k) = \Psi(f_k) \circ \cdots \circ \Psi(f_2) \circ \Psi(f_1).$$

The permutation $\Psi^k(f_1, \ldots, f_k)$ is called a "balanced Feistel scheme with k rounds" or shortly Ψ^k. When $f_1, \ldots, f_k$ are randomly and independently chosen in $\mathscr{F}_n$, then $\Psi^k(f_1, \ldots, f_k)$ is called a "random Feistel scheme with k rounds" or a "Luby-Rackoff construction with k rounds."

J. Patarin et al., *Guide to Feistel Ciphers*, Information Security and Cryptography,
https://doi.org/10.1007/978-3-031-99850-8_17

Remark 17.1 In this chapter, for convenience, balanced Feistel are denoted by Ψ^k instead of Ψ^r since r will be used for the number of independent equations in R.

Definition 17.1 *(Definition of H for Ψ^k)* When $[L_i, R_i], [S_i, T_i], 1 \le i \le q$, is a given sequence of $2q$ values of $\{0,1\}^{2n}$, we will denote by $H_k(L, R, S, T)$ or in short by H_k, or simply by H, the number of k-tuples of functions $(f_1, \dots f_k)$ of F_n^k such that:

$$\forall i,\ 1 \le i \le q,\ \Psi^k(f_1, \dots, f_k)[L_i, R_i] = [S_i, T_i]$$

We will analyze the properties of these H values in order to obtain our security results.

Let $[L_i, R_i], [S_i, T_i], 1 \le i \le q$, be a given sequence of $2q$ values of $\{0,1\}^{2n}$. Let r be the number of independent equalities $R_i = R_j, i \ne j$, and let s be the number of independent equalities $S_i = S_j, i \ne j$.

Theorem 17.1 *The exact formula for H_1 (i.e., for Ψ^1) is:*

$$H_1 = 0 \text{ if } (C) \text{ is not satisfied}$$

$$H_1 = \frac{|\mathcal{F}_n|}{2^{nq}} \cdot 2^{nr} \text{ if } (C) \text{ is satisfied}$$

where (C) is this set of conditions:

1. $\forall i,\ 1 \le i \le q,\ R_i = S_i$
2. $\forall i, j\ 1 \le i \le q,\ 1 \le j \le q,\ R_i = R_j \Rightarrow T_i \oplus L_i = T_j \oplus L_j$.

Proof For one round, we have $\Psi^1([L_i, R_i]) = [S_i, Y_i] \Leftrightarrow S_i = R_i$ and $T_i = L_i \oplus f_1(R_i)$. Therefore, if (C) is not satisfied, $H_1 = 0$. Now if (C) is satisfied, then f_1 is fixed on exactly $q - r$ points by $f_1(R_i) = T_i \oplus L_i$, and we obtain Theorem 17.1 as claimed. □

Theorem 17.2 *The exact formula for H_2 (i.e., for Ψ^2) is:*

$$H_2 = 0 \text{ if } (C) \text{ is not satisfied}$$

$$H_2 = \frac{|\mathcal{F}_n|^2}{2^{2nq}} \cdot 2^{n(r+s)} \text{ if } (C) \text{ is satisfied}$$

where (C) is this set of conditions:

1. $\forall i, j\ 1 \le i \le q,\ 1 \le j \le q,\ R_i = R_j \Rightarrow L_i \oplus L_j = S_i \oplus S_j$
2. $\forall i, j\ 1 \le i \le q,\ 1 \le j \le q,\ S_i = S_j \Rightarrow R_i \oplus R_j = T_i \oplus T_j$.

Proof For two rounds we have $\psi^2([L_i, R_i]) = [S_i, T_i] \Leftrightarrow S_i = L_i \oplus f_1(R_i)$ and $T_i = R_i \oplus f_2(S_i)$. Therefore if (C) is not satisfied, $H_2 = 0$. Now if (C) is satisfied then (f_1, f_2) is fixed on exactly $2q - r - s$ points, and we obtain Theorem 17.2 as claimed. □

Definition 17.2 (*Framework for Ψ^3*) For 3 rounds, Ψ^3, we define a "framework" as a set of equations $X_i = X_j$. We will say that two frameworks are equal if they imply exactly the same set of equations in X.

Theorem 17.3 *The exact formula for H_3 (i.e., for Ψ^3) is:*

$$H_3 = \frac{|\mathcal{F}_n|^3 \cdot 2^{n(r+s)}}{2^{3nq}} \sum_{\substack{\text{all frameworks } \mathcal{F} \\ \text{that satisfy } (F1)}} 2^{nx} [\, \text{Number of } X_i \text{ satisfying } (C1)]$$

where:

- *x is the number of independent equalities $X_i = X_j$ for a framework $\mathcal{F}$.*
- *$(F1) : X_i = X_j$ is in $\mathcal{F} \Rightarrow S_i \oplus S_j = R_i \oplus R_j$*
- $(C_1) : \begin{cases} R_i = R_j \Rightarrow X_i \oplus X_j = L_i \oplus L_j \\ S_i = S_j \Rightarrow X_i \oplus X_j = T_i \oplus T_j \\ \text{The only equations } X_i = X_j, i < j, \text{ are exactly those implied by } \mathcal{F}. \end{cases}$

Proof We write $\Psi^3 = \Psi \circ \Psi^2$ with $\Psi^2([L_i, R_i]) = [X_i, S_i]$ and $\Psi([X_i, S_i]) = [S_i, T_i]$. For Ψ^2, we obtain from Theorem 17.2, $2^{n(r+x)} \frac{|\mathcal{F}_n|^2}{2^{2nq}}$ solutions when $(C1)$ is satisfied. For Ψ, we obtain from Theorem 17.1, $2^{ns} \frac{|\mathcal{F}_n|}{2^{nq}}$ solutions when $(C1)$ is satisfied. Thus, we obtain Theorem 17.3 as claimed. □

Definition 17.3 (*Framework for Ψ^4*) For 4 rounds, Ψ^4, let us define a "framework" as a set of equations $X_i = X_j$ or $Y_i = Y_j$. We will say that two frameworks are equal if they imply exactly the same set of equalities in X and Y. For a framework $\mathcal{F}$, we denote by x the number of independent equalities $X_i = X_j$, and by y the number of independent equalities $Y_i = Y_j$.

Theorem 17.4 *The exact formula for H_4 (i.e., for Ψ^4) is:*

$$H_4 = \frac{|\mathcal{F}_n|^4 \cdot 2^{n(r+s)}}{2^{4nq}} \sum_{\text{all frameworks } \mathcal{F}} 2^{n(x+y)} [\, \text{Number of } X_i \text{ satisfying } (C1)]$$

$$\cdot [\, \text{Number of } Y_i \text{ satisfying } (C2)]$$

where

$$(C_1) : \begin{cases} R_i = R_j \Rightarrow X_i \oplus X_j = L_i \oplus L_j \\ Y_i = Y_j \text{ is in } \mathcal{F} \Rightarrow X_i \oplus X_j = S_i \oplus S_j \\ \text{The only equations } X_i = X_j, i < j, \text{ are exactly those implied by } \mathcal{F}. \end{cases}$$

$$(C_2) : \begin{cases} S_i = S_j \Rightarrow Y_i \oplus Y_j = T_i \oplus T_j \\ X_i = X_j \text{ is in } \mathcal{F} \Rightarrow Y_i \oplus Y_j = R_i \oplus R_j \\ \text{The only equations } Y_i = Y_j, i < j, \text{ are exactly those implied by } \mathcal{F}. \end{cases}$$

Proof We write $\psi^4 = \Psi \circ \Psi^3$ with $\Psi^3([L_i, R_i]) = [Y_i, S_i]$ and $\Psi([Y_i, S_i]) = [S_i, T_i]$, and we sum over all possible Y. Then from Theorems 17.1 and 17.3, we obtain Theorem 17.4. □

Definition 17.4 (*Framework for Ψ^5*) For 5 rounds, Ψ^5, a "framework" is a set of equations $X_i = X_j$ or $Y_i = Y_j$, or $Z_i = Z_j$. We will say that two frameworks are equal if they imply exactly the same set of equalities in X, Y, and Z. For a framework $\mathcal{F}$, we denote by x the number of independent equalities $X_i = X_j$, by y the number of independent equalities $Y_i = Y_j$, and by z the number of independent equalities $Z_i = Z_j$.

Theorem 17.5 *The exact formula for H_5 (i.e., for Ψ^5) is:*

$$H_5 = \frac{|\mathcal{F}_n|^5 \cdot 2^{n(r+s)}}{2^{5nq}} \sum_{\text{all frameworks } \mathcal{F}} 2^{n(x+y+z)}[\text{ Number of } X_i, Z_i \text{ satisfying } (C1)]$$

$$\cdot[\text{ Number of } Y_i \text{ satisfying } (C2)]$$

where

$$(C_1) : \begin{cases} R_i = R_j \Rightarrow X_i \oplus X_j = L_i \oplus L_j \\ Y_i = Y_j \text{ is in } \mathcal{F} \Rightarrow X_i \oplus X_j = Z_i \oplus Z_j \\ S_i = S_j \Rightarrow Z_i \oplus Z_j = T_i \oplus T_j \\ \text{The only equations } X_i = X_j, i < j, \text{ are exactly those implied by } \mathcal{F}. \\ \text{The only equations } Z_i = Z_j, i < j, \text{ are exactly those implied by } \mathcal{F}. \end{cases}$$

$$(C_2) : \begin{cases} X_i = X_j \text{ is in } \mathcal{F} \Rightarrow Y_i \oplus Y_j = R_i \oplus R_j \\ Z_i = Z_j \text{ is in } \mathcal{F} \Rightarrow Y_i \oplus Y_j = S_i \oplus S_j \\ \text{The only equations } Y_i = Y_j, i < j, \text{ are exactly those implied by } \mathcal{F}. \end{cases}$$

Proof We write $\Psi^5 = \Psi \circ \Psi^4$ with $\Psi^4([L_i, R_i]) = [Z_i, S_i]$ and $\Psi([Z_i, S_i]) = [S_i, T_i]$, and we sum over all possible Z. Then from Theorems 17.1 and 17.4, we obtain Theorem 17.5. □

Definition 17.5 (*Framework for* Ψ^6) For 6 rounds, Ψ^6, a "framework" is a set of equations $X_i = X_j$ or $Y_i = Y_j$, $Z_i = Z_j$, or $A_i = A_j$. We will say that two frameworks are equal if they imply exactly the same set of equalities in X, Y, Z, and A. For a framework $\mathcal{F}$, we denote by x (respectively, y, z, or a) the number of independent equalities $X_i = X_j$ (respectively $Y_i = Y_j$, $Z_i = Z_j$, or $A_i = A_j$).

Theorem 17.6 *The exact formula for* H_6 *(i.e., for* Ψ^6*) is:*

$$H_6 = \frac{|\mathcal{F}_n|^6 \cdot 2^{n(r+s)}}{2^{6nq}} \sum_{\textit{all frameworks } \mathcal{F}} 2^{n(x+y+z+a)} \cdot [\textit{ Number of } X_i, Z_i \textit{ satisfying } (C1)] \cdot [\textit{ Number of } Y_i, A_i \textit{ satisfying } (C2)]$$

where

$$(C_1): \begin{cases} R_i = R_j \Rightarrow X_i \oplus X_j = L_i \oplus L_j \\ Y_i = Y_j \textit{ is in } \mathcal{F} \Rightarrow X_i \oplus X_j = Z_i \oplus Z_j \\ A_i = A_j \textit{ is in } \mathcal{F} \Rightarrow Z_i \oplus Z_j = S_i \oplus S_j \\ \textit{The only equations } X_i = X_j, i < j, \textit{ are exactly those implied by } \mathcal{F}. \\ \textit{The only equations } Z_i = Z_j, i < j, \textit{ are exactly those implied by } \mathcal{F}. \end{cases}$$

$$(C_2): \begin{cases} X_i = X_j \textit{ is in } \mathcal{F} \Rightarrow Y_i \oplus Y_j = R_i \oplus R_j \\ Z_i = Z_j \textit{ is in } \mathcal{F} \Rightarrow Y_i \oplus Y_j = A_i \oplus A_j \\ S_i = S_i \Rightarrow A_i \oplus A_j = T_i \oplus T_j \\ \textit{The only equations } Y_i = Y_j, i < j, \textit{ are exactly those implied by } \mathcal{F}. \\ \textit{The only equations } A_i = A_j, i < j, \textit{ are exactly those implied by } \mathcal{F}. \end{cases}$$

Proof We write $\Psi^6 = \Psi \circ \Psi^5$ with $\Psi^5([L_i, R_i]) = [A_i, S_i]$ and $\Psi([A_i, S_i]) = [S_i, T_i]$, and we sum over all possible A. Then from Theorems 17.1 and 17.5, we obtain Theorem 17.6. □

17.2 KPA Security for Ψ^4

Notations for 4 Rounds

- We will denote by $[L_i, R_i]$, $1 \le i \le q$, the q plaintexts. These plaintexts can be assumed to be pairwise distinct, i.e., $i \neq j \Rightarrow L_i \neq L_j$ or $R_i \neq R_j$.
- $[R_i, X_i]$ is the output after one round, i.e.,

$$\forall i, 1 \le i \le q, X_i = L_i \oplus f_1(R_i).$$

- $[X_i, Y_i]$ is the output after two rounds, i.e.,

$$\forall i, 1 \leq i \leq q, Y_i = R_i \oplus f_2(X_i) = R_i \oplus f_2(L_i \oplus f_1(R_i)).$$

- $[Y_i, S_i]$ is the output after three rounds, i.e.,

$$\forall i, 1 \leq i \leq q, S_i = X_i \oplus f_3(Y_i) = L_i \oplus f_1(R_i) \oplus f_3(Y_i).$$

- $[S_i, T_i]$ is the output after 4 rounds, i.e.,

$$\forall i, 1 \leq i \leq q, T_i = Y_i \oplus f_4(S_i).$$

Definition 17.6 Let $\mathscr{F}$ be a framework. We will denote by Weight ($\mathscr{F}$) the number of variables (X_i, Y_i), $1 \leq i \leq q, X_i \in \{0,1\}^n, Y_i \in \{0,1\}^n$, that satisfy $\mathscr{F}$, i.e., such that the equalities $X_i = X_j$ or $Y_i = Y_i$ are exactly those in $\mathscr{F}$.

Theorem 17.7 *If we denote by x the number of independent equalities $X_i = X_j$ of a framework $\mathscr{F}$ and by y the number of independent equalities of $\mathscr{F}$, we have: Weight ($\mathscr{F}$) $= J_{q-x} \cdot J_{q-y}$ where J_i denotes $J_i = 2^n(2^n-1)\dots(2^n-i+1)$.*

Proof We have fixed x values X_i from the other values, and all the other variables X_j must be pairwise distinct. Therefore for $(X_1, X_2, \dots, X_q)$ we have exactly J_{q-x} possibilities. Similarly for $(Y_1, Y_2, \dots, Y_q)$ we have exactly J_{q-y} possibilities. □

The results on Ψ^4 for proving KPA security in $\frac{2^n}{n^2}$ will be based on this Theorem.

Theorem 17.8 *For random values $[L_i, R_i], [S_i, T_i]$, $1 \leq i \leq q$ such that $[L_i, R_i]$, $1 \leq i \leq q$ are pairwise distinct, with probability $\geq 1-\beta$ we have:*

1. *The number H of $(f_1, f_2, f_3, f_4) \in F_n^4$ such that $\Psi^4(f_1, f_2, f_3, f_4)[L_i, R_i] = [S_i, T_i]$ satisfies $H \geq \frac{|F_n^4|}{2^{nq}}(1-\alpha)$.*
2. *α and β can be chosen $\ll 1$ when $q \ll \frac{2^n}{n^2}$. (Moreover we will obtain explicit values for α and β: $\alpha = \frac{2q}{2^n}$, $\beta = \frac{2q}{2^n}$ when $q \leq \frac{2^n}{12n^2}$).*

Proof To prove Theorem 17.8, we will evaluate H from the formula given by Theorem 17.4, i.e., with the sum on all frameworks $\mathscr{F}$. When a framework $\mathscr{F}$ is fixed, 5 cases can occur:

Case 1. A contradiction appears by linearity in the equations generated by $(C1)$. In this case we will say that we have a "circle" of equalities in $R, Y_{\mathscr{F}}$ that generates a "circle" of equalities in X by $(C1)$.

Case 2. A contradiction appears by linearity in the equations generated by $(C2)$. In this case we will say that we have a "circle" of equalities in $S, X_{\mathscr{F}}$ that generates a "circle" of equations in Y by $(C2)$.

Case 3. No contradiction occurs by linearity but $\xi_X q$ is not $\ll \frac{2^n}{n}$, where ξ_X denotes the maximum number of X_j variables fixed in the equations $(C1)$ when one variable

X_i is fixed. In this case we will say that we have a big "line" of equalities $R, Y_{\mathscr{F}}$ that generates a big "line" of equalities in X by $(C1)$.

Case 4. No contradiction occurs by linearity but $\xi_Y q$ is not $\ll \frac{2^n}{n}$, where ξ_Y denotes the maximum number of Y_j variables fixed in the equations $(C1)$ when one variable Y_i is fixed. In this case we will say that we have a big "line" of equalities $S, X_{\mathscr{F}}$ that generates a big "line" of equalities in Y by $(C2)$.

Case 5. No contradiction occurs by linearity and $\xi_X q \ll \frac{2^n}{n}$ and $\xi_Y q \ll \frac{2^n}{n}$.

In order to prove Theorem 17.8 we will first prove that Cases 1, 2, 3, 4 appear with a negligible probability when the values $[L_i, R_i]$, $[S_i, T_i]$ are randomly chosen, and when $\mathscr{F}$ is randomly chosen with a distribution of probability proportional to $Weight(\mathscr{F})$. That is what we will do now.

Circles in $R, Y_{\mathscr{F}}$

Definition 17.7 We will say that we have a "circle in $R, Y_{\mathscr{F}}$" if there are k indices $i_1, \ldots, i_k$ with $k \geq 3$ and such that:

1. $i_1, i_2, \ldots, i_{k-1}$ are pairwise distinct and $i_k = i_1$.
2. $\forall \lambda,\ 1 \leq \lambda \leq k-2$ we have at least one of these conditions: $R_{i_\lambda} = R_{i_{\lambda+1}}$ or $Y_{i_\lambda} = Y_{i_{\lambda+1}}$ is one of the equalities of $\mathscr{F}$.

Example 17.1 If $R_1 = R_2$ and $Y_1 = Y_2$ is one of the equalities of $\mathscr{F}$, we have a circle in $R, Y_{\mathscr{F}}$.

Clearly, if we have a circle in $R, Y_{\mathscr{F}}$, from it we can generate a "minimum circle" in $R, Y_{\mathscr{F}}$, i.e., keeping only one equation $R_{i_1} = R_{i_l}$ per line of equations $R_{i_1} = R_{i_2} = R_{i_3} = \ldots = R_{i_l}$, and keeping only one equation $Y_{i_1} = Y_{i_l}$ per line of equations $Y_{i_1} = Y_{i_2} = Y_{i_3} = \ldots = Y_{i_l}$. Therefore, we have a circle in $R, Y_{\mathscr{F}}$ if and only if there is an even integer μ and there are μ pairwise distinct indices $i_1, \ldots, i_\mu$ with $(R_{i_1} = R_{i_2}), (Y_{i_2} = Y_{i_3}$ is in $\mathscr{F}), (R_{i_3} = R_{i_4}) \ldots (Y_{i_\mu} = Y_{i_1}$ is in $\mathscr{F})$. μ will be called the length of the circle.

Lemma 17.1 *When the values $[L_i, R_i]$ are randomly chosen, pairwise distinct, and when the framework $\mathscr{F}$ is randomly chosen, with a distribution of probability proportional to $Weight(\mathscr{F})$, then the probability to have a circle in $R, Y_{\mathscr{F}}$ is $\leq \frac{q^2}{2^{2n}(1-\frac{q^2}{2 \cdot 2^{2n}})}$.*

Proof When we will say that $\mathscr{F}$ is "randomly chosen" it will always mean with a distribution of probability proportional to $Weight(\mathscr{F})$

First Property.

First, we can notice that with probability $\geq 1 - \frac{q^2}{2 \cdot 2^{2n}}$ we can assume that the $[L_i, R_i]$ values, for $1 \leq i \leq q$, are random, without considering the fact that they are pairwise distinct, since the probability to distinguish these two distributions is $\leq \frac{q^2}{2 \cdot 2^{2n}}$. (This is the Functions/Permutations switching lemma on permutations on $2n$ bits.).

Circle of length 2

To have a circle of length 2, we must find two indices i and j, $i < j$ such that $(R_i = R_j)$ and ($Y_i = Y_j$ is in $\mathscr{F}$). For $(i, j), i < j$, we have $\frac{q(q-1)}{2}$ possibilities. Now when i and j are fixed, the probability to have $R_i = R_j$ is $\frac{1}{2^n}$ if the R_i values are random, and the probability to have $Y_i = Y_j$ is $\frac{1}{2^n}$ since $\mathscr{F}$ is randomly chosen (i.e., generated with a distribution proportional to $Weight(\mathscr{F})$). Therefore the probability to have $(R_i = R_j)$ and ($Y_i = Y_j$ is in $\mathscr{F}$ is $\leq \frac{q(q-1)}{2 \cdot 2^n}$ if the R_i values are randomly chosen.

Circles of length μ, μ even

To have a circle of length μ, we must find μ pairwise distinct indices $i_1, i_2, \ldots, i_\mu$ such that $(R_{i_1} = R_{i_l})$, ($Y_{i_1} = Y_{i_2}$ is in $\mathscr{F}$), $(R_{i_3} = R_{i_4}) \ldots (Y_{i_\mu} = Y_{i_1}$ is in $\mathscr{F}$). If the R_i values are randomly chosen, and if $\mathscr{F}$ is randomly chosen (i.e., with $Weight(\mathscr{F})$ distribution), then this probability is $\leq \frac{q^\mu}{\mu! 2^{\mu n}}$. Therefore, by using the First property above, we see that the probability to have a circle in $R, Y_\mathscr{F}$ is $\leq \frac{q^2}{2 \cdot 2^{2n}} + \sum_{i=1}^{+\infty} \frac{q^{2i}}{(2i)! 2^{2in}} \leq \frac{q^2}{2^{2n}(1 - \frac{q^2}{2 \cdot 2^{2n}})}$ as claimed. □

Lines in $R, Y_\mathscr{F}$

Definition 17.8 We will say that we have a "line in $R, Y_\mathscr{F}$" of length θ if there are $\theta + 1$ pairwise distinct indices $i_1, i_2, \ldots, i_{\theta+1}$ such that $\forall \lambda,\ 1 \leq \lambda \leq \theta$, we have at least one of these two conditions: $R_{i_\lambda} = R_{i_{\lambda+1}}$ or ($Y_{i_\lambda} = Y_{i_{\lambda+1}}$ is one of the equalities of $\mathscr{F}$).

Example 17.2 If $R_1 = R_2$, $R_2 = R_3$, ($Y_3 = Y_4$ is in $\mathscr{F}$), and $R_4 = R_5$, then we have a line of length 4 in $R, Y_\mathscr{F}$.

Lemma 17.2 *$\forall \theta \in \mathbb{N}$, when the values $[L_i, R_i]$ are randomly chosen, pairwise distinct, and when the framework $\mathscr{F}$ is randomly chosen (i.e., with a distribution proportional to $Weight(\mathscr{F})$), then the probability to have a line of length $\geq \theta$ in $R, Y_\mathscr{F}$ is $\leq \frac{q^{\theta+1} \cdot 2^\theta}{2^{n\theta}} + \frac{q^2}{2 \cdot 2^{2n}}$.*

Proof The proof is easy, we proceed as we did for the circles. The term $q^{\theta+1}$ comes from the possibilities for the $\theta + 1$ indices $i_1, i_2, \ldots, i_{\theta+1}$, the term 2^θ comes from the fact that from i_λ to $i_{\lambda+1}$ we can have 2 possibilities: one equation in R, or one equation in Y, $1 \leq \lambda \leq \theta$, and the term $\frac{q^2}{2 \cdot 2^{2n}}$ comes from the property seen in the proof of Lemma 17.1. □

Lemma 17.3 *When the values $[L_i, R_i]$ are randomly chosen, pairwise distinct, the values $[S_i, T_i]$ are randomly chosen, and $\mathscr{F}$ is randomly chosen (i.e., with a distribution of probability in $Weight(\mathscr{F})$), then the probability p to have a circle in $R, Y_\mathscr{F}$, or to have a circle in $S, X_\mathscr{F}$, or to have a line in $R, Y_\mathscr{F}$ of length $\geq \theta$, or to have a line in $S, X_\mathscr{F}$ f length $\geq \theta$ satisfies:*

$$p \leq \frac{3q^2}{2^{2n}(1 - \frac{q^2}{2 \cdot 2^{2n}})} + \frac{2q^{\theta+1} \cdot 2^\theta}{2^{n\theta}}$$

Moreover, if $\theta \geq n$ and $8 \leq q \leq \frac{2^n}{8}$, we have $p \leq \frac{4q^2}{2^{2n}}$ and $\frac{2q^{\theta+1}.2^\theta}{2^{n\theta}} \leq \frac{2q}{2^{2n}}$. (Therefore, we can assume that we have no line of length $\geq n$ if $q \ll 2^n$).

Proof The probability to have a circle in $R, Y_{\mathscr{F}}$ is $\leq \frac{q^2}{2^{2n}(1-\frac{q^2}{2 \cdot 2^{2n}})}$. (cf Lemma 17.1). Similarly, the probability to have a circle in $S, X_{\mathscr{F}}$ is $\leq \frac{q^2}{2^{2n}(1-\frac{q^2}{2 \cdot 2^{2n}})}$. (symmetry of the hypothesis). The probability to have a line in $R, Y_{\mathscr{F}}$ of length $\geq \theta$ is $\leq \frac{q^{\theta+1}.2^\theta}{2^{n\theta}} + \frac{q^2}{2 \cdot 2^{2n}}$ (cf Lemma 17.2). Similarly, for a line in $S, X_{\mathscr{F}}$ of length $\geq \theta$. Therefore $p \leq \frac{3q^2}{2^{2n}(1-\frac{q^2}{2 \cdot 2^{2n}})} + \frac{2q^{\theta+1}.2^\theta}{2^{n\theta}}$ as claimed.

Moreover, if $\theta \geq n$ and $q \leq \frac{2^n}{8}$, we have $\frac{2q^{\theta+1}.2^\theta}{2^{n\theta}} \leq 2q \cdot (\frac{2q}{2^n})^\theta \leq 2q(\frac{1}{4})^\theta \leq \frac{2q}{2^{2n}}$. Then $p \leq \frac{3q^2}{2^{2n}(1-\frac{1}{128})} + \frac{2q}{2^{2n}}$ and this is $\leq \frac{4q^2}{2^{2n}}$ if $q \geq 8$ as claimed. □

From now on, we assume $8 \leq q \leq \frac{2^n}{8}$. We take $\theta = n$, and we will say that a framework $\mathscr{F}$ is in "Case 5" if $\mathscr{F}$ has no circle in $R, Y_{\mathscr{F}}$, no circle in $S, X_{\mathscr{F}}$, no line of length $\geq n$ in $R, Y_{\mathscr{F}}$ and no line of length $\geq n$ in $S, X_{\mathscr{F}}$. Then,

Lemma 17.4 *When the values $[L_i, R_i]$ are randomly chosen, pairwise distinct, and the values $[S_i, T_i]$ are randomly chosen, then the probability p that*

$$\left[\sum_{\textit{all frameworks } \mathscr{F} \textit{ of Case 5}} Weight(\mathscr{F}) \leq (2^{2nq})(1 - \frac{2q}{2^n})\right] satisfy$$

$p \leq \frac{2q}{2^n}$.

Proof This comes immediately from Lemma 17.3 and

$$\sum_{\text{all frameworks } \mathscr{F}} Weight(\mathscr{F}) = \text{Number of all sequences } (X_i, Y_i),\ 1 \leq i \leq q$$

$$= 2^{2nq}$$

□

Proof of Theorem 17.8. Let $\mathscr{F}$ be a framework of Case 5. We will use this Theorem of Chap. 16:

Theorem 17.9 *"Theorem $P_i \oplus P_j$" for any ξ_{max}.*

Let (A) be a set of equations $P_i \oplus P_j = \lambda_k$, with p variables such that:

1. *There are no circle in P in the equations (A).*
2. *There are no more than ξ_{max} indices in the same block.*
3. *By linearity from (A), it is not possible to generate an equation $P_i = P_j$ with $i \neq j$.*

Then, if $p \le \sqrt{N}$ *or* $\sqrt{N} \ge \xi_{\max}^2 n + \xi_{\max}$, *and* $1 \le p \le 2^n/12\xi_{\max}^2$, *we have* $H_p(A) \ge J_p$.

Then, from this theorem, if $(12q)n^2 \le 2^n$, and $n \ge 30$, since $\xi_X \le n$, we have at least $\frac{J_{q-x}}{2^{n(r+y)}}$ solutions $(X_1, \dots, X_q)$ that satisfy $(C1)$, and at least $\frac{J_{q-y}}{2^{n(s+x)}}$ solutions $(Y_1, \dots, Y_q)$ that satisfy $(C2)$. Since $Weight(\mathcal{F}) = J_{q-x} \cdot J_{q-y}$ we see that the number of $(X_1, \dots, X_q), (Y_1, \dots, Y_q)$ solutions of $(C1)$ and $(C2)$ is $\ge \frac{Weight(\mathcal{F})}{2^{n(r+s)}.2^{n(x+y)}}$. Therefore, from Lemma 17.3 and Theorem 17.9 we obtain: if $8 \le q \le \frac{2^n}{12n^2}$, then when the $[L_i, R_i]$ are randomly chosen pairwise distinct, and when the $[S_i, T_i]$ are randomly chosen, we have with a probability $\ge 1 - \frac{2q}{2^n}$ that

$$H \ge \frac{|\mathcal{F}_n|^4}{2^{4nq}} \sum_{\text{all frameworks } \mathcal{F} \text{ of Case 5}} Weight(\mathcal{F})$$

$$H \ge \frac{|\mathcal{F}_n|^4}{2^{2nq}}\left(1 - \frac{2q}{2^n}\right)$$

Therefore we have proved Theorem 17.8 with $\alpha = \frac{2q}{2^n}$, $\beta = \frac{2q}{2^n}$ and $8 \le q \le \frac{2^n}{12n^2}$. □

Example 17.3 Let $\mathcal{F}$ be this framework of equalities: $(Y_2 = Y_3)$ and $(X_2 = X_4 = X_5)$. Let assume $R_1 = R_2$. Then for $(C1)$ we have

$$\begin{cases} X_1 \oplus X_2 = L_1 \oplus L_2 \\ X_2 \oplus X_3 = S_2 \oplus S_3 \end{cases}$$

And for $(C2)$ we have

$$\begin{cases} Y_2 \oplus Y_4 = R_2 \oplus R_4 \\ Y_2 \oplus Y_5 = R_2 \oplus R_5 \end{cases}$$

We have here $\ge \frac{J_{q-2}}{2^{2n}}$ solutions for $X_1, \dots, X_q$ and $\ge \frac{J_{q-1}}{2^{2n}}$ solutions for $Y_1, \dots, Y_q$, and here $r = 1, s = 0, x = 2, y = 1$.

Theorem 17.10 *If* $q \le \frac{2^n}{12n^2}$, *then for every KPA with q (random) known plaintexts we have:* $\mathcal{A}^{PRF} \le \frac{4q}{2^n}$ *where* $\mathcal{A}$ *denotes the advantage to distinguish* Ψ^4 *from a random function* $f \in_R \mathcal{F}_{2n}$, *and* $\mathcal{A}^{PRP} \le \frac{4q}{2^n} + \frac{q^2}{2 \cdot 2^{2n}}$ *where* $\mathcal{A}^{PRP}$ *denotes the advantage to distinguish* Ψ^4 *from a random permutation* $f \in_R \mathcal{P}_{2n}$.

Proof This comes immediately from Theorems 3.1, 17.8 and the classical pseudo-random function/pseudo-random permutation switching lemma (i.e., First property in the proof of Lemma 17.3). □

17.3 CPA Security for Ψ^5

Theorem 17.11 *If $q \leq \frac{2^n}{12n^2}$ and $n \geq 30$, then for every CPA with q adaptive chosen plaintexts, we have: $\mathcal{A}^{PRF} \leq \frac{5q}{2^n}$ and $\mathcal{A}^{PRR} \leq \frac{5q}{2^n} + \frac{q^2}{2.2^{2n}}$. Where $\mathcal{A}^{PRF}$ denotes the advantage to distinguish Ψ^5 from $f \in_R \mathcal{F}_{2n}$, and $\mathcal{A}^{PRP}$ denotes the advantage to distinguish Ψ^5 from $f \in_R \mathcal{P}_{2n}$.*

The results on Ψ^5 for proving CPA security will be based on this Theorem:

Theorem 17.12 *There are some values $\alpha > 0$ and $\beta > 0$ and there is a subset $E \subset (\{0,1\}^{2n})^q$ such that:*
1. $|E| \geq (1-\beta)2^{2nq}$
2. For all sequences $[L_i, R_i]$, $1 \leq i \leq q$ of pairwise distinct element of $\{0,1\}^{2n}$ and for all sequences $[S_i, T_i]$, $1 \leq i \leq q$, of E we have: $H \geq \frac{|\mathcal{F}_n|^5}{2^{2nq}}(1-\alpha)$ where H denotes the number of $(f_1, f_2, f_3, f_4, f_5) \in \mathcal{F}_n^5$ such that $\forall i$, $1 \leq i \leq q$, $\Psi^5(f_1, f_2, f_3, f_4, f_5)[L_i, R_i] = [S_i, T_i]$.
3. α and β can be chosen $\ll 1$ when $q \ll \frac{2^n}{n}$. (Moreover we will obtain explicit values for α and β: $\alpha = \frac{4q}{2^n}$, $\beta = \frac{q}{2^n}$ when $q \leq \frac{2^n}{12n^2}$).

Remark 17.2 It is equivalent to speak of such a subset E, or to say that when $[S_i, T_i]$, $1 \leq i \leq q$ are randomly chosen, the probability to have 2 is $\geq 1-\beta$.

To prove Theorem 17.12, we will still use the formula given for H_4 (i.e., for Ψ^4) by Theorem 17.4, but we will perform one more round at the beginning. More precisely, we write $\Psi^5 = \Psi^4 \circ \Psi^1$.
The q chosen plaintexts are $[L_i, R_i]$, $1 \leq i \leq q$.
After one round they become $[R_i, X'_i]$, $1 \leq i \leq q$, with $X'_i = L_i \oplus f_1(R_i)$.
After two rounds they become $[X'_i, X_i]$, $1 \leq i \leq q$.
After three rounds they become $[X_i, Y_i]$, $1 \leq i \leq q$.
After four rounds they become $[Y_i, S_i]$, $1 \leq i \leq q$.
Finally after 5 rounds they become $[S_i, T_i]$, $1 \leq i \leq q$.
We have: $H_5([L_i, R_i, S_i, T_i], 1 \leq i \leq q) = \sum_{f_1 \in \mathcal{F}_n} (H_4([R_i, X'_i, S_i, T_i], 1 \leq i \leq q))$ with $\forall i$, $1 \leq i \leq q$, $X'_i = L_i \oplus f_1(R_i)$. When a framework $\mathcal{F}$ for Ψ^4 is fixed (in $\Psi^5 = \Psi^4 \circ \Psi^1$), 5 cases occur:

Case 1. A contradiction appears by linearity in the equations generated by $(C1)$. In this case we have a "circle" of equalities in X', $Y_{\mathcal{F}}$ that generates a "circle" in X by $(C1)$.

Case 2. A contradiction appears by linearity in the equations generated by $(C2)$. In this case we have a "circle" of equalities in S, $X_{\mathcal{F}}$ that generates a "circle" in Y by $(C2)$.

Case 3. No contradiction appears by linearity but $\xi_X > n$. In this case, we have a line of equalities in X', $Y_{\mathcal{F}}$ of length $> n$ that generates a line of equalities in X by $(C1)$ of length $> n$.

Case 4. No contradiction appears by linearity but $\xi_Y > n$. In this case, we have a line of equalities in $S, X_{\mathcal{F}}$ of length $> n$ that generates a line of equalities in Y by $(C2)$ of length $> n$.

Case 5. No contradiction occurs by linearity and $\xi_X \leq n$ and $\xi_Y \leq n$
To prove Theorem 17.12 we will first prove that Cases 1, 2, 3, 4 appear with a negligible probability when the $[S_i, T_i]$ variables are randomly chosen, when f_1 is randomly chosen in $\mathcal{F}_n$, and when $\mathcal{F}$ is randomly chosen (this means as usual with a distribution of probability proportional to $Weight(\mathcal{F})$). This is what we will do now.

Circles in $X', Y_{\mathcal{F}}$.

Lemma 17.5 *$\forall \lambda > 0$, for all pairwise distinct $[L_i, R_i]$, $1 \leq i \leq q$, when f_1 is randomly chosen in $\mathcal{F}_n$ we have with probability $\geq 1 - \frac{1}{\lambda}$ that the number N of (i, j), $i \neq j$ such that $X'_i = X'_j$ satisfies: $N \leq \frac{\lambda q(q-1)}{2^n}$.*

Proof This comes immediately from the following lemma. □

Lemma 17.6 *For all pairwise distinct $[L_i, R_i]$, $1 \leq i \leq q$ (of course pairwise distinct means $i \neq j \Rightarrow L_i \neq L_j$ or $R_i \neq R_j$) the number of (f_1, i, j) such that $X'_i = X'_j$, $i \neq j$ is $\leq |\mathcal{F}_n| \frac{q(q-1)}{2^n}$.*

Proof $X'_i = X'_j$ means $L_i \oplus f_1(R_i) = L_j \oplus f_1(R_j)$. This implies $R_i \neq R_j$ (because $L_i = L_j$ and $R_i = R_j$ implies $i = j$). Thus, when (i, j) is fixed, the number of f_1 such that $X'_i = X'_j$ is exactly $\frac{|\mathcal{F}_n|}{2^n}$ if $R_i \neq R_j$, and is exactly 0 if $R_i = R_j$. Therefore, since we have at most $q(q-1)$ values (i, j), $i \neq j$, $R_i \neq R_j$, the total number of (f_1, i, j) such that $X'_i = X'_j$ is $\leq \frac{|F_n| q(q-1)}{2^n}$ as claimed. □

Lemma 17.7 *For all $\lambda > 0$, for all values $[L_i, R_i]$, $1 \leq i \leq q$ (no matter how cleverly chosen they are), when $f_1 \in_R \mathcal{F}_n$, and $\mathcal{F}$ is randomly chosen, the probability p to have a circle in $X', Y_{\mathcal{F}}$ satisfies:*

$$p \leq \frac{1}{\lambda} + \frac{q^2}{2 \cdot 2^{2n}} + \frac{\lambda^2 q^4}{2^{4n}(1 - \frac{\lambda q^2}{2^{2n}})}$$

For $\lambda = \frac{2^n}{q}$ this gives:

$$p \leq \frac{q}{2^n} + \frac{q^2}{2 \cdot 2^{2n}} + \frac{q^2}{2^{2n}(1 - \frac{q}{2^n})}$$

Therefore, if $q \leq \frac{2^n}{4}$, we have: $p \leq \frac{2q}{2^n}$.

Proof **Circles of length 2.**
To have a circle of length 2, we must find two indices i and j, $i < j$, such that: $(X'_i = X'_j)$ and ($Y_i = Y_j$ is in $\mathcal{F}$). For $(i, j), i < j$, we have $\frac{q(q-1)}{2}$ possibilities. Now when i and j are fixed, the probability to have $X'_i = X'_j$, i.e., to have $L_i \oplus f_1(R_i) =$

$L_j \oplus f_1(R_j)$, is $\frac{1}{2^n}$ when $f_1 \in_R \mathcal{F}_n$, since here we must have $R_i \neq R_j$ (because $R_i = R_j$ and $L_i = L_j$ imply $i = j$). Similarly, the probability to have ($Y_i = Y_j$ is in $\mathcal{F}$) when $\mathcal{F}$ is randomly chosen (i.e., with a distribution in $Weight(\mathcal{F})$) is $\frac{1}{2^n}$. Therefore the probability to have ($X'_i = X'_j$) and ($Y_i = Y_j$ is in $\mathcal{F}$) is $\leq \frac{q(q-1)}{2 \cdot 2^{2n}}$.

Circles of length μ, $\mu \geq 4$, μ even

To have a circle of lengthμ, we must find μ pairwise distinct indices $i_1, i_2, \dots, i_\mu$ such that ($X'_{i_1} = X'_{i_2}$), ($Y_{i_2} = Y_{i_3}$ is in $\mathcal{F}$), ($X'_{i_3} = X'_{i_4}$), …, ($Y_{i_\mu} = Y_{i_1}$ is in $\mathcal{F}$). From the lemmas above, we know that $\forall \lambda > 0$ we have a probability $\geq 1 - \frac{1}{\lambda}$ that for $(i_1, i_2, \dots, i_\mu)$ we have at most $\left(\frac{\lambda q^2}{2^n}\right)^{\frac{\mu}{2}}$ possibilities, when f_1 is randomly chosen in $\mathcal{F}_n$. Now when $(i_1, i_2, \dots, i_\mu)$ are fixed, when $\mathcal{F}$ is randomly chosen, we have a probability $\frac{1}{2^{\frac{\mu}{2}n}}$ to have ($Y_{i_2} = Y_{i_3}$ is in $\mathcal{F}$), …, ($Y_{i_\mu} = Y_{i_1}$ is in $\mathcal{F}$). Therefore, $\forall \lambda > 0$ the probability to have a circle of length ≥ 4 is $\leq \frac{1}{\lambda} + \sum_{\mu=4,\, \mu \text{ even}}^{=\infty} \frac{\lambda^{\frac{\mu}{2}} q^\mu}{2^{n\mu}} \leq \frac{1}{\lambda} + \frac{\lambda^2 q^4}{2^{4n}(1 - \frac{\lambda q^2}{2^{2n}})}$. Since we have seen that the probability to have a circle of length 2 is $\leq \frac{q^2}{2 \cdot 2^{2n}}$ we obtain Lemma 17.7. □

Lines in X', $Y_{\mathcal{F}}$

Lemma 17.8 *$\forall \lambda > 0$, $\forall k \in \mathbb{N}^*$, for all pairwise distinct $[L_i, R_i]$, $1 \leq i \leq q$, when f_1 is randomly chosen in $\mathcal{F}_n$ we have a probability $\geq 1 - \frac{1}{\lambda}$ that the number N of $(i_1, i_2, \dots, i_k)$ such that $(i_1, i_2, \dots, i_k)$ are pairwise distinct and $X'_{i_1} = X'_{i_2} = \dots = X'_{i_k}$ satisfies: $N \leq \frac{\lambda q^k}{2^{(k-1)n}}$.*

Remark 17.3 Lemma 17.2 was a special case of Lemma 17.8 with $k = 2$.

Proof This result comes essentially from the following lemma. □

Lemma 17.9 *For all pairwise distinct $[L_i, R_i]$, $1 \leq i \leq q$, the number of $(f_1, i_1, i_2, \dots, i_k)$ such that $i_1, i_2, \dots, i_k$ are pairwise distinct and $X'_{i_1} = X'_{i_2} = \dots = X'_{i_k}$ is $\leq |\mathcal{F}_n| \frac{q^k}{2^{(k-1)n}}$.*

Proof $X'_{i_1} = X'_{i_2} = \dots = X'_{i_k}$ means $L_{i_1} \oplus f_1(R_{i_1}) = L_{i_2} \oplus f_1(R_{i_2}) = \dots = L_{i_k} \oplus f_1(R_{i_k})$ (1). Since $i_1, i_2, \dots, i_k$ are pairwise distinct, this implies here $R_{i_1}, R_{i_2}, \dots, R_{i_k}$ are pairwise distinct (because $L_i = L_j$ and $R_i = R_j \Rightarrow i = j$). Thus, when $i_1, i_2, \dots, i_k$ are fixed, the number of f_1 that satisfy (1) is exactly $\frac{|\mathcal{F}_n|}{2^{(k-1)n}}$ if $R_{i_1}, R_{i_2}, \dots R_{i_k}$ are pairwise distinct and exactly 0 if $R_{i_1}, R_{i_2}, \dots R_{i_k}$ are not pairwise distinct. Since we have at most q^k values $(i_1, i_2, \dots, i_k)$, we obtain Lemma 17.9. □

Lemma 17.10 $\forall\lambda > 0,\ \forall\theta \in \mathbb{N}^*$, *for all values* $[L_i, R_i],\ 1 \le i \le q$ *(no matter how cleverly chosen they are), when* $f_1 \in_R \mathcal{F}_n$, *and* $\mathcal{F}$ *is randomly chosen, the probability* p *to have a line of length* $\ge \theta$ *in* $X', Y_{\mathcal{F}}$ *satisfies:*

$$p \le \frac{1}{\lambda} + 2^{\theta}\frac{\lambda^{\frac{\theta+1}{2}} q^{\theta+1}}{2^{n\theta}}$$

For $\lambda = \frac{2^n}{q}$, *we see that if* $q \le \frac{2^n}{12}$, *and* $\theta \ge n$, *we have:* $p \le \frac{2q}{2^n}$.

Proof The proof is easy from Lemma 17.8 since from Lemma 17.8, with $\lambda \ge 1$, we see that when θ is fixed the larger value for p is obtained when equalities in X' and equalities in $Y_{\mathcal{F}}$ alternate: $X'_{i_1} = X'_{i_2}, (Y_{i_2} = Y_{i_3}$ is in $\mathcal{F})$ etc. The coefficient 2^θ comes from the fact that for each indice i_λ, between i_λ and $i_{\lambda+1}$ we can have either an equality in X', or an equality in $Y_{\mathcal{F}}$. Now for $\lambda = \frac{2^n}{q}$, $\frac{1}{\lambda} + 2^{\theta}\frac{\lambda^{\frac{\theta+1}{2}} q^{\theta+1}}{2^{n\theta}} = \frac{q}{2^n} + \sqrt{2^n}\sqrt{q}\left(\frac{2\sqrt{q}}{\sqrt{2^n}}\right)^{\theta}$. If $q \le \frac{2^n}{12}$ and $\theta \ge n$, this gives $p \le \frac{q}{2^n} + \frac{\sqrt{2^n}\sqrt{q}}{2^{2n}}$ therefore we can write with simply $p \le \frac{2q}{2^n}$ as claimed. □

From now on, we assume $q \le \frac{2^n}{12}$. We take $\theta = n$ and we will say that a framework $\mathcal{F}$ is in "Case 5" if $\mathcal{F}$ has no circle in $X', Y_{\mathcal{F}}$, no circle in $S, X_{\mathcal{F}}$, no line of length $\ge n$ in $X', Y_{\mathcal{F}}$ and no line of length $\ge n$ in $S, X_{\mathcal{F}}$.

Lemma 17.11 *If* $q \le \frac{2^n}{12}$, *for all pairwise distinct values* $[L_i, R_i],\ 1 \le i \le q$ *when the values* $[S_i, T_i]$ *are randomly chosen,* $1 \le i \le q$, *the probability* p *that*

$$\left[\sum_{f_1\in\mathcal{F}_n}\ \sum_{\text{all frameworks } \mathcal{F} \text{ of Case 5}} Weight(\mathcal{F}) \le |\mathcal{F}_n|\cdot 2^{nq}\left(1-\frac{4q}{2^n}\right)\right]$$

satisfies: $p \le \frac{q}{2^n}$.

Proof This comes from Lemmas 17.8, 17.7 and 17.10. The term in $p \le \frac{q}{2^n}$ comes from Lemma 17.4, i.e., the circles and the lines in $S, X_{\mathcal{F}}$. These circles and lines in $S, X_{\mathcal{F}}$ are analyzed for Ψ^5 exactly as we did for Ψ^4 (only the first round has changed, not the last round). The term $|\mathcal{F}_n|\cdot 2^{nq}(1-\frac{4q}{2^n})$ comes from Lemma 17.7 (with $\frac{-2q}{2^n}$) and Lemma 17.10 (another $\frac{-2q}{2^n}$), and the fact that $\sum_{\text{all frameworks } \mathcal{F}} Weight(\mathcal{F}) = 2^{2nq}$. □

Proof From Lemma 17.11 and the theorem "$P_i \oplus P_j$" Theorem 17.9, we see that if $q \le \frac{2^n}{12n^2}$, then for all pairwise distinct $[L_i, R_i],\ 1 \le i \le q$, we have a probability $\ge 1 - \frac{q}{2^n}$ that $H \ge \frac{|\mathcal{F}_n|^5}{2^{2nq}}(1-\frac{4q}{2^n})$ when the $[S_i, T_i]$ values, $1 \le i \le q$, are randomly chosen. Therefore, we have proved Theorem 17.12 with $\alpha = \frac{4q}{2^n}$ and $\beta = \frac{q}{2^n}$ when $q \le \frac{2^n}{12n^2}$. □

Theorem 17.13 *If $q \le \frac{2^n}{12n^2}$ and $n \ge 30$, then for every CPA with q adaptive chosen plaintexts, we have: $\mathscr{A}^{PRF} \le \frac{5q}{2^n}$ and $\mathscr{A}^{PRP} \le \frac{5q}{2^n} + \frac{q^2}{2 \cdot 2^{2n}}$. Where $\mathscr{A}^{PRF}$ denotes the advantage to distinguish Ψ^5 from $f \in_R \mathscr{F}_{2n}$, and $\mathscr{A}^{PRP}$ denotes the advantage to distinguish Ψ^5 from $f \in_R \mathscr{P}_{2n}$.*

Proof This comes immediately from Theorems 3.4, 17.12, and the classical pseudo-random functions/pseudo-random permutations switching lemma. □

17.4 CCA Security for Ψ^5

It is possible to prove the CCA security of Ψ^5 when $q \ll \frac{2^n}{n^2}$. However, the proof is noticeably harder for Ψ^5 than for Ψ^6. Indeed, for Ψ^6, we have better properties on the H coefficients. For example, there is no "hole" when $q \ll 2^n$, where holes are defined in Problem 17.3. Thus, for Ψ^5, it is necessary to distinguish direct and inverse queries, which makes the analysis trickier. Such a proof can be obtained by introducing "standard systems of equalities," which are systems of equalities such that the number of blocks with $\xi = i$ is much smaller than the number of blocks with $\xi = i + 1$ for all $i \ge 2$. However, in this book, we focus on the simpler proof for Ψ^6.

17.5 CCA Security for Ψ^6

We write $\Psi^6 = \Psi^1 \circ \Psi^4 \circ \Psi^1$, and we proceed with the $[S_i, T_i]$ values of Ψ^6 exactly as we did in the previous sections with the $[L_i, R_i]$ values of Ψ^5. Then for Ψ^6 we obtain:

Theorem 17.14 *For all pairwise distinct $[L_i, R_i]$, $1 \le i \le q$ and for all pairwise distinct $[S_i, T_i]$, $1 \le i \le q$ the number H of $(f_1, f_2, f_3, f_4, f_5, f_6) \in \mathscr{F}_n^6$ such that $\forall i,\ 1 \le i \le q$,*

$$\Psi^6(f_1, f_2, f_3, f_4, f_5, f_6)[L_i, R_i] = [S_i, T_i]$$

satisfies $H \ge \frac{|\mathscr{F}_n|^6}{2^{2nq}}(1-\alpha)$ where α can be chosen $\ll 1$ when $q \ll \frac{2^n}{n^2}$. More precisely, we can choose $\alpha = \frac{8q}{2^n}$ if $q \le \frac{2^n}{12n^2}$.

Proof We write $\Psi^6 = \Psi^1 \circ \Psi^4 \circ \Psi^1$. We use these notations:
plaintexts: $[L_i, R_i]$, $1 \le i \le q$
1 round: $[R_i, X_i']$ with $X_i' = L_i \oplus f_1(R_i)$
2 rounds: $[X_i', X_i]$
3 rounds: $[X_i, Y_i]$

4 rounds: $[Y_i, Y_i']$ with $Y_i' = T_i \oplus f_6(S_i)$
5 rounds: $[Y_i', S_i]$
6 rounds: $[S_i, T_i]$
The analysis of the circle and lines in $Y', X_{\mathscr{F}}$ can now be done exactly as we did the analysis of the circles end lines in $X', Y_{\mathscr{F}}$ for Ψ^5 (symmetry of the hypothesis). Therefore, if $q \leq \frac{2^n}{12}$, for all pairwise distinct values $[L_i, R_i]$, $1 \leq i \leq q$ and for all pairwise distinct values $[S_i, T_i]$, $1 \leq i \leq q$:

$$\sum_{f_1 \in \mathscr{F}_n} \sum_{f_6 \in F_n} \sum_{\text{all frameworks } \mathscr{F} \text{ of Case 5}} Weight(\mathscr{F}) \geq |F_n|^2 \cdot 2^{2qn}\left(1 - \frac{8q}{2^n}\right) \quad (1)$$

(same proof as for Lemma 17.11). Thus, if $q \leq \frac{2^n}{12n^2}$ and $n \geq 30$, we can use the theorem $P_i \oplus P_j$ with $\xi_{max} \leq n$ of Chap. 16, and (1) gives Theorem 17.14 with $\alpha = \frac{8q}{2^n}$. □

Theorem 17.15 *If $q \leq \frac{2^n}{12n^2}$, then for every CCA with q adaptive chosen plaintexts or chosen ciphertexts, we have: $\mathscr{A}^{PRP} \leq \frac{8q}{2^n} + \frac{q^2}{2 \cdot 2^{2n}}$ where $\mathscr{A}^{PRP}$ denote the advantage to distinguish Ψ^6 from $f \in_R \mathscr{P}_{2n}$.*

Proof This comes immediately from Theorems 17.14 and 3.5. □

17.6 Security Results on Ψ^k, $k \geq 6$, with the Composition Theorem

In Theorem 17.14, we have seen that when $q \leq \frac{2^n}{12n}$, then H is always greater than or equal to $\frac{|\mathscr{F}_n|^6}{2^{2nq}}(1 - \frac{9q}{2^n})$. (There are no "holes" where H can be much smaller than its average value). Thus we are able to use the simple theorem of composition of Chap. 3. Then, for Ψ^{6r}, we have, when $q \leq \frac{2^n}{12}$:

$$\mathscr{A}^{CCA} \leq \left(\frac{8q}{2^n} + \frac{q^2}{2 \cdot 2^{2n}}\right)^r$$

where $\mathscr{A}^{CCA}$ denotes the advantage to distinguish Ψ^{6r} from $f \in_R \mathscr{P}_{2n}$. We do not improve the bound on q here, but for a fixed q, $\mathscr{A}^{CCA}$ decreases when r increases.

17.7 Results from Mirror Theory Compared with Results from Coupling on Ψ^k

With the coupling technique (cf Chap. 14), we have obtained proofs of security like this:

$$\begin{array}{ll} \text{NCPA:} & 3 \text{ rounds} \leftarrow q^{n/2} \\ & 5 \text{ rounds} \leftarrow q^{2n/3} \\ & 7 \text{ rounds} \leftarrow q^{3n/4} \\ & \text{etc.} \end{array}$$

$$\begin{array}{ll} \text{CCA:} & 5 \text{ rounds} \leftarrow q^{n/2} \\ & 7 \text{ rounds} \leftarrow q^{2n/3} \\ & 9 \text{ rounds} \leftarrow q^{3n/4} \\ & \text{etc.} \end{array}$$

For example, 3 rounds $\leftarrow q^{n/2}$ means that when $q^2 \ll 2^n$, the advantage to distinguish Ψ^3 from a random permutation is negligible.

With Mirror theory and the H-coefficient technique, we have obtained proofs of security like this:

$$\begin{array}{ll} \text{KPA:} & 4 \text{ rounds} \leftarrow q^{n} \\ \text{CCA:} & 3 \text{ rounds} \leftarrow q^{n/2} \\ & 5 \text{ rounds} \leftarrow q^{n} \end{array}$$

We can notice many differences between the two techniques:

1. With the Mirror Theory/H-coefficient technique, we have direct proofs in CCA, while with the coupling technique, we have better bounds in NCPA than in CCA.
2. In Mirror theory, we can prove the optimal bound q^n after a finite number of rounds: 5, while with the coupling technique, the optimal bound is only reached asymptotically when the number of rounds tends to infinity.
3. With Mirror theory, our composition theorem does not change the bound on q, while with the coupling technique, we improve this bound when we compose more rounds.
4. The proofs with Mirror theory look more difficult than the proof with the coupling technique. However, if instead of proofs of security in q^n in the theorem $P_i \oplus P_j$, we stop at a bound $q^{\frac{t}{t+1}n}$, for example with $t = 4$, then the proofs are relatively easy.
5. It is also possible to combine both methods. For example, we can obtain a proof for Ψ^6 on $q^{\frac{t}{t+1}n}$ with $t = 4$ from Mirror theory, and then improve this bound on q by using the coupling technique from Ψ^6 (instead of starting from Ψ^3).

Problems

17.1 From the formula H with framework, (i.e., Theorem 17.6), show that Ψ^6 is not homogeneous. The definition of homogeneous was given in Chap. 4.

17.2 Show that for $r \geq 6$, Ψ^r is not homogeneous.

17.3 We will say the there is a "hole" of length q in Ψ^r if there are some values $[L_1, R_1], [L_2, R_2], \ldots, [L_q, R_q], [S_1, T_1], [S_2, T_2], \ldots, [S_q, T_q]$ such that $\forall i,\ 1 \leq i \leq q$, the $[L_i, R_i]$ are pairwise distinct, the $[S_i, T_i]$ are pairwise distinct but $H([L_1, R_1], \ldots, [S_q, T_q]) \ll \tilde{H}$ where $\tilde{H}$ is the mean value of H on q inputs/outputs.

Show that:

1. For Ψ^1, there is a hole of length 1.
2. For Ψ^2 and Ψ^3, there is a hole of length 2.
3. For Ψ^4 and Ψ^5, there is a hole of length $\sqrt{2^n}$.
4. For Ψ^r, $r \geq 6$, there is no hole of length much less than 2^n.

Chapter 18
Indifferentiability

Abstract Indifferentiability is a stronger notion than indistinguishability which considers the case where the adversary has oracle access to the inner round functions. It allows to rigorously formalize the fact that a block cipher "behaves" as an ideal cipher. It is known that at least six rounds of balanced Feistel ciphers are necessary to achieve this security notion. Currently, the lowest number of rounds known to be sufficient to achieve the notion is eight.

18.1 Introduction

Previous chapters have focused on the indistinguishability setting: the adversary is given access to a black-box permutation, and tries to distinguish between the block cipher with a random secret key, or a uniformly random permutation. In this chapter, we will consider a different, stronger security setting, which does not involve any secret key. At first view this might seem a bit puzzling: which useful security property of a block cipher could one define that does not involve a secret key?

Actually, there are many such situations. For example, block ciphers are sometimes used to construct hash functions, as in the so-called Davies–Meyer construction: Given a block cipher E with key space and message space $\{0, 1\}^n$, one can define a compression function from $2n$ bits to n bits as $h(a, b) = E_a(b) \oplus b$; this compression function can then be turned into a full-fledged hash function using a chaining mechanism such as the Merkle–Damgaard construction. Since typical security goals for hash functions (such as collision resistance or pre-image resistance) do not involve secret keys, we see that the attacker in that case is in a completely different (and stronger) position compared to the standard indistinguishability security experiment. In particular, the adversary is allowed to "play with" the key input in order to find collisions for instance.

In this context, we ideally would require from the block cipher E to behave as an "ideal cipher", i.e., a block cipher drawn uniformly at random from the set of all block ciphers with the same key and message spaces. This would ensure that the block cipher has no "strange" property that the attacker would use to break, say, collision resistance of the Davies–Meyer compression function built from E.

J. Patarin et al., *Guide to Feistel Ciphers*, Information Security and Cryptography,
https://doi.org/10.1007/978-3-031-99850-8_18

The indifferentiability notion allows exactly to formalize this strong requirement of "behaving as an ideal cipher". For this, we need to model the round functions as random oracles. More precisely, we consider that the round functions are of the form $F_i(k, x) = H(\langle i \rangle \| k \| x)$ where H is a hash function modeled as a random oracle, $\langle i \rangle$ is the bit-string representation of integer i indexing the round, k is the key, and x is the round input. This implies that the resulting cipher behaves independently for each key k in the key space, and so we can focus on the case where the key space is a singleton (in which case an ideal cipher is simply a single random permutation). The extension to the case of an arbitrary key space is straightforward. Hence, in all the following, we will simply consider the r-round Feistel construction with un-keyed round functions $F_i : \{0, 1\}^n \to \{0, 1\}^n$ modeled as random function oracles.

18.2 Formal Definition of Indifferentiability

We focus on the case of balanced Feistel ciphers, but the definition of indifferentiability can be generalized to any construction of a block cipher from an ideal primitive.

Definition 18.1 The Feistel cipher $\Psi^r[\mathbf{F}]$ based on random round functions $\mathbf{F} = (F_1, \ldots, F_r)$ is said $(q_D, q_S, t_S, \varepsilon)$-indifferentiable from a random permutation if there exists a simulator S with oracle access to a random permutation oracle P such that for any distinguisher D making at most q_D queries, the simulator S makes at most q_S queries, runs in time t_S, and the advantage of D in distinguishing $(\Psi^r[\mathbf{F}], \mathbf{F})$ from (P, S^P), defined as

$$\left| \Pr\left[\mathsf{D}^{(\Psi^r[\mathbf{F}],\mathbf{F})} = 1\right] - \Pr\left[\mathsf{D}^{(P,\mathsf{S}^P)} = 1\right] \right|,$$

is at most ε.

Hence, the role of the simulator is to make the random permutation P "look like" an r-round Feistel cipher by simulating round functions that are coherent with the random permutation P, without deviating too much from uniformly random round functions. We stress that the simulator is not allowed to know the queries made by the distinguisher to the random permutation P. See Fig. 18.1 for an illustration.

Indifferentiability comes with a useful composition theorem which states, informally, that if Ψ^r is indifferentiable from an ideal cipher, then it can be used in any cryptosystem which is provably secure in the ideal cipher model, and the resulting cryptosystem will be secure in the random oracle model (i.e., assuming the round functions of the Feistel cipher are random oracles). In other words, any cryptographic functionality which is provably achievable in the ideal cipher model is also achievable in the random oracle model (the other direction also holds, as shown by Coron et al. (2005).

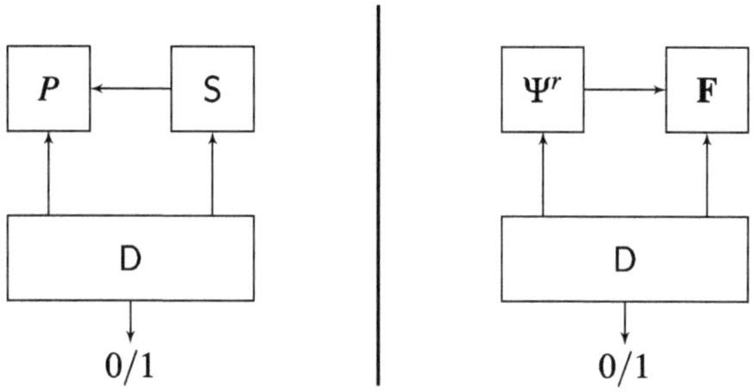

Fig. 18.1 The indifferentiability notion. On the left, the distinguisher interacts with (P, S^P), where P is a random permutation oracle and S is a simulator. On the right, the distinguisher interacts with $(\Psi^r[\mathbf{F}], \mathbf{F})$, where $\mathbf{F} = (F_1, \ldots, F_r)$ is a tuple of uniformly random round functions

18.3 Five Rounds of Balanced Feistel is Not Indifferentiable from an Ideal Cipher

In this section, we present an attack against the indifferentiability of the 5-round Feistel construction, which shows that at least six rounds are necessary to achieve indifferentiability from an ideal cipher.

The distinguisher D proceeds as follows (see Fig. 18.2). It chooses an arbitrary value Z_{13}, two arbitrary values Y_{14} et Y_{23}, and queries $F_3(Y_{14})$ and $F_3(Y_{23})$. It then computes

$$\begin{cases} X_{12} = Z_{13} \oplus F_3(Y_{14}) \\ X_{34} = Z_{13} \oplus F_3(Y_{23}). \end{cases}$$

Notations are chosen such that input round values sharing a common index correspond to the same input-output pair of the Feistel scheme: we say they constitute a chain. For example, (X_{12}, Y_{14}, Z_{13}) constitute a chain since $X_{12} = Z_{13} \oplus F_3(Y_{14})$.

The distinguisher then queries $F_2(X_{12})$ and $F_2(X_{34})$ and computes

$$\begin{cases} R_1 = Y_{14} \oplus F_2(X_{12}) \\ R_2 = Y_{23} \oplus F_2(X_{12}) \\ R_3 = Y_{23} \oplus F_2(X_{34}) \\ R_4 = Y_{14} \oplus F_2(X_{34}). \end{cases}$$

Note that necessarily $R_1 \oplus R_2 \oplus R_3 \oplus R_4 = 0$.

Then the distinguisher queries $F_1(R_1)$, $F_1(R_2)$, $F_1(R_3)$, and $F_1(R_4)$ and computes

$$\begin{cases} L_1 = X_{12} \oplus F_1(R_1) \\ L_2 = X_{12} \oplus F_1(R_2) \\ L_3 = X_{34} \oplus F_1(R_3) \\ L_4 = X_{34} \oplus F_1(R_4). \end{cases}$$

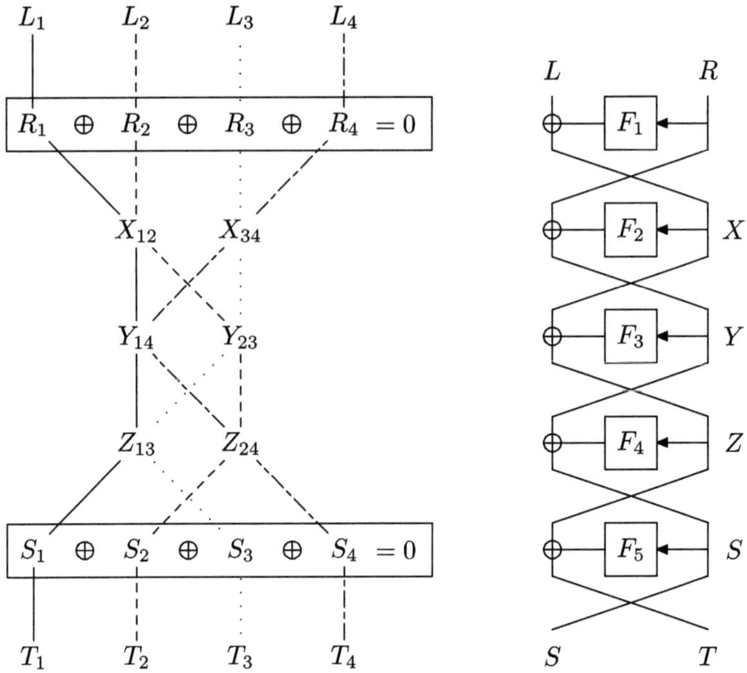

Fig. 18.2 The five-round distinguishing attack. The lines with four distinct patterns on the left side represent the computation paths in the Feistel construction for each input/output (L_i, R_i), (S_i, T_i) involved in the attack

Finally the distinguisher makes the P-queries $(S_1, T_1) = P(0, (L_1, R_1))$, $(S_2, T_2) = P(0(L_2, R_2))$, $(S_3, T_3) = P(0, (L_3, R_3))$ and $(S_4, T_4) = P(0, (L_4, R_4))$. If $S_1 \oplus S_2 \oplus S_3 \oplus S_4 = 0$, it returns 1, otherwise it returns 0.

First, one can easily verify that D always returns 1 when it interacts with $(\Psi^5[\mathbf{F}], \mathbf{F})$. Indeed, denote $Z_{24} = X_{12} \oplus F_3(Y_{23})$ the input value to F_4 associated with (L_2, R_2). Since $X_{12} \oplus F_3(Y_{14}) = X_{34} \oplus F_3(Y_{23}) = Z_{13}$, then $Z_{24} = X_{34} \oplus F_3(Y_{14})$, so that Z_{24} is also the input value to F_4 associated with (L_4, R_4). It follows that

$$\begin{cases} S_1 = Y_{14} \oplus F_4(Z_{13}) \\ S_2 = Y_{23} \oplus F_4(Z_{24}) \\ S_3 = Y_{23} \oplus F_4(Z_{13}) \\ S_4 = Y_{14} \oplus F_4(Z_{24}) \ , \end{cases}$$

and the relation $S_1 \oplus S_2 \oplus S_3 \oplus S_4 = 0$ is always verified.

On the contrary, when interacting with (P, S^P), it returns 1 only with negligible probability. Indeed, considering the union of D and S as a single machine making a polynomial number of queries to the random permutation P, it can find four input/output pairs $(S_i, T_i) = P(0, (L_i, R_i))$ satisfying $R_1 \oplus R_2 \oplus R_3 \oplus R_4 = 0$ and $S_1 \oplus S_2 \oplus S_3 \oplus S_4 = 0$ only with negligible probability.

18.4 Positive Results

Now that we know that at least six rounds are necessary for the indifferentiability of the Feistel construction, the question is how many rounds are sufficient (if the notion is achievable at all). The first proof that it is indeed possible to achieve indifferentiability within a constant number of rounds was given by Holenstein et al. (2011), namely for 14 rounds. The proof of this important result is out of the scope of this book. The main point is to design an adequate simulator. This is achieved using a *chain detection/completion* mechanism. In a nutshell, when the distinguisher makes queries to specific consecutive round functions, the simulator detects a "chain" that it will complete, adapting two round functions values to ensure consistency with the random permutation. Completing a chain is likely to create other chains, a significant challenge is to show that this recursive completion mechanism does not result in an uncontrolled "chain reaction" leading to an exponential running time of the simulator. Another challenge is to show that the simulator can always adapt chains and will never be trapped into an inconsistency because it must adapt two different chains with distinct round function values. We refer to Holenstein et al. (2011); Coron et al. (2016) for more details.

18.5 Further Reading

The indifferentiability notion has been first introduced by Maurer et al. (2004) and subsequently used by Coron et al. (2005) to show that a random oracle can be constructed (in the indifferentiability sense) from an ideal cipher. The attack against the 5-round Feistel construction was given (Coron et al. 2008). In the same paper, a proof of the indifferentiability of the 6-round Feistel construction was claimed but this proof was later found to be flawed. The proof that the 14-round Feistel construction is indifferentiable from an ideal cipher was given in Holenstein et al. (2011); see also Coron et al. (2016). The number of rounds was recently decreased to ten rounds by Dai and Steinberger (2015), Dachman-Soled et al. (2016), and then to eight rounds by Dai and Steinberger (2016).

References

Coron, J. S., Y. Dodis, C. Malinaud, and P. Puniya. 2005. Merkle-Damgård revisited: how to construct a hash function. In *Advances in cryptology—CRYPTO 2005*, ed. V. Shoup. Volume 3621 of LNCS, 430–448. Berlin: Springer.

Coron, J. S., J. Patarin, and Y. Seurin. 2008. The random oracle model and the ideal cipher model are equivalent. In *Advances in cryptology—CRYPTO 2008*, ed. D. Wagner. Volume 5157 of LNCS, 1–20. Berlin: Springer.

Coron, J., T. Holenstein, R. Künzler, J. Patarin, Y. Seurin, and S. Tessaro. 2016. How to build an ideal cipher: the indifferentiability of the feistel construction. *Journal of Cryptology* 29 (1): 61–114.

Dachman-Soled, D., J. Katz, and A. Thiruvengadam. 2016. 10-round feistel is indifferentiable from an ideal cipher. In *Advances in Cryptology—EUROCRYPT 2016 (Proceedings, Part II)*, ed. A. Fischlin and J. Coron. Volume 9666 of LNCS, 649–678. Berlin: Springer Full version available at http://eprint.iacr.org/2015/876

Dai, Y., and J. Steinberger. 2015. Feistel networks: indifferentiability at 10 rounds. In *IACR cryptology ePrint Archive, Report 2015/874, 2015*. Available at http://eprint.iacr.org/2015/874

Dai, Y., and J. Steinberger. 2016. Feistel Networks: Indifferentiability at 8 Rounds. In *CRYPTO 2016*, to appear, 2016. Available at http://eprint.iacr.org/2015/874.

Holenstein, T., R. Künzler, and S. Tessaro. 2011. The equivalence of the random oracle model and the ideal cipher model, revisited. In *Symposium on theory of computing—STOC 2011*, ed. L. Fortnow and S. P. Vadhan, 89–98. Full version available at http://arxiv.org/abs/1011.1264.

Maurer, U. M., R. Renner, and C. Holenstein. 2004. Indifferentiability, impossibility results on reductions, and applications to the random oracle methodology. In *Theory of cryptography conference—TCC 2004*, ed. M. Naor. Volume 2951 of LNCS, 21–39. Berlin: Springer.

Solutions

Problems of Chap. 2

2.1 Let $G = \Psi^3(f, f, f)$. From Theorem 2.1, we see that $G^{-1} = \sigma \circ G \circ \sigma$. Therefore, we have this attack in CPA with $q = 2$: ask for $G([L_1, R_1]) = [S_1, T_1]$, ask for $G^([T_1, S_1])$, and check if this is $[R_1, L_1]$. Since this attack, with only two queries, distinguishes G from a random permutation with a probability close to 1, G is not secure in CPA. Similarly, with $F = \Psi^7(f_1, f_2, f_3, f_4, f_3, f_2, f_1)$, we have $F^{-1} = \sigma \circ F \circ \sigma$, and the same attack works.

2.2 On G there is an "attack by the middle." We can find the key k_1 and k_2 after only about 2^{40} computations (this is realistic at present) and 2^{40} in memory. Moreover, we need only $q = 2$ queries if F_1 and F_2 inputs and outputs have more than 40 bits (if not see Problem 2.3), and we need only KPA. Some time/memory trade-off also exists, for example, with 2^{50} computations and 2^{30} in memory.

We give quickly the idea of the attack by the middle. Let (a, b) be one of the plaintext/ciphertext that we have. Let t be the "middle value," i.e., the value such that: $F_1(a) = t$ and $F_2(t) = b$.

Step 1. We build a table T like this: for each key k_1 we compute $t = F_{1,k_1}(a)$ and we store k_1 at address. t. Step 1 requires about 2^{40} computations and 2^{40} in memory.

Step 2. For each k_2, we compute $F_{2,k_2}^{-1}(b) = t$ and we look at the address t of Table T if there is a value k_1. If yes, we check if this (k_1, k_2) also works on another plaintext/ciphertext that we have. If yes, we have found (k_1, k_2) with a high probability.

Remark A.1 If the computation of G was very slow (for example, if it requires more than 2^{40} computations), then this attack would require more than 2^{40} computations (see Problem 2.4).

J. Patarin et al., *Guide to Feistel Ciphers*, Information Security and Cryptography,
https://doi.org/10.1007/978-3-031-99850-8

2.3 Here an attacker will not try to find the value of the secret key, but he will use the fact that the input and the output space is quite small. Let us assume, for example, in KPA that he has 2^{30} plaintext/ciphertext pairs (a_i, b_i), $1 \leq i \leq 30$. Then he can store them in a Table T (the values a_i at the address b_i) and each time a value b is sent, he looks at the address b if he knows the corresponding plaintext a. In this example, for each b he has a non-negligible probability (here $2^{\frac{1}{10}}$) to know a.

2.4 Let G be a permutation that we will design, with a key k of 50 bits. We can proceed like this.

1. From k (of 50 bits), we generate a key K of 256 bits by using a very slow function. For example, the computation of K from k would require 2^{40} computations.
2. Now we can use a secure classical cryptographic cipher with this key K of 256 bits (for example, a strong Feistel cipher).

To perform an exhaustive search on k (or K) from known plaintext/ciphertext pairs requires here to perform $2^{40} \cdot 2^{50} = 2^{90}$ computations, despite the fact that the"official" key for this algorithm has only 50 bits.

Notice that the user has to compute K from k only once, and then can use a fast cipher. Therefore, the cipher is almost as efficient as a standard cipher.

Remark A.2 This idea was given to us by Jean-Jacques Quisquater.

Problems of Chap. 4

4.1 Let $([L_i, R_i], [S_i, T_i])$, $1 \leq i \leq q$ be the q input/output pairs in KPA. Since the attacker is limited by the number of queries, but not by the number of computations, he can try all the possibilities for $(f_1, \ldots, f_k)$ such that $\forall i$, $1 \leq i \leq q$, $\Psi^k(f_1, \ldots, f_j)([L_i, R_i]) = [S_i, T_i]$. We have a space of $|\mathscr{F}_n|^k$ possible $(f_1, \ldots, f_k)$, and for each i, $1 \leq i \leq q$, $\Psi^k(f_1, \ldots, f_j)([L_i, R_i]) = [S_i, T_i]$ divide this space of possible keys by a factor about $\frac{1}{2^{2n}}$. Therefore, we will have only one possibility with a probability near 1, when $(2^{2n})^q \geq |\mathscr{F}_n|^k$. This means: $2^{2nq} \geq \left((2^n)^{2^n}\right)^k$, $2nq \geq n \cdot k \cdot 2^n$, $q \geq \frac{k \cdot 2^n}{2}$.

4.2 Let G be a permutation generator such that for a key k, there exists another key k' such that $G_k^{-1} = G_{k'}$. Then $G_{k'} \circ G_k \circ G_{k'} \circ G_k \circ \ldots \circ G_{k'} \circ G_k =$ Identity, and this is less secure than $G^1 = G$. This shows that the security of G^r can decrease when r increases if the keys are not independent.

4.3 If G generates functions from $\{0, 1\}^n \to \{0, 1\}^n$ that are not permutations, then the number of collisions in G^r will increase when r increases, and G can be a worse function generator when r increases. This is, for example, what appears when we iterate Butterfly ciphers or Benes ciphers (cf. Patarin 2008).

4.4
Solution 1 with $q = 2$.

If $\Psi^4[L_1, R_1] = [S_1, T_1]$ and $\Psi^4[L_2, R_2] = [S_2, T_2]$, and $R_1 = R_2$, $L_1 \neq L_2$, then the probability that $S_1 \oplus S_2 = L_1 \oplus L_2$ is about twice what it would be with a truly random permutation of $\mathscr{P}_{2n}$ (instead of Ψ^4). This result shows that the security bound given by Luby and Rackoff for Ψ^4 in a chosen plaintext attack is tight (the attack requires $q \simeq \sqrt{2^n}$ messages to ensure $S_i \oplus S_j = L_i \oplus L_j$).

This result also shows that Ψ^4 is not homogeneous, and the non-homogeneity property appears with only two (very special) messages.

Solution 2 with $q = 4$.

Let $R_1 = R_3$, $R_2 = R_4 = R_1 \oplus \alpha$, $S_1 = S_2$, $S_3 = S_4 = S_1 \oplus \alpha$, $L_1 = L_2$, $L_3 = L_4 = L_1 \oplus \alpha$, $T_1 = T_3$, $T_2 = T_4 = T_1 \oplus \alpha$.

Then the value H for Ψ^4 with these R, L, S, T is at least about $\frac{|\mathscr{F}_n|^4}{2^{6n}}$ (instead of about $\frac{|\mathscr{F}_n|^4}{2^{8n}}$ as expected if it was homogeneous).

4.5 If $\Psi^5[L_1, R_1] = [S_1, T_1]$ and $\Psi^5[L_2, R_2] = [S_2, T_2]$, and if $R_1 = R_2$ and $L_1 \neq L_2$, then the probability that $S_1 = S_2$ and $L_1 \oplus L_2 = T_1 \oplus T_2$ is about twice what it would be with a truly random permutation of B_{2n} (instead of Ψ^5). Therefore Ψ^5 is not homogeneous, and the non-homogeneity property appears with only two (very special) messages.

Remark A.3 However, since here we have two equations and two indices ($S_i = S_j$ and $L_i \oplus L_j = T_i \oplus T_j$), this non-homogeneity property would require about $q = 2^n$ messages in a chosen plaintext attack (instead of the $\sqrt{2^n}$ messages above for Ψ^4).

Problems of Chap. 6

6.1 Here $n = 50$ since we want to generate a pseudo-random permutation from 100 bits to 100 bits with a balanced Feistel cipher. At present (2016), a number of at least 2^{100} computations for the attacker are recommended for high security. From Table 6.1, we see that for 4 rounds, we have an attack with 2^{25} computations. For 5 rounds, we have an attack with 2^{50} computations. For $k \geq 6$, the lower bound on the number of computations. For the best known generic attacks is 2^{100}. Therefore, we must recommend to choose $k \geq 6$. Moreover, in practical design, most of the time, the round functions are not pseudo-random. Thus it might be wise to use an even larger value for k in order to have a security margin.

6.2 Here, in KPA, with 2^{32} messages (i.e., the whole codebook), the adversary will have all the table of corresponding plaintext/ciphertext pairs. In many cryptographic applications, this will be enough to break the system. However, if we assume that knowing the table is not a problem, but it should require at least 2^{100} computations to distinguish this table from the table of a truly random permutation on 32 bits, here $n = 16$, and from Table 6.1, we will recommend to choose $k \geq 11$ (since $7 \times 16 \geq 100$).

Problems of Chap. 9

9.1 There are only four functions from 1 bit to 1 bit, and all of them are linear: $f_1 : x_i \mapsto 0$, $f_2 : x_i \mapsto 1$, $f_3 : x_i \mapsto x_i$, and $f_4 : x_i \mapsto x_i \oplus 1$, where $x_i = 0$ or 1. Therefore with this "extremely Expanding Feistel Cipher," every round is linear and the cipher is linear, thus not secure. In KPA, the secret linear coefficient can easily be found by Gaussian reductions.

Problems of Chap. 12

12.1 In NCPA, we can choose two messages with the same bits, except for bit number 2. Then, for the outputs of these two messages, only S-box number 2 will have a different input, and therefore the 2 outputs of 32 bits will be the same, except for at most 4 bits, and we know the position of these 4 bits. Similarly, in KPA, it is also easy to distinguish the round functions from pseudo-random functions from 32 bits to 32 bits. For example, with about eight messages, we have a large probability that at least one S-box has the same 6-bit inputs for two of these messages, and therefore the four corresponding output bits will be the same.

12.2 In A, all the transformation are linear (i.e., affine). Therefore, A is not secure, and it is easier to attack A than DES. For example, in KPA, with Gaussian reduction, we will easily find the matrix M and the vector V such that $A(X) = MX + V$.

12.3 In Davies' attack, the number m of plaintext/ciphertext pairs that are needed for the attack to succeed is determined by probability theory (typically the difference between two mean values should be larger than a standard deviation in order to distinguish the information from noise). Thus m will not be smaller if we use CPA instead of KPA, and the complexity of the attack will be almost the same in both cases.

12.4 For the same reason as the previous problem, the complexity of the linear cryptanalysis of DES is almost the same in KPA and CPA.

Problems of Chap. 16

16.1 We will use this H-coefficient theorem for CPA, directly obtained from Theorem 3.4 of Chap. 3.

Theorem A.1 *Let α and β be real numbers, $\alpha > 0$, and $\beta > 0$. Let E be a subset of $(\{0,1\}^n)^q$ such that $|E| \geq (1-\beta)2^{nq}$.*
If
1. For all sequences a_i, $1 \leq i \leq q$, of pairwise distinct elements of $\{0,1\}^n$ and for all sequences b_i, (not necessary distinct), $1 \leq i \leq q$, of E we have $H \geq \frac{|\mathcal{P}_n|}{2^{nq}}(1-\alpha)$.
Then
2. For every CPA with q chosen plaintexts we have: $p \leq \alpha + \beta$ where $p = \mathcal{A}^{PRF}$ denotes the advantage to distinguish $f(x\|0) \oplus f(x\|1)$ when $f \in_R \mathcal{P}_n$ from a random function $g : \{0,1\}^{n-1} \to \{0,1\}^n$.

How to get Theorem 16.3 **from Theorem** A.1
In order to get Theorem 16.3 from Theorem A.1, a sufficient condition is to prove that for "most" (since we need β small) sequences of values b_i, $1 \leq i \leq q, b_i \in \{0,1\}^n$, we have: the number H of $f \in \mathcal{P}_n$ such that $\forall i,\ 1 \leq i \leq q,\ f(a_i\|0) \oplus f(a_i\|1) = b_i$ satisfies $H \geq \frac{|\mathcal{P}_n|}{2^{nq}}(1-\alpha)$ for a small value α (more precisely with $\alpha \ll O(\frac{q}{2^n})$ or $\alpha \ll O(\frac{q}{n \cdot 2^n})$). This is what we will do in the next sections. For E, we will take

$$E \stackrel{\text{def}}{=} \left\{(b_1, \ldots, b_q) \in (\{0,1\}^n)^q,\ \text{such that } \forall i,\ 1 \leq i \leq q,\ b_i \neq 0\right\}.$$

Since $(2^n-1)^q = 2^{nq}\left(1-\frac{1}{2^n}\right)^q \geq 2^{nq}\left(1-\frac{q}{2^n}\right)$, we will take $\beta = \frac{q}{2^n}$.
Let h be the number of sequences P_i, $1 \leq i \leq 2q$, $P_i \in \{0,1\}^n$ such that:

1. The P_i are pairwise distinct, $1 \leq i \leq 2q$.
2. $\forall i,\ 1 \leq i \leq q,\ P_{2i-1} \oplus P_{2i} = b_i$.

We see that $H = h \cdot \frac{|\mathcal{P}_n|}{2^n(2^n-1)\ldots(2^n-2q+1)}$, since when the P_i are fixed, then f is fixed on exactly $2q$ pairwise distinct points by $\forall i,\ 1 \leq i \leq q,\ f(a_i\|0) = P_{2i-1}$ and $f(a_i\|1) = P_{2i}$. Therefore we see that to prove Theorem 16.3, we want to prove this property:

For all (most would be enough but we will prove for all) sequences of values b_i, $1 \leq i \leq q, b_i \in \{0,1\}^n, b_i \neq 0$, we have: the number h of sequences P_i, $1 \leq i \leq 2q$; $P_i \in \{0,1\}^n$ such that the P_i are pairwise distinct and $\forall i,\ 1 \leq i \leq q,\ P_{2i-1} \oplus P_{2i} = b_i$ satisfies $h \geq \frac{2^n(2^n-1)\ldots(2^n-2q+1)}{2^{nq}}(1-\alpha)$ for a small value α.

This is exactly what we have proved in Chap. 16 (cf Theorem $P_i \oplus P_j$ with $\xi_{max} = 2$) and moreover with $\alpha = 0$, when $q \leq \frac{2^n}{48}$ and $n \geq 30$ from Theorem 16.1. Thus, $\mathcal{A}^{PRF} \leq \frac{q}{2^n}$ (here $\alpha = 0$ and $\beta = \frac{q}{2^n}$).

Remark A.4 We can notice that to distinguish $f(x\|0) \oplus f(x\|1)$ from a random function, securities in KPA, NCPA, and CPA are exactly the same: specific choices of the values a_i have no influence, i.e., there is no "clever" way to choose the a_i, or random values a_i are as useful as other choices. However many values a_i can

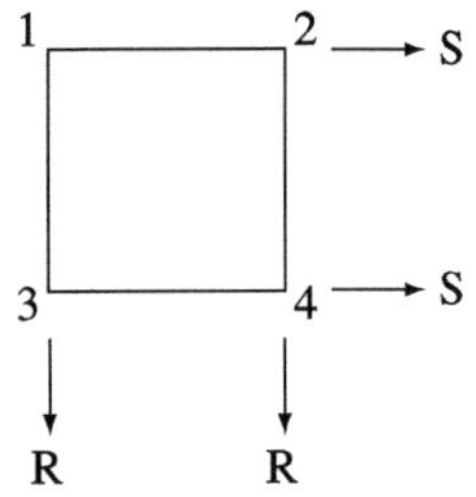

Fig. A.1 A representation of the equations $S_1 = S_2$, $S_3 = S_4$, $R_1 = R_3$, and $R_2 = R_4$

still give information since the distribution of the b_i is not perfectly random and the security in $O(\frac{q}{2^n})$ or $O(\frac{q}{n \cdot 2^n})$ was not obvious.

16.2 For different values x_i, all the values $b_i = f(x_i\|0) \oplus f(x_i\|1)$ are pairwise distinct since f is a permutation. This is not the case for $b_i = g(x_i)$, where f is a random function of $\{0,1\}^{n-1} \to \{0,1\}^n$. Therefore, to distinguish $F(x) = f(x\|0) \oplus f(x\|1)$ from g, we consider the following distinguisher D:

1. D chooses q different values $x_1, \ldots, x_q$ and obtain $b_1, \ldots, b_q$.
2. Then D tests if: $\exists i,\ 1 \le i \le q,\ b_i = 0$. In that case, D outputs 1, otherwise, D outputs 0.

The probability that D outputs 1 on F is 0 and the probability that D outputs 1 on g is $1 - \left(1 - \frac{1}{2^n}\right)^q$. Therefore $\mathscr{A}_{\mathsf{D}} = 1 - \left(1 - \frac{1}{2^n}\right)^q \simeq \frac{q}{2^n}$. This shows that the bound in $O(\frac{q}{2^n})$ found in Theorem 15.21 is not far to be tight.

Problems of Chap. 17

17.1
Solution 1 with $q = 4$.
Let $\Psi^6[L_i, R_i] = [S_i, T_i]$ for $i = 1, 2, 3, 4$.

If $R_1 = R_3, R_2 = R_4 \neq R_1, S_1 = S_2, S_3 = S_4 \neq S_1, L_1 \oplus L_3 = L_2 \oplus L_4 = S_1 \oplus S_3 \neq 0$, and $T_1 \oplus T_2 = T_3 \oplus T_4 = R_1 \oplus R_2 \neq 0$, then we will see that H is at least about $2 \cdot \frac{|\mathscr{F}_n|^6}{2^{8n}}$, instead of $\frac{|\mathscr{F}_n|^6}{2^{8n}}$ as expected if it was homogeneous. Therefore, Ψ^6 is not homogeneous (Fig. A.1).

Proof We know that the exact value of H (cf Theorem 17.6) is

$$H = \sum_{(X,A,Y,Z) \text{ satisfying } (C)} \frac{|\mathscr{F}_n|^6}{2^{6mn}} \cdot 2^{n(r+s+x+y+p+m)},$$

with (C) being the following set of conditions:

$$\forall i, j, 1 \le i \le q, 1 \le j \le q, i \ne j \quad \begin{cases} R_i = R_j \Rightarrow X_i \oplus L_i = X_j \oplus L_j \\ S_i = S_j \Rightarrow A_i \oplus T_i = A_j \oplus T_j \\ X_i = X_j \Rightarrow Y_i \oplus R_i = Y_j \oplus R_j \\ A_i = A_j \Rightarrow Z_i \oplus S_i = Z_j \oplus S_j \\ Y_i = Y_j \Rightarrow X_i \oplus Z_i = X_j \oplus Z_j \\ Z_i = Z_j \Rightarrow Y_i \oplus A_i = Y_j \oplus A_j, \end{cases}$$

and with q being the number of independent equations $R_i = R_j$, $i \ne j$, s is the number of independent equations $S_i = S_j$, $i \ne j$, etc. up to m being the number of independent equations $Z_i = Z_j$, $i \ne j$.

We will consider two special sets of values for (X, A, Y, Z).

First possible set.

Let X_1, A_1, Z_1, Y_1 have any value (thus we have 2^{4n} possible values here), and let $X_1 = X_2$, $X_3 = X_4 = X_1 \oplus L_1 \oplus L_3$, $A_1 = A_3$, $A_2 = A_4 = A_1 \oplus T_1 \oplus T_2$, $Z_1 = Z_2$, $Z_3 = Z_4 = Z_1 \oplus S_1 \oplus S_3$, $Y_1 \oplus Y_3$, and $Y_2 \oplus Y_4 = Y_1 \oplus R_1 \oplus R_2$.

It is easy to see that for these values all the conditions (C) are satisfied:

$$\begin{aligned}
R_1 = R_3 &\Rightarrow X_1 \oplus L_1 = X_3 \oplus L_3 \text{ (by definition of } X_3)\\
R_1 = R_3 &\Rightarrow X_1 \oplus L_1 = X_3 \oplus L_3 \text{ (by definition of } X_3)\\
R_2 = R_4 &\Rightarrow X_2 \oplus L_2 = X_4 \oplus L_4 \text{ (since } L_2 \oplus L_4 = L_1 \oplus L_3)\\
S_1 = S_2 &\Rightarrow A_1 \oplus T_1 = A_2 \oplus T_2 \text{ (by definition of } A_2)\\
S_3 = S_4 &\Rightarrow A_3 \oplus T_3 = A_4 \oplus T_4 \text{ (since } T_1 \oplus T_2 = T_3 \oplus T_4)\\
X_1 = X_2 &\Rightarrow Y_1 \oplus R_1 = Y_2 \oplus R_2 \text{ (by definition of } Y_2)\\
X_3 = X_4 &\Rightarrow Y_3 \oplus R_3 = Y_4 \oplus R_4 \text{ (since } R_1 \oplus R_2 = R_3 \oplus R_4)\\
A_1 = A_3 &\Rightarrow Z_1 \oplus S_1 = Z_3 \oplus S_3 \text{ (by definition of } Z_3)\\
A_2 = A_4 &\Rightarrow Z_2 \oplus S_2 = Z_4 \oplus S_4 \text{ (since } S_2 \oplus S_4 = S_1 \oplus S_3)\\
Y_1 = Y_3 &\Rightarrow X_1 \oplus Z_1 = X_3 \oplus Z_3 \text{ (since } L_1 \oplus L_3 = S_1 \oplus S_3)\\
Y_2 = Y_4 &\Rightarrow X_2 \oplus Z_2 = X_4 \oplus Z_4 \text{ (since } L_1 \oplus L3 = S_1 \oplus S_3)\\
Z_1 = Z_2 &\Rightarrow Y_1 \oplus A_1 = Y_2 \oplus A_2 \text{ (since } R_1 \oplus R_2 = T_1 \oplus T_2)\\
Z_3 = Z_4 &\Rightarrow Y_3 \oplus A_3 = Y_4 \oplus A_4 \text{ (since } R_1 \oplus R_2 = T_1 \oplus T_2).
\end{aligned}$$

Only from these X, A, Y, Z we see that:

$$H \ge 2^{4n} \cdot \frac{|\mathcal{F}_n|^6}{2^{24n}} \cdot 2^{n(2+2+2+2+2+2)} = \frac{|F_n|^6}{2^{8n}}.$$

Note. Here we have $r = 2$ equalities in R and $s = 2$ equalities in S, and we have found variables X, A, Y, Z that satisfy all the equations (C) by introducing only $\mu = 4$ equations with nonzero constants (i.e., $X_3 = X_1 \oplus L_1 \oplus L_3$, $A_2 = A_1 \oplus T_1 \oplus T_2$, $Z_3 = Z_1 \oplus S_1 \oplus S_3$, and $Y_2 = Y_1 \oplus R_1 \oplus R_2$). Since all the equations of (C) are satisfied with $\mu \le r + s$ it will give a proof of non-homogeneity.

Second possible set.

There is also the "usual" set, i.e., the values X, A, Y, Z that we have used in the proof that Ψ^6 is super-pseudo-random (these values introduce no equalities in the X, A, Y, Z variables, so this second set is entirely disjoint from the first set).

Here, we have

- X_1 has 2^n possibilities.
- X_2 has $(2^n - 2)$ possibilities (because $X_2 \neq X_1$ and $X_2 \neq X_1 \oplus L_1 \oplus L_3$ and since here $L_1 \oplus L_3 = L_2 \oplus L_4$ these two inequalities will imply $X_2 \oplus L_2 \oplus L_4 \neq X_1$ and $X_2 \oplus L_2 \oplus L_4 \neq X_1 \oplus L_1 \oplus L_3$).
- $X_3 = X_1 \oplus L_1 \oplus L_3$, $X_4 = X_2 \oplus L_2 \oplus L_4$.
- A_1 has 2^n possibilities, $A_2 = A_1 \oplus T_1 \oplus T_2$.
- A_3 has $(2^n - 2)$ possibilities (because $A_3 \neq Y_1$ and $A_3 \neq A_1 \oplus T_1 \oplus T_2$ and since here we have $T_1 \oplus T_2 = T_3 \oplus T_4$ these two inequalities will imply $A_3 \oplus T_3 \oplus T_4 \neq A_1$ and $A_3 \oplus T_3 \oplus T_4 \neq A_1 \oplus T_1 \oplus T_2$).
- Y_1 has 2^n possibilities, and Y_2 has $2^n - 1$ possibilities (because $Y_2 \neq Y_1$). Similarly Y_3 has $2^n - 2$ possibilities (because $Y_3 \neq Y_1$ and $Y_3 \neq Y_2$) and Y_4 has $2^n - 3$ possibilities (because $Y_4 \neq Y_1$ and $Y_4 \neq Y_2$ and $Y_4 \neq Y_3$).
- For the same reason Z_1, Z_2, Z_3, and Z_4 have, respectively, 2^n, $2^n - 1$, $2^n - 2$, and $2^n - 3$ possibilities.

Only from these X, A, Y, Z we see that

$$H \geq 2^{4n} \cdot (2^n - 1)^2 \cdot (2^n - 2)^4 \cdot (2^n - 3)^2 \cdot \frac{|\mathcal{F}_n|^6}{2^{24n}} \cdot 2^{n(2+2)} \approx \frac{|\mathcal{F}_n|^6}{2^{8n}}.$$

Therefore by combining the first and the second sets, we have $H \geq$ about $2\frac{|\mathcal{F}_n|^6}{2^{8n}}$, as claimed (instead of $H \approx \frac{|\mathcal{F}_n|^6}{2^{8n}}$ if Ψ^6 was homogeneous). □

Remark A.5 Since $L_4 = L_1 \oplus L_2 \oplus L_3$ and $R_4 = R_2$, the index 4 is fixed, the indices 1, 2, and 3 are fixed.

In fact we have here 3 indices 1, 2, and 3 and at least 4 equations on these indices that we cannot impose with a plaintext/ciphertext attack: $T_1 \oplus T_2 = R_1 \oplus R_2$, $L_1 \oplus L_3 = S_1 \oplus S_3$, $S_{4(1,2,3)} = S_3$, $T_{4(1,2,3)} = T_1 \oplus T_2 \oplus T_3$.

Thus, this example shows that Ψ^6 is not homogeneous, but it does not give a cryptographic attack when $q < 2^n$.

Remark A.6 It is sometimes interesting to see if there is an attack when $2^n < q \ll 2^{2n}$, when this attack requires $\ll 2^{2n}$ computations.

However here when the index 1 is fixed (we have m possibilities for it), the index 2 is also "in a way" fixed, since $S_2 = S_1$ and $T_2 \oplus R_2 = T_1 \oplus R_1$ (because on average when 1 is fixed there will be about only one index 2 such that these two equations are satisfied). Similarly, when the index 1 is fixed, the index 3 is "in a way" fixed, since $R_3 = R_1$ and $S_3 \oplus L_3 = S_1 \oplus L_1$. So in fact, when 1 is fixed, 2, 3, and 4 are fixed. But there are still two exceptional equations: $S_{4(1)} = S_3$ and $T_{4(1)} = T_1 \oplus T_{2(1)} \oplus T_{3(1)}$,

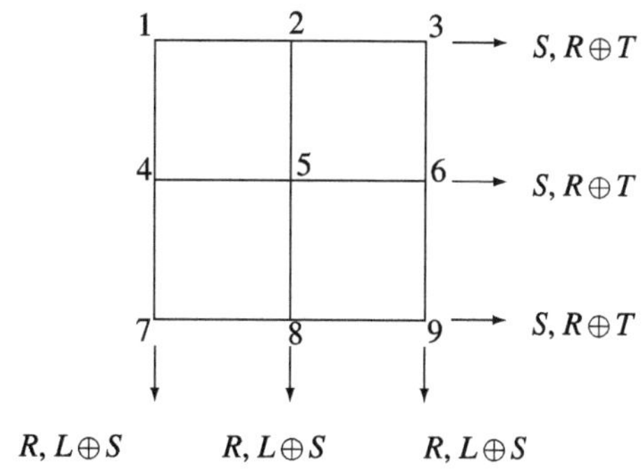

Fig. A.2 A representation of the 24 equations in S, L, R, T

and when $q \ll 2^{2n}$ the probability that these equations occur is negligible. Therefore this example 1 does not give an attack even when $2^n < q \ll 2^{2n}$.

Solution 2 with $q = 9$.
Let $\Psi^6[L_i, R_i] = [S_i, T_i]$ for $1 \le i \le 9$. We study the values of H when

$$\begin{cases} R_1 = R_4 = R_7 \\ R_2 = R_5 = R_8 \\ R_3 = R_6 = R_9 \\ L_1 \oplus S_1 = L_4 \oplus S_4 = L_7 \oplus S_7 \\ L_2 \oplus S_2 = L_5 \oplus S_5 = L_8 \oplus S_8 \\ L_3 \oplus S_3 = L_6 \oplus S_6 = L_9 \oplus S_9 \end{cases} \quad \text{and} \quad \begin{cases} S_1 = S_2 = S_3 \\ S_4 = S_5 = S_6 \\ S_7 = S_8 = S_9 \\ R_1 \oplus T_1 = R_2 \oplus T_2 = R_3 \oplus T_3 \\ R_4 \oplus T_4 = R_5 \oplus T_5 = R_6 \oplus T_6 \\ R_7 \oplus T_7 = R_8 \oplus T_8 = R_9 \oplus T_9. \end{cases}$$

All these relations are represented in Fig. A.2. We also assume that $R_1 \neq R_2$, $R_1 \neq R_3$, $R_2 \neq R_3$, $S_1 \neq S_4$, $S_1 \neq S_7$, and $S_4 \neq S_7$.

Then—as we will see below—for such L, R, S, T values, the value of H is at least $\frac{|\mathcal{F}_n|^6}{2^{14n}}$, instead of $\frac{|\mathcal{F}_n|^6}{2^{18n}}$ as expected if it was homogeneous. Therefore, Ψ^6 is not homogeneous.

Proof Let $\alpha = R_1 \oplus R_2$, $\beta = R_1 \oplus R_3$, $\alpha' = S_1 \oplus S_4$, and $\beta' = S_1 \oplus S_7$ (by definition we have $\alpha \neq 0$, $\beta \neq 0$, $\alpha' \neq 0$, and $\beta' \neq 0$). We consider (X, A, Y, Z) values such that:

$$\begin{cases} X_1 = X_2 = X_3 \\ X_4 = X_5 = X_6 = X_1 \oplus \alpha' \\ X_7 = X_8 = X_9 = X_1 \oplus \beta' \\ A_1 = A_4 = A_7 \\ A_2 = A_5 = A_8 = A_1 \oplus \alpha \\ A_3 = A_6 = A_9 = A_1 \oplus \beta \end{cases} \quad \text{and} \quad \begin{cases} Z_1 = Z_2 = Z_3 \\ Z_4 = Z_5 = Z_6 = Z_1 \oplus \alpha' \\ Z_7 = Z_8 = Z_9 = Z_1 \oplus \beta' \\ Y_1 = Y_4 = Y_7 \\ Y_2 = Y_5 = Y_8 = Y_1 \oplus \alpha \\ Y_3 = Y_6 = Y_9 = Y_1 \oplus \beta. \end{cases}$$

It is easy to verify that for there values all the conditions (C) are satisfied (these conditions were explicitly written for Ψ^6 in solution 1):

$$
\begin{array}{l}
R_1 = R_4 \Rightarrow X_1 \oplus L_1 = X_4 \oplus L_4 \text{ (because } \alpha' = L_1 \oplus L_4 = S_1 \oplus S_4)\\
R_1 = R_7 \Rightarrow X_1 \oplus L_1 = X_7 \oplus L_7 \text{ (because } \beta' = S_1 \oplus S_7 = L_1 \oplus L_7)\\
R_2 = R_5 \Rightarrow X_2 \oplus L_2 = X_5 \oplus L_5 \text{ (because } \alpha' = S_1 \oplus S_4 = S_2 \oplus S_5 = L_2 \oplus L_5)\\
R_2 = R_8 \Rightarrow X_2 \oplus L_2 = X_8 \oplus L_8 \text{ (because } \beta' = S_1 \oplus S_7 = S_2 \oplus S_8 = L_2 \oplus L_8)\\
R_3 = R_6 \Rightarrow X_3 \oplus L_3 = X_6 \oplus L_6 \text{ (because } \alpha' = S_1 \oplus S_4 = S_3 \oplus S_6 = L_3 \oplus L_6)\\
R_3 = R_9 \Rightarrow X_3 \oplus L_3 = X_9 \oplus L_9 \text{ (because } \beta' = S_1 \oplus S_7 = S_3 \oplus S_9 = L_3 \oplus L_9)\\
S_1 = S_2 \Rightarrow A_1 \oplus T_1 = A_2 \oplus T_2 \text{ (because } \alpha = R_1 \oplus R_2 = T_1 \oplus T_2)\\
S_1 = S_3 \Rightarrow A_1 \oplus T_1 = A_3 \oplus T_3 \text{ (because } \beta = R_1 \oplus R_3 = T_1 \oplus T_3)\\
S_4 = S_5 \Rightarrow A_4 \oplus T_4 = A_5 \oplus T_5 \text{ (because } \alpha = R_1 \oplus R_2 = R_4 \oplus R_5 = T_4 \oplus T_5)\\
S_4 = S_6 \Rightarrow A_4 \oplus T_4 = A_6 \oplus T_6 \text{ (because } \beta = R_1 \oplus R_3 = R_4 \oplus R_6 = T_4 \oplus T_6)\\
S_7 = S_8 \Rightarrow A_7 \oplus T_7 = A_8 \oplus T_8 \text{ (because } \alpha = R_1 \oplus R_2 = R_7 \oplus R_8 = T_7 \oplus T_8)\\
S_7 = S_9 \Rightarrow A_7 \oplus T_7 = A_9 \oplus T_9 \text{ (because } \beta = R_1 \oplus R_3 = R_7 \oplus R_9 = T_7 \oplus T_9)\\
X_1 = X_2 \Rightarrow Y_1 \oplus R_1 = Y_2 \oplus R_2 \text{ (because } \alpha = R_1 \oplus R_2)\\
X_1 = X_3 \Rightarrow Y_1 \oplus R_1 = Y_3 \oplus R_3 \text{ (because } \beta = R_1 \oplus R_3)\\
X_4 = X_5 \Rightarrow Y_4 \oplus R_4 = Y_5 \oplus R_5 \text{ (because } \alpha = R_1 \oplus R_2 = R_4 \oplus R_5)\\
X_4 = X_6 \Rightarrow Y_4 \oplus R_4 = Y_6 \oplus R_6 \text{ (because } \beta = R_1 \oplus R_3 = R_4 \oplus R_6)\\
X_7 = X_8 \Rightarrow Y_7 \oplus R_7 = Y_8 \oplus R_8 \text{ (because } \alpha = R_1 \oplus R_2 = R_7 \oplus R_8)\\
X_7 = X_9 \Rightarrow Y_7 \oplus R_7 = Y_9 \oplus R_9 \text{ (because } \beta = R_1 \oplus R_3 = R_7 \oplus R_9)\\
A_1 = A_4 \Rightarrow Z_1 \oplus S_1 = Z_4 \oplus S_4 \text{ (because } \alpha' = S_1 \oplus S_4)\\
A_1 = A_7 \Rightarrow Z_1 \oplus S_1 = Z_7 \oplus S_7 \text{ (because } \beta' = S_1 \oplus S_7)\\
A_2 = A_5 \Rightarrow Z_2 \oplus S_2 = Z_5 \oplus S_5 \text{ (because } \alpha' = S_1 \oplus S_4 = S_2 \oplus S_5)\\
A_2 = A_8 \Rightarrow Z_2 \oplus S_2 = Z_8 \oplus S_8 \text{ (because } \beta' = S_1 \oplus S_7 = S_2 \oplus S_8)\\
A_3 = A_6 \Rightarrow Z_3 \oplus S_3 = Z_6 \oplus S_6 \text{ (because } \alpha' = S_1 \oplus S_4 = S_3 \oplus S_6)\\
A_3 = A_9 \Rightarrow Z_3 \oplus S_3 = Z_9 \oplus S_9 \text{ (because } \beta' = S_1 \oplus S_7 = S_3 \oplus S_9)\\
Z_1 = Z_2 \Rightarrow Y_1 \oplus A_1 = Y_2 \oplus A_2 \text{ (because } Y_1 \oplus Y_2 = \alpha = A_1 \oplus A_2)\\
Z_1 = Z_3 \Rightarrow Y_1 \oplus A_1 = Y_3 \oplus A_3 \text{ (because } Y_1 \oplus Y_3 = \beta = A_1 \oplus A_3)\\
Z_4 = Z_5 \Rightarrow Y_4 \oplus A_4 = Y_5 \oplus A_5 \text{ (because } Y_4 \oplus Y_5 = \alpha = A_4 \oplus A_5)\\
Z_4 = Z_6 \Rightarrow Y_4 \oplus A_4 = Y_6 \oplus A_6 \text{ (because } Y_4 \oplus Y_6 = \beta = A_4 \oplus A_6)\\
Z_7 = Z_8 \Rightarrow Y_7 \oplus A_7 = Y_8 \oplus A_8 \text{ (because } Y_7 \oplus Y_8 = \alpha = A_7 \oplus A_8)\\
Z_7 = Z_9 \Rightarrow Y_7 \oplus A_7 = Y_9 \oplus A_9 \text{ (because } Y_7 \oplus Y_9 = \beta = A_7 \oplus A_9)\\
Y_1 = Y_4 \Rightarrow X_1 \oplus Z_1 = X_4 \oplus Z_4 \text{ (because } Z_1 \oplus Z_4 = \alpha' = X_1 \oplus X_4)\\
Y_1 = Y_7 \Rightarrow X_1 \oplus Z_1 = X_7 \oplus Z_7 \text{ (because } Z_1 \oplus Z_7 = \beta' = X_1 \oplus X_7)\\
Y_2 = Y_5 \Rightarrow X_2 \oplus Z_2 = X_5 \oplus Z_5 \text{ (because } Z_2 \oplus Z_5 = \alpha' = X_2 \oplus X_5)\\
Y_2 = Y_8 \Rightarrow X_2 \oplus Z_2 = X_8 \oplus Z_8 \text{ (because } Z_2 \oplus Z_8 = \beta' = X_2 \oplus X_8)\\
Y_3 = Y_6 \Rightarrow X_3 \oplus Z_3 = X_6 \oplus Z_6 \text{ (because } Z_3 \oplus Z_6 = \alpha' = X_3 \oplus X_6)\\
Y_3 = Y_9 \Rightarrow X_3 \oplus Z_3 = X_9 \oplus Z_9 \text{ (because } Z_3 \oplus Z_9 = \beta' = X_3 \oplus X_9).
\end{array}
$$

Therefore, from the exact value of H and by considering only such (X, A, Y, Z), we have

$$H \geq 2^{4n} \cdot \frac{|\mathscr{F}_n|^6}{2^{54n}} \cdot 2^{n(6+6+6+6+6+6)} = \frac{|\mathscr{F}_n|^6}{2^{14n}},$$

as claimed (instead of $H \simeq \frac{|\mathscr{F}_n|^6}{2^{18n}}$ if Ψ^6 was homogeneous).

17.2
Solution 1 with $q = (k/2)^2$. For simplicity, we assume that k is even (the proof is very similar when k is odd). Let $k = 2\lambda$. Let $\Psi^k[L_i, R_i] = [S_i, T_i]$ for $1 \le i \le m$. We essentially generalize to Ψ^k the construction given in solution 2 for Ψ^6.

The exact value of H is

$$H = \sum_{(X^{(1)},\ldots,X^{(k-2)}) \text{ satisfying } (C)} \frac{|\mathcal{F}_n|^k}{2^{knq}} \cdot 2^{n(r+s+x^{(1)}+\ldots+x^{(k-2)})},$$

where the $X^{(1)}$, ..., $X^{(k-2)}$ variables are the intermediate round variables, and where (C) denotes the conditions on the equalities (*i.e.*, $R_i = R_j \Rightarrow X_i^{(1)} \oplus L_i = X_j^{(1)} \oplus L_j$, etc.). The proof of this formula is not difficult.

We take $q = \lambda^2 (= \frac{k^2}{4})$.

We study the value H when $L_i, R_i, S_i, T_i, 1 \le i \le m$ satisfy the equalities illustrated in Fig. A.3. (For simplicity, we do not write these equalities explicitly.)

We will consider values $X^{(1)}, \ldots, X^{(k-2)}$ such that:

1. In Fig. A.3 the $\oplus$ of two elements on the same line $= 0$ for $X^{(1)}, X^{(3)}, \ldots, X^{(k-3)}$.
2. In Fig. A.3 the $\oplus$ of two elements on the same column $= 0$ for $X^{(2)}, X^{(4)}, \ldots, X^{(k-2)}$.
3. We have the $(k-2) \cdot (\lambda - 1)$ equalities with nonzero constant needed to satisfy all the (C) conditions.

Then, in the exact formula given above for H, for these $X^{(1)}, \ldots, X^{(k-2)}$ we have: $r = \lambda(\lambda-1)$, $s = \lambda(\lambda-1)$, $X^{(1)} = \lambda(\lambda-1)$, ... , $X^{(k-2)} = \lambda(\lambda-1)$. And we have $2^{(k-2)n}$ possibilities for $X^{(1)}, \ldots, X^{(k-2)}$. Then:

$$H \ge 2^{(k-2)n} \cdot \frac{|\mathcal{F}_n|^k}{2^{knq}} \cdot 2^{nk\lambda(\lambda-1)},$$

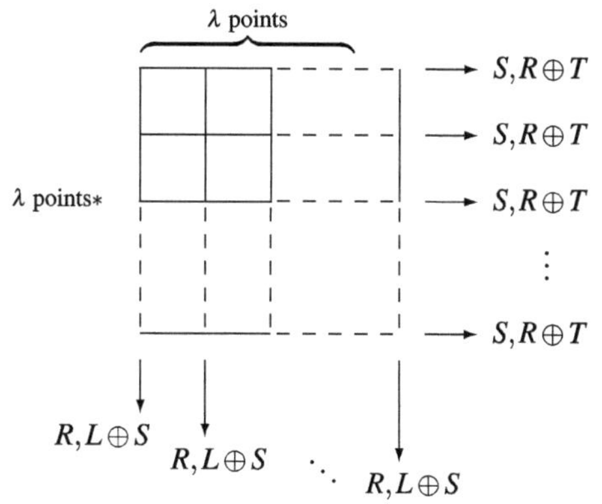

Fig. A.3 Modeling the $4 \cdot \lambda(\lambda - 1)$ equations in S, L, R, T

so that, with $q = \lambda^2 = \frac{k^2}{4}$,

$$H \geq 2^{(k-2)n} \cdot \frac{|\mathcal{F}_n|^k}{2^{2qn}}$$

(instead of $\frac{|\mathcal{F}_n|^k}{2^{2nq}}$ if Ψ^k was homogeneous). Therefore, Ψ^k is not homogeneous, as claimed.

Note:

Here we have $2\lambda(\lambda - 1)$ equalities $R_i = R_j$ or $S_i = S_k$, and only $(k-2)(\lambda - 1)$ equalities with nonzero constants have been used to satisfy all the conditions (C). So the deviation from the average is about $2^{(k-2)n}$.

Solution 2 with $q = (k/2 - 1)^2$.
If we take $\lambda = \frac{k}{2} - 1$ (instead of $\lambda = \frac{k}{2}$), then we will have still $2\lambda(\lambda - 1)$ equalities in $S_i = S_j$ or $R_i = R_j, i \neq j$, and only $(k-2)(\lambda - 1)$ equalities with nonzero constants to satisfy all the conditions (C). Here the obtained value of H will be about twice the average value. (This attack needs less points: $(\frac{k}{2} - 1)^2$ instead of $(\frac{k}{2})^2$, but the deviation from the average is less important.)

17.3

1. **1 round.**
   For Ψ^1, we can choose $[L_1, R_1], [Su_1, T_1]$ with $R_1 \neq S_1$. Then $H = 0 \ll \tilde{H} = \frac{|\mathcal{F}_n|}{2^{2n}}$.
2. **2 rounds.**
   For Ψ^2, we can choose $[L_1, R_1], [L_2, R_2], [S_1, T_1], [S_2, T_2]$ with $R_1 = R_2$ and $L_1 \oplus L_2 \neq S_1 \oplus S_2$. Then $H = 0 \ll \tilde{H} = \frac{|\mathcal{F}_n|^2}{2^{2n}(2^{2n}-1)}$.
   3 rounds.
   For Ψ^3, we can choose $[L_1, R_1], [L_2, R_2], [S_1, T_1], [S_2, T_2]$ with $R_1 = R_2$, $S_1 = S_2$ and $L_1 \oplus L_2 \neq T_1 \oplus T_2$. Then $H = 0 \ll \tilde{H} = \frac{|\mathcal{F}_n|^3}{2^{2n}(2^{2n}-1)}$.
3. **4 rounds.**
   For Ψ^4, with $q \simeq \sqrt{2^n}$, we can choose all the R_i with the same value, all the S_i pairwise distinct, and the property: $\forall i, j,\ 1 \leq i \leq q,\ 1 \leq j \leq q,\ S_i \oplus S_j \neq L_i \oplus L_j$. For example, the first $\frac{n}{2}$ bits of the S_i values are always 0 and the last $\frac{n}{2}$ bits of the L_i values are always 0. Since all the R_i values are equal, then all the L_i values are pairwise distinct (because we want pairwise distinct $[L_i, R_i]$) and all the X_i values are pairwise distinct (because $R_i = R_j \Rightarrow X_i \oplus X_j = L_i \oplus L_j$. Moreover here all the Y_i values are also pairwise distinct because $Y_i = Y_j \Rightarrow X_i \oplus X_j = S_i \oplus S_j \Rightarrow L_i \oplus L_j = S_i \oplus S_j$, but we always have here $L_i \oplus L_j \neq S_i \oplus S_j$.

We know (cf Theorem 17.4) that the exact value for H is

$$H_4 = \frac{|\mathscr{F}_n|^4 \cdot 2^{n(r+s)}}{2^{4nq}} \sum_{\textit{all frameworks } \mathscr{F}} 2^{n(x+y)} [\textit{ Number of } X_i \textit{ satisfying } (C1)]$$

$$\cdot [\textit{ Number of } Y_i \textit{ satisfying } (C2)].$$

Here it gives

$$H_4 = \frac{|\mathscr{F}_n|^4}{2^{4nq}} \cdot 2^{n(q-1)} \cdot 2^0 (2^n)(2^n)(2^n - 1) \ldots (2^n - q + 1)$$

$$H_4 == \frac{|\mathscr{F}_n|^4}{2^{2nq}} \left(1 - \frac{1}{2^n}\right)\left(1 - \frac{2}{2^n}\right) \ldots \left(1 - \frac{q-1}{2^n}\right) \ll \frac{|F_n|^4}{2^{2nq}}$$

when $q \simeq \sqrt{2^n}$. However $\tilde{H}_4 = \frac{|\mathscr{F}_n|^4}{(2^n)(2^n-1)\ldots(2^n-q+1)} \simeq \frac{|\mathscr{F}_n|^4}{2^{2nq}}$. Therefore here we have $H_4 \ll \tilde{H}_4$, i.e., a "hole" of length $\sqrt{2^n}$.

5 rounds.

For Ψ^5, with $q \simeq \sqrt{2^n}$, we can choose all the R_i with the same value, all the S_i with the same value and the property: $\forall i, j,\ 1 \leq i \leq q,\ 1 \leq j \leq q, T_i \oplus T_j \neq L_i \oplus L_j$. For example, the first $\frac{n}{2}$ bits of the L_i values are always 0 and the last $\frac{n}{2}$ bits of the T_i values are always 0. Since all the R_i values are equal, then all the L_i values are pairwise distinct (because we want pairwise distinct $[L_i, R_i]$) and all the X_i values are pairwise distinct (because $R_i = R_j \Rightarrow X_i \oplus X_j = L_i \oplus L_j$. Similarly, since all the S_i values are equal, then all the T_i values are distinct (because we want pairwise distinct $[S_i, T_i]$) and all the Z_i values are pairwise distinct (because $S_i = S_j \Rightarrow Z_i \oplus Z_j = T_i \oplus T_j$). Moreover all the Y_i values are also pairwise distinct, because $Y_i = Y_j \Rightarrow X_i \oplus X_j = Z_i \oplus Z_j \Rightarrow L_i \oplus L_j = T_i \oplus T_j$, but we always have: $L_i \oplus L_j \neq T_i \oplus T_j$.

We know (cf Theorem 17.5) that the exact formula for H is:

$$H_5 = \frac{|\mathscr{F}_n|^5 \cdot 2^{n(r+s)}}{2^{5nq}} \sum_{\textit{all frameworks } \mathscr{F}} 2^{n(x+y+z)}$$

$$\cdot [\textit{ Number of } X_i, Z_i \textit{ satisfying } (C1)]$$
$$\cdot [\textit{ Number of } Y_i \textit{ satisfying } (C2)].$$

Here we have only one framework (all the X_in pairwise distinct, Y_i pairwise distinct, Z_i pairwise distinct) with $r = q - 1, s = q - 1, x = y = z = 0$, [Number of X_i satisfying $(C1)] = 2^n$, [Number of Z_i satisfying $(C1)] = 2^n$, and [Number of Y_i satisfying $(C2)] = 2^n(2^n - 1) \ldots (2^n - q + 1)$. We obtain

$$H_5 = \frac{|\mathscr{F}_n|^5}{2^{2nq}} \cdot \left(1 - \frac{1}{2^n}\right)\left(1 - \frac{2}{2^n}\right) \ldots \left(1 - \frac{q-1}{2^n}\right) \ll \frac{|F_n|^5}{2^{2nq}}$$

when $q \ll \sqrt{2^n}$. However $\tilde{H}_5 = \frac{|\mathcal{F}_n|^5}{(2^n)(2^n-1)...(2^n-q+1)} \simeq \frac{|\mathcal{F}_n|^5}{2^{2nq}}$. Therefore here we have $H_5 \ll \tilde{H}_5$, i.e., a "hole" of length $\sqrt{2^n}$.

Remark A.7 This result is not in contradiction with the act that Ψ^5 is CCA secure when $q \ll 2^n$ because it is not possible in a CCA attack with q queries to obtain $R_1 = R_2 = \ldots = R_m$ and $S_1 = S_2 = \ldots = S_m$ with $m \simeq \sqrt{2^n}$.

4. **6 rounds.**
   As seen in Chap. 17, for 6 rounds, we always have $H \geq \tilde{H}(1-\epsilon)$ when $q \ll 2^n$. Therefore we have no hole of length $q \ll 2^n$ for Ψ^6. Similarly for Ψ^r, $r \geq 6$, since then $\Psi^r = \Psi^{k-6} \circ \Psi^6$ with no hole in Ψ^6.

Reference

Patarin, J. 2008. A proof of security in $O(2^n)$ for the benes schemes. In *Progress in cryptology—AFRICACRYPT '08*, vol. 5023, ed. S. Vaudenay. Lecture notes in computer science, 209–220. Berlin: Springer.

MIX
Papier aus verantwortungsvollen Quellen
Paper from responsible sources
FSC® C105338
FSC
www.fsc.org

If you have any concerns about our products,
you can contact us on
ProductSafety@springernature.com

In case Publisher is established outside the EU,
the EU authorized representative is:
Springer Nature Customer Service Center GmbH
Europaplatz 3, 69115 Heidelberg, Germany

Printed by Libri Plureos GmbH
in Hamburg, Germany